Astronomy and Astrophysics Olympiad

Volume 2: Problems and Solutions

Mihail Sandu

Romanian Physics Olympiad Committee, Romania

World Scientific

W JERSEY • LONDON • SINGAPORE • BEIJING • SHANGHAI • HONG KONG • TAIPEI • CHENNAI • TOKYO

Published by

World Scientific Publishing Europe Ltd.
57 Shelton Street, Covent Garden, London WC2H 9HE
Head office: 5 Toh Tuck Link, Singapore 596224
USA office: 27 Warren Street, Suite 401-402, Hackensack, NJ 07601

British Library Cataloguing-in-Publication Data
A catalogue record for this book is available from the British Library.

Cover image credit: NASA

Translation from Romanian by Daniela Berciu

ASTRONOMY AND ASTROPHYSICS OLYMPIAD
Volume 2: Problems and Solutions

ISBN 978-1-80061-709-4 (hardcover)
ISBN 978-1-80061-693-6 (paperback)
ISBN 978-1-80061-694-3 (ebook for institutions)
ISBN 978-1-80061-695-0 (ebook for individuals)

For any available supplementary material, please visit
https://www.worldscientific.com/worldscibooks/10.1142/Q0499#t=suppl

Desk Editors: Kannan Krishnan/Gabriel Rawlinson/Shi Ying Koe

Typeset by Stallion Press
Email: enquiries@stallionpress.com

Astronomy and Astrophysics Olympiad

Volume 2: Problems and Solutions

Foreword

Astronomy and Astrophysics are among the oldest sciences, yet they continue to undergo rapid development. Over time, the focus has shifted from simple observations of celestial motions to detailed investigations of galaxies, black holes and the cosmic microwave background. Today, progress in these fields is shaped by powerful telescopes, large-scale surveys, and theoretical models that push the limits of our current knowledge.

Such progress depends not only on professional research but also on the cultivation of new generations of scientists. Competitions like the Astronomy Olympiad and the Astronomy and Astrophysics Olympiad play an important role in this process. By challenging high school students to engage with advanced concepts and problem-solving techniques, these events help develop the critical skills and curiosity essential for a career in science.

Despite their educational value, there remains a lack of comprehensive resources tailored specifically to these Olympiads. The present collection of three volumes, containing original problems crafted by Professor Mihail Sandu, addresses this need in an outstanding manner, offering students and teachers alike a wide range of tools critically needed for preparing for such competitions.

Professor Sandu is a highly respected educator with decades of experience at both high school and university levels. His contributions to academic competitions are remarkable. Among other achievements, he prepared all theory problems for the 2014 International Olympiad on Astronomy and Astrophysics, as well as all practical test problems for the 2019 International Astronomy Olympiad, both held in Romania. More recently, he played a key role in the first Junior International Olympiad on Astronomy and Astrophysics in 2022. He has published nearly 100 books in Romanian and English, including textbooks and problem collections in Physics and Astronomy that have guided generations of students.

I have had the privilege of knowing Professor Sandu personally. His dedication, clarity of thought, and genuine enthusiasm for teaching are evident both in his presence at competitions and in his writing. Over more than five decades, he has supported and inspired countless students, exemplifying a rare commitment to education and mentorship. I am confident that the present collection will prove to be an invaluable resource for students preparing for Olympiads, and that it will continue to inspire future generations of astronomers and astrophysicists.

Andrei Constantin
Royal Society Dorothy Hodgkin Fellow,
Department of Physics, University of Oxford,
and Physics & Mathematics Tutor
at Mansfield College Oxford

Contents

Problem 1

Earth and Venus

Drawing (a) in Figure 1.1 corresponds to a certain moment (considered the starting moment for the proposed problem) at which the center of the Earth and the center of Venus are in the same direction as the center of the Sun, in the direction of the star σ, the distance between the Earth and the center of Venus being minimal (inferior conjunction of Venus).

The two arrows, one on the surface of Venus (white arrow) and the other on the surface of the Earth (black arrow), mark points on the equator of each of the two planets, as shown in the drawing in Figure 1.2, at the start of the lower conjunction of Venus, the two

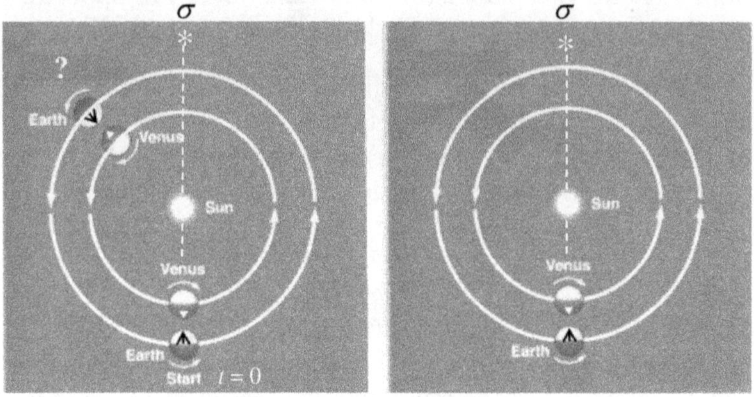

(a) (b)

Fig. 1.1

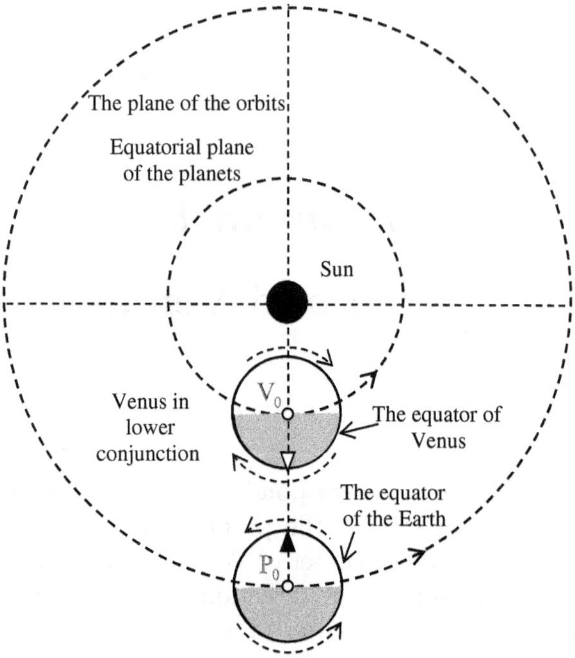

Fig. 1.2

points being face to face, in the direction that passes through the center of the Sun and the centers of the two planets. The rotation directions of the two planets, around their axes, are opposite. The orbits of the two planets around the Sun are coplanar circles.

a) *Determine* the time after which Venus will again be in the lower conjunction for the first time, as indicated by the sequence marked with (?) in drawing (a) in Figure 1.1. *Argue* whether, at the time of this lower conjunction, the two arrows, marked on the equator of each planet, will be again face to face in the direction determined by the centers of the two planets and the center of the Sun.

b) *Determine* the time interval after which Venus will return to the lower conjunction when the two planets expose the same two equatorial points at the minimum distance at the time of launch, as shown in drawing (b) in Figure 1.1. The decimal numbers $a.07$ and $b.06$ can be considered integers: $a.07 = a; b.06 = b$.

c) *Determine* the time after which Venus will be in the upper conjunction for the first time after the initial lower conjunction, when the distance between the center of the Earth and the center of Venus is maximum, as shown in drawing (b) in Figure 1.3, when the center of the Earth has returned to its original position in the direction of the center of the Sun and the star σ. *Analize* the possibility that in this position, the planet Venus will have reversed its luminous hemisphere. *Determine* the time interval relative to the time of the first upper conjunction of Venus after which the next upper conjunction of Venus will occur.

Given: The period of the Earth's sidereal revolution is $T_{\text{sidereal, Earth}} = 365$ terrestrial days; the period of the sidereal revolution of Venus is $T_{\text{sidereal, Venus}} = 224$ terrestrial days; the duration of a complete rotation of Venus around its axis is $\tau_V = 243$ terrestrial days; the duration of a complete rotation of the Earth around its own axis is $\tau_E = 1$ terrestrial day; the orbital speed of the Earth is $v_E = 30\,\text{km/s}$; the orbital speed of Venus is $v_V = 35\,\text{km/s}$.

It will be considered that the whole process takes place without the variation of its angular momentums and without the variation of the angular momentums of the revolution of the two planets. The movement of the Sun relative to the star σ is neglected.

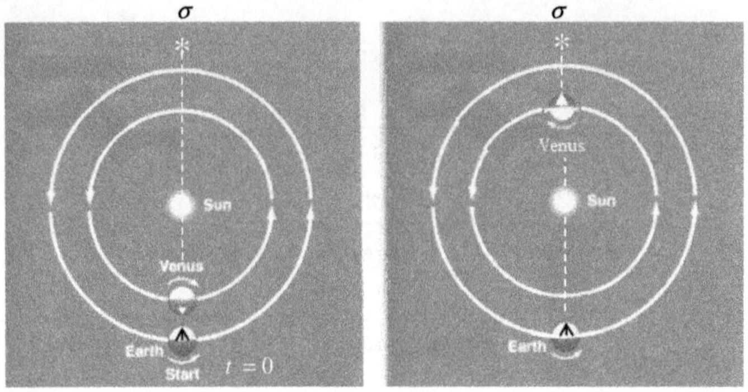

(a) (b)

Fig. 1.3

Solution

a) The duration of a complete rotation of Venus around its axis (the duration of the day on Venus) is $\tau_V = 243$ terrestrial days.

The duration of a complete rotation of the Earth around its axis is $\tau_E = 1$ terrestrial day.

The duration of a complete rotation of Venus around the Sun (the duration of the year on Venus) is 224 terrestrial days. This duration represents the period of the sidereal revolution of Venus, $T_{\text{sidereal, Venus}} = 224$ terrestrial days. The average speed of Venus in its orbit around the Sun is $v_{\text{Venus}} = 35\,\text{km/s}$. The radius of Venus's orbit around the Sun is $R_{\text{Venus}} = 6050\,\text{km}$.

The duration of a complete rotation of the Earth around the Sun (the length of the year on Earth) is 365 terrestrial days. This duration represents the period of the Earth's sidereal revolution, which is $T_{\text{sidereal, Earth}} = 365$ terrestrial days. The average speed of the Earth in its orbit around the Sun is $v_{\text{Earth}} = 30\,\text{km/s}$. The radius of the Earth's orbit around the Sun is $R_{\text{Earth}} = 6370\,\text{km}$.

The *period of synodic revolution*, T_{synodic}, of a planet is the time interval after which the planet's center returns to the same position relative to the center of the Earth (the time interval between two consecutive identical configurations of the planet).

The relationship between the sidereal and synodic periods for the inner planets Venus and Earth is obtained using the drawing in Figure 1.4, which shows:

$$\alpha_V = 2\pi + \alpha_E;$$

$$(\omega_V - \omega_E)T_{\text{synodic, V}} = 2\pi;$$

$$\left(\frac{2\pi}{T_{\text{sidereal,V}}} - \frac{2\pi}{T_{\text{sidereal, E}}}\right) \cdot T_{\text{synodic, V}} = 2\pi;$$

$$\frac{2\pi}{T_{\text{sidereal, V}}} - \frac{2\pi}{T_{\text{sidereal, E}}} = \frac{2\pi}{T_{\text{synodic, V}}};$$

$$\frac{1}{T_{\text{sidereal, V}}} - \frac{1}{T_{\text{sidereal, E}}} = \frac{1}{T_{\text{synodic, V}}};$$

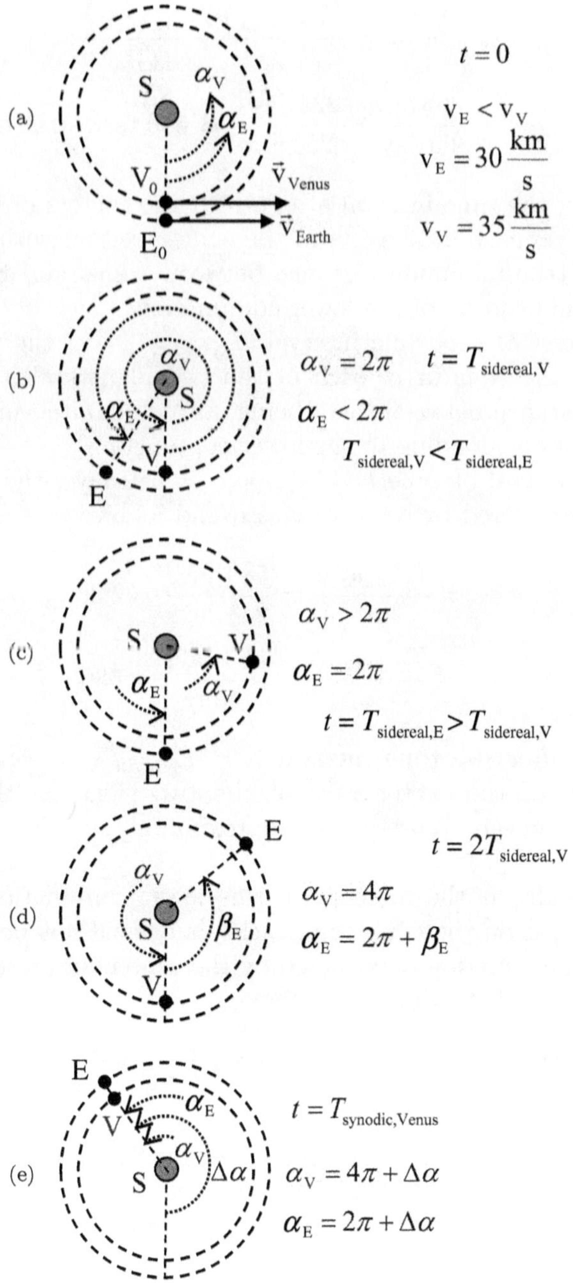
(a)
S
α_V
α_E
V_0
E_0
$\vec{v}_{Venus}$
$\vec{v}_{Earth}$
$t = 0$
$v_E < v_V$
$v_E = 30 \frac{km}{s}$
$v_V = 35 \frac{km}{s}$
(b)
α_V
S
α_E
V
E
$\alpha_V = 2\pi$ $t = T_{sidereal,V}$
$\alpha_E < 2\pi$
$T_{sidereal,V} < T_{sidereal,E}$
(c)
S
V
α_E
α_V
E
$\alpha_V > 2\pi$
$\alpha_E = 2\pi$
$t = T_{sidereal,E} > T_{sidereal,V}$
(d)
E
α_V
S
β_E
V
$t = 2T_{sidereal,V}$
$\alpha_V = 4\pi$
$\alpha_E = 2\pi + \beta_E$
(e)
E
V
α_E
α_V
S
$\Delta\alpha$
$t = T_{synodic,Venus}$
$\alpha_V = 4\pi + \Delta\alpha$
$\alpha_E = 2\pi + \Delta\alpha$

Fig. 1.4

$$T_{\text{synodic, V}} = \frac{T_{\text{sidereal, E}} T_{\text{sidereal, V}}}{T_{\text{sidereal, E}} - T_{\text{sidereal, V}}};$$

$$T_{\text{synodic,V}} = \frac{365\,\text{days} \cdot 224\,\text{days}}{365\,\text{days} - 224\,\text{days}} \approx 580 \text{ terrestrial days},$$

representing the time interval after which the centers of the planets Earth and Venus will be again in the same position with respect to each other (the minimum distance between Venus and Earth, with Venus in the position of the lower conjunction).

At the end of this time interval, $T_{\text{synodic, V}}$, will the two arrows marked on the equator of each of the two planets also be face to face, the distance between them being, as at the initial moment, the minimum (corresponding to the lower conjunction)?

In the interval of time $t = T_{\text{synodic,V}} = 580\,\text{days}$, the number of rotations completed by each planet around its own axis is:

$$n_{\text{V}} = \frac{T_{\text{synodic, V}}}{\tau_{\text{V}}} = \frac{580\,\text{days}}{243\,\text{days}} = 2.38;$$

$$n_{\text{E}} = \frac{T_{\text{synodic, V}}}{\tau_{\text{E}}} = \frac{580\,\text{days}}{1\,\text{day}} = 580,$$

such that, after the time interval $t = T_{\text{synodic, V}} = 580\,\text{days}$, the orientations of the vector rays of the two signs on the equator of each planet are represented approximately in the drawing in Figure 1.5.

As a result, at the moment of this lower conjunction, the two arrows marked on the equator of each planet will not be again face to face in the direction determined by the centers of the two planets and the center of the Sun.

b) If $T_{\text{synodic, V}}$ is the time interval after which, for the first time since the starting moment, the centers of the two planets will be at a minimum distance again (obviously in the direction of the Sun's center), then the centers of the two planets will return to a minimum distance, but in the same positions with respect to the center of the Sun, as at the initial moment, after a time interval

$$t = N \cdot T_{\text{synodic, V}},$$

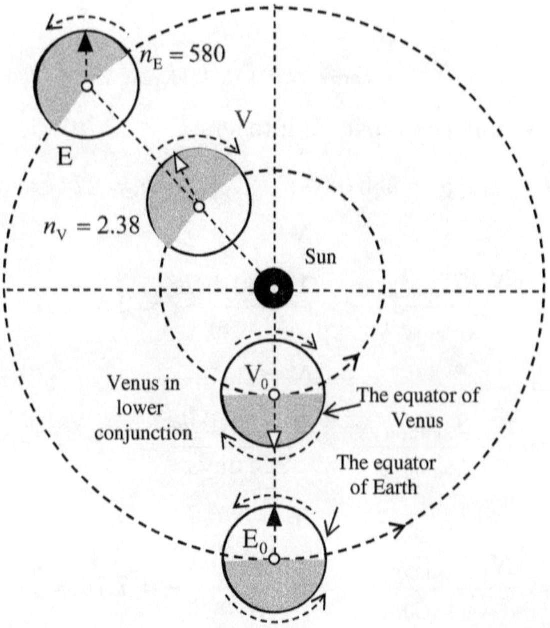

Fig. 1.5

where N is an integer (the number of synod cycles after which the centers of the two planets and the center of the Sun are in the same position with respect to the star σ, as at the initial moment).

The number of rotations performed by each planet around the Sun, from the initial moment until the return of the centers of the two planets to the initial arrangement, in relation to the Sun, remaining permanently in the direction of the star σ, in the time frame t, must be an integer, which is calculated as follows:

$$N_V = \frac{t}{T_{\text{sidereal, V}}} = N\frac{T_{\text{synodic, V}}}{T_{\text{sidereal, V}}};$$

$$N_P = \frac{t}{T_{\text{sidereal, E}}} = N\frac{T_{\text{synodic, V}}}{T_{\text{sidereal, E}}};$$

such that

$$N \cdot T_{\text{synodic, V}} = N_V \cdot T_{\text{sidereal, V}} = N_E \cdot T_{\text{sidereal, E}},$$

where both N and N_V, as well as N_E, must be integers.

Under these conditions:

$$N \cdot T_{\text{synodic, V}} = N_{\text{V}} T_{\text{sidereal, V}},$$

where both N and N_{V} must be integers;

$$T_{\text{synodic, V}} = 580\,\text{days}; \quad T_{\text{sidereal, V}} = 224\,\text{days};$$

$$N = 1;$$

$$N_{\text{V}} = \frac{N \cdot T_{\text{sinodic, V}}}{T_{\text{sydereal, V}}} = \frac{1 \cdot 580\,\text{days}}{224\,\text{days}} = 2.58 \neq \text{integer};$$

$$N = 2;$$

$$N_{\text{V}} = \frac{N \cdot T_{\text{synodic, V}}}{T_{\text{sidereal, V}}} = \frac{2 \cdot 580\,\text{days}}{224\,\text{days}} = 5.17 \neq \text{integer};$$

$$N = 3;$$

$$N_{\text{V}} = \frac{N \cdot T_{\text{synodic, V}}}{T_{\text{sidereal, V}}} = \frac{3 \cdot 580\,\text{days}}{224\,\text{days}} = 7.76 \neq \text{integer};$$

$$N = 4;$$

$$N_{\text{V}} = \frac{N \cdot T_{\text{synodic, V}}}{T_{\text{sidereal, V}}} = \frac{4 \cdot 580\,\text{days}}{224\,\text{days}} = 10.35 \neq \text{integer};$$

$$N = 5;$$

$$N_{\text{V}} = \frac{N \cdot T_{\text{synodic, V}}}{T_{\text{sidereal, V}}} = \frac{5 \cdot 580\,\text{days}}{224\,\text{days}} \approx 12.94 \neq \text{integer};$$

$$N = 6;$$

$$N_{\text{V}} = \frac{N \cdot T_{\text{synodic, V}}}{T_{\text{sidereal, V}}} = \frac{6 \cdot 580\,\text{days}}{224\,\text{days}} \approx 15.53 \neq \text{integer};$$

$$N = 7;$$

$$N_{\text{V}} = \frac{N \cdot T_{\text{synodic, V}}}{T_{\text{sidereal, V}}} = \frac{7 \cdot 580\,\text{days}}{224\,\text{days}} \approx 18.12 \neq \text{integer};$$

$$N = 8;$$

$$N_{\text{V}} = \frac{N \cdot T_{\text{synodic, V}}}{T_{\text{sidereal, V}}} = \frac{8 \cdot 580\,\text{days}}{224\,\text{days}} \approx 20.71 \neq \text{integer};$$

$$N = 9;$$

$$N_V = \frac{N \cdot T_{\text{synodic, V}}}{T_{\text{sidereal, V}}} = \frac{9 \cdot 580\,\text{days}}{224\,\text{days}} \approx 23.30 \neq \text{integer};$$

$$N = 10;$$

$$N_V = \frac{N \cdot T_{\text{synodic, V}}}{T_{\text{sidereal, V}}} = \frac{10 \cdot 580\,\text{days}}{224\,\text{days}} \approx 25.89 \neq \text{integer};$$

$$N = 11;$$

$$N_V = \frac{N \cdot T_{\text{synodic, V}}}{T_{\text{sidereal, V}}} = \frac{11 \cdot 580\,\text{days}}{224\,\text{days}} \approx 28.48 \neq \text{integer};$$

$$N = 12;$$

$$N_V = \frac{N \cdot T_{\text{synodic, V}}}{T_{\text{sidereal, V}}} = \frac{12 \cdot 580\,\text{days}}{224\,\text{days}} \approx 31.07 \approx \text{integer}.$$

For Earth's rotations:

$$N_E = N\frac{T_{\text{synodic, V}}}{T_{\text{sidereal, E}}};$$

$$T_{\text{synodic, V}} = 580\,\text{days}; \quad T_{\text{sidereal, E}} = 365\,\text{days};$$

$$N = 1;$$

$$N_E = \frac{N \cdot T_{\text{synodic, V}}}{T_{\text{sidereal, E}}} = \frac{1 \cdot 580\,\text{days}}{365\,\text{days}} = 1.58 \neq \text{integer};$$

$$N = 2;$$

$$N_E = \frac{N \cdot T_{\text{synodic, V}}}{T_{\text{sidereal, E}}} = \frac{2 \cdot 580\,\text{days}}{365\,\text{days}} = 3.17 \neq \text{integer};$$

$$N = 3;$$

$$N_E = \frac{N \cdot T_{\text{synodic, V}}}{T_{\text{sidereal, E}}} = \frac{3 \cdot 580\,\text{days}}{365\,\text{days}} = 4.76 \neq \text{integer};$$

$$N = 4;$$

$$N_E = \frac{N \cdot T_{\text{synodic, V}}}{T_{\text{sidereal, E}}} = \frac{4 \cdot 580\,\text{days}}{365\,\text{days}} = 6.35 \neq \text{integer};$$

$$N = 5;$$

$$N_E = \frac{N \cdot T_{\text{synodic, V}}}{T_{\text{sidereal, E}}} = \frac{5 \cdot 580\,\text{days}}{365\,\text{days}} \approx 7.94 \neq \text{integer};$$

$$N = 6;$$

$$N_{\mathrm{E}} = \frac{N \cdot T_{\text{synodic, V}}}{T_{\text{sidereal, E}}} = \frac{6 \cdot 580\,\text{days}}{365\,\text{days}} \approx 9.53 \neq \text{integer};$$

$$N = 7;$$

$$N_{\mathrm{E}} = \frac{N \cdot T_{\text{synodic, V}}}{T_{\text{sidereal, E}}} = \frac{7 \cdot 580\,\text{days}}{365\,\text{days}} \approx 11.12 \neq \text{integer};$$

$$N = 8;$$

$$N_{\mathrm{E}} = \frac{N \cdot T_{\text{synodic, V}}}{T_{\text{sidereal, E}}} = \frac{8 \cdot 580\,\text{days}}{365\,\text{days}} \approx 12.71 \neq \text{integer};$$

$$N = 9;$$

$$N_{\mathrm{E}} = \frac{N \cdot T_{\text{synodic, V}}}{T_{\text{sidereal, E}}} = \frac{9 \cdot 580\,\text{days}}{365\,\text{days}} \approx 14.30 \neq \text{integer};$$

$$N = 10;$$

$$N_{\mathrm{E}} = \frac{N \cdot T_{\text{synodic, V}}}{T_{\text{sidereal, E}}} = \frac{10 \cdot 580\,\text{days}}{365\,\text{days}} \approx 15.89 \neq \text{integer};$$

$$N = 11;$$

$$N_{\mathrm{E}} = \frac{N \cdot T_{\text{synodic, V}}}{T_{\text{sidereal, E}}} = \frac{11 \cdot 580\,\text{days}}{365\,\text{days}} \approx 17.47 \neq \text{integer};$$

$$N = 12;$$

$$N_{\mathrm{E}} = \frac{N \cdot T_{\text{synodic, V}}}{T_{\text{sidereal, E}}} = \frac{12 \cdot 580\,\text{days}}{365\,\text{days}} \approx 19.06 \approx \text{integer}.$$

This results in:

$$t = N \cdot T_{\text{synodic, V}} = 12 \cdot 580\,\text{days};$$

$$t = 6960\,\text{days} \approx 19\,\text{years},$$

representing the time interval after which the initial arrangement of the Earth–Venus–Sun system, in relation to the star σ, with Venus in the lower conjunction, will be repeated after 19 years.

It is easy to see that the value of $t = 6960\,\text{days}$ is the least common multiple of the values $T_{\text{sidereal, V}} = 224\,\text{days}$ and $T_{\text{sidereal, E}} = 365\,\text{days}$:

$$\frac{6960}{224} \approx 31; \quad \frac{6960}{365} \approx 19.$$

c) Following the evolution of the two planets from the moment of the lower conjunction of Venus ($t = 0$) to the moment of the upper conjunction of Venus, when the Earth returns to the initial position with respect to the Sun ($t > T_{\text{sidereal, E}}$), as represented in the sequences in Figure 1.6, it results that:

$$t > T_{\text{sidereal, E}}; \quad \Delta\alpha + \Delta\beta = 2\pi;$$

$$\alpha_{\text{V}} = 4\pi + \Delta\alpha + \Delta\beta + \pi; \quad \alpha_{\text{V}} = 7\pi;$$

$$\alpha_{\text{V}} = \omega_{\text{V}} t = \frac{2\pi}{T_{\text{sidereal, V}}} t;$$

$$\alpha_{\text{E}} = 2\pi + \Delta\beta + \Delta\alpha; \quad \alpha_{\text{E}} = 4\pi;$$

$$\alpha_{\text{E}} = \omega_{\text{E}} t = \frac{2\pi}{T_{\text{sidereal, E}}} t;$$

$$\alpha_{\text{V}} = \alpha_{\text{E}} + 3\pi;$$

$$\frac{2\pi}{T_{\text{sidereal, V}}} t = \frac{2\pi}{T_{\text{sidereal, E}}} t + 3\pi; 2\pi \left(\frac{1}{T_{\text{sidereal, V}}} - \frac{1}{T_{\text{sidereal, E}}} \right) t = 3\pi;$$

$$t = \frac{3}{2} \cdot \frac{T_{\text{E}} T_{\text{V}}}{T_{\text{E}} - T_{\text{V}}} = \frac{3}{2} T_{\text{synodic, V}} > T_{\text{synodic, V}};$$

$$t = \frac{3}{2} \cdot \frac{365\,\text{days} \cdot 224\,\text{days}}{365\,\text{days} - 224\,\text{days}} = \frac{3}{2} \cdot 580\,\text{days} \approx 870\,\text{days},$$

representing the time after which Venus passes from the lower conjunction to the upper conjunction.

During the time interval $t = 870\,\text{days}$, the number of rotations performed by each planet around its own axis is:

$$n_{\text{V}} = \frac{t}{\tau_{\text{V}}} = \frac{870\,\text{days}}{243\,\text{days}} \approx 3.58;$$

$$n_{\text{E}} = \frac{t}{\tau_{\text{E}}} = \frac{870\,\text{days}}{1\,\text{day}} = 870,$$

so that, over the time $t = 870\,\text{days}$, the orientations of the vector rays of the two signs on the equator of each planet are those represented approximately in the drawing in Figure 1.7.

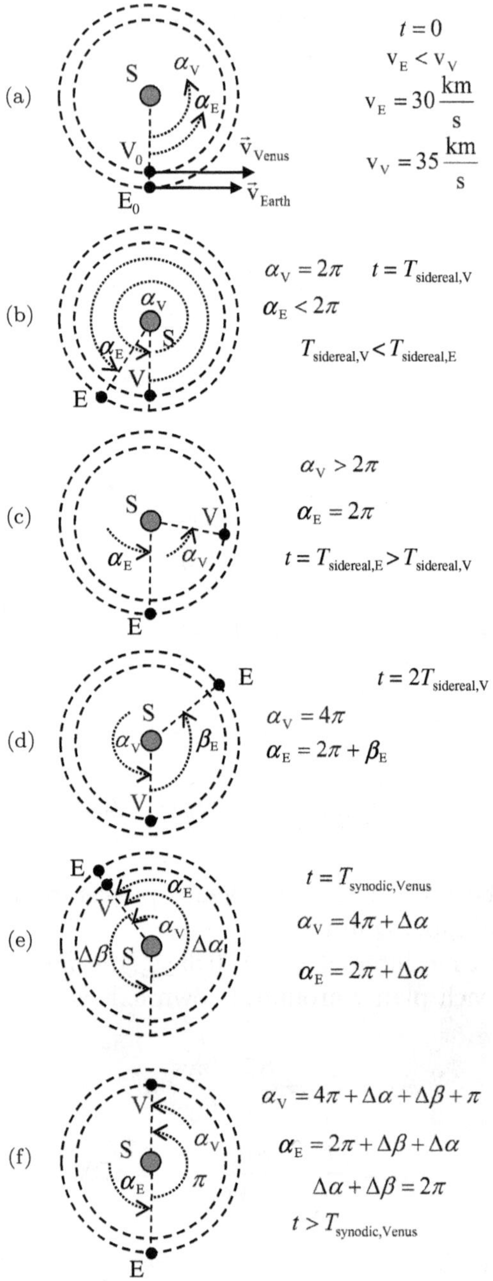

(a)
S
α_V
α_E
V_0
E_0
$\vec{v}_{Venus}$
$\vec{v}_{Earth}$
$t = 0$
$v_E < v_V$
$v_E = 30\,\frac{km}{s}$
$v_V = 35\,\frac{km}{s}$
(b)
$\alpha_V = 2\pi$ $t = T_{sidereal,V}$
$\alpha_E < 2\pi$
$T_{sidereal,V} < T_{sidereal,E}$
(c)
$\alpha_V > 2\pi$
$\alpha_E = 2\pi$
$t = T_{sidereal,E} > T_{sidereal,V}$
(d)
$t = 2T_{sidereal,V}$
$\alpha_V = 4\pi$
$\alpha_E = 2\pi + \beta_E$
(e)
$t = T_{synodic,Venus}$
$\alpha_V = 4\pi + \Delta\alpha$
$\alpha_E = 2\pi + \Delta\alpha$
(f)
$\alpha_V = 4\pi + \Delta\alpha + \Delta\beta + \pi$
$\alpha_E = 2\pi + \Delta\beta + \Delta\alpha$
$\Delta\alpha + \Delta\beta = 2\pi$
$t > T_{synodic,Venus}$

Fig. 1.6

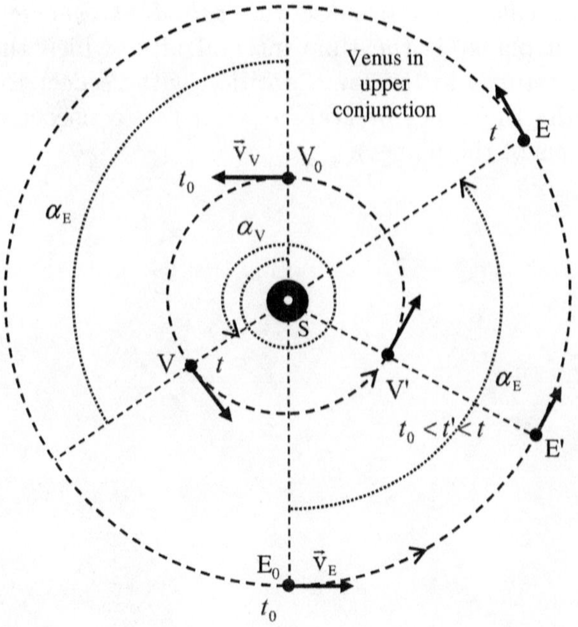

Fig. 1.7

According to the notation in Figure 1.7, for the calculation of the duration of the time interval between two consecutive upper conjunctions of Venus, the result is:

$$\alpha_V = \omega_V t = \frac{2\pi}{T_{\text{sidereal, V}}} t;$$

$$\alpha_E = \omega_E t = \frac{2\pi}{T_{\text{sidereal, E}}} t;$$

$$\alpha_V = \alpha_E + 2\pi; \quad \alpha_V - \alpha_E = 2\pi;$$

$$\frac{2\pi}{T_{\text{sidereal, V}}} t - \frac{2\pi}{T_{\text{sidereal, E}}} t = 2\pi;$$

$$t = \frac{T_{\text{sid, V}} T_{\text{sid, E}}}{T_{\text{sid, E}} - T_{\text{sid, V}}} = T_{\text{synodic, V}} = \frac{365\,\text{days} \cdot 224\,\text{days}}{365\,\text{days} - 224\,\text{days}}$$
$$\approx 580\ \text{terrestrial days}.$$

This can be calculated because the *period of synodic revolution*, T_{synodic}, of a planet is the time interval after which the center of that planet returns to the same position with respect to the center of the Earth (the time interval between two consecutive identical configurations of the planet).

Problem 2

The Thermal Kelvin–Helmholtz Time of a Star

The time required for a stellar configuration to contract, from the initial state of the interstellar spherical cloud with infinite distribution to the current state of hydrostatic equilibrium, when the radius and mass of the star are, respectively, R and M and the brightness of the star remains constant, represents the thermal Kelvin–Helmholtz time, t_{KH}, of the star.

a) *Determine* the thermal Kelvin–Helmholtz time for a spherical star, t_{KH}, knowing that the gravitational potential energy of a star, E_{pg}, as well as its internal energy, U, in a state of hydrostatic echilibrium are given by:

$$E_{\mathrm{pg}} = -\frac{3}{5}\frac{KM^2}{R}; \quad U = \frac{3}{10} \cdot \frac{KM^2}{R}.$$

b) *Determine* the thermal Kelvin–Helmholtz time for the Sun, $t_{\mathrm{KH,S}}$, knowing: the radius of the Sun, $R_{\mathrm{S}} = 6.96 \cdot 10^5\,\mathrm{km}$; the mass of the Sun, $M_{\mathrm{S}} = 1.99 \cdot 10^{30}\,\mathrm{kg}$; the brightness of the Sun, $L_{\mathrm{S}} = 3.86 \cdot 10^{26}\,\mathrm{W}$; the constant of the gravitational attraction, $G = 6.67 \cdot 10^{-11}\,\mathrm{Nm^2\,kg^{-2}}$.

Solution

a)

$$E_{\text{pg}} = -\frac{3}{5}\frac{GM^2}{R};$$

$$U = \frac{3}{10} \cdot \frac{GM^2}{R} = -\frac{1}{2} \cdot \left(-\frac{3}{5} \cdot \frac{GM^2}{R}\right) = -\frac{1}{2} \cdot E_{\text{pg}},$$

such that:

$$2U = \frac{3}{5} \cdot \frac{GM^2}{R},$$

$$2U + E_{\text{pg}} = \frac{3}{5} \cdot \frac{GM^2}{R} - \frac{3}{5} \cdot \frac{GM^2}{R};$$

$$2U + E_{\text{pg}} = 0,$$

representing the virial theorem;

$$E = U + E_{\text{pg}} = \text{constant},$$

representing the total energy of a star, which, according to the law of conservation of energy, is constant.

From the virial theorem and knowing that

$$E_{\text{pg}} = -2U,$$

the result is that the total energy of the star is:

$$E = U - 2U = -U;$$

$$E = -\frac{3}{10} \cdot \frac{GM^2}{R} = \frac{1}{2}\left(-\frac{3}{5} \cdot \frac{GM^2}{R}\right) = \frac{1}{2} \cdot E_{\text{pg}}; \quad E_{\text{pg}} = -\frac{3}{5}\frac{GM^2}{R};$$

$$E = -\frac{3}{10}\frac{GM^2}{R}.$$

The total energy of the star decreases during gravitational contraction when the radius of the star decreases in its evolution. The energy thus released, radiated into space, is equal to half the potential gravitational energy.

The time in which the star contracts from the infinite initial extension to the radius corresponding to the hydrostatic equilibrium, the brightness of the star remaining constant, is the Kelvin time of that star, t_{KH}.

If $R_0 \to \infty$ is the initial radius of the dust cloud from which, by gravitational contraction, the star was formed, which now, under conditions of hydrostatic equilibrium, has the radius R, then the variation of the total energy of the star, as a result of its gravitational contraction, is:

$$\Delta E = E_{\mathrm{initial}} - E_{\mathrm{final}} = -\frac{3}{10} \cdot \frac{GM^2}{R_0} - \left(-\frac{3}{10} \cdot \frac{GM^2}{R} \right);$$

$$R_0 \to \infty;$$

$$\Delta E = \frac{3}{10} \cdot \frac{GM^2}{R}.$$

Considering the definition of the brightness of the star (the rate of decrease of the total energy of the star), the result is:

$$L = \frac{\Delta E}{t_{\mathrm{KH}}}; \quad t_{\mathrm{KH}} = \frac{\Delta E}{L};$$

$$t_{\mathrm{KH}} = \frac{3}{10} \cdot \frac{GM^2}{RL}.$$

b)

$$t_{\mathrm{KH,S}} = \frac{3}{10} \cdot \frac{GM_{\mathrm{S}}^2}{R_{\mathrm{S}} L_{\mathrm{S}}} \approx 55 \cdot 10^{13}\ s \approx 1.74 \cdot 10^7 \text{ years}.$$

Problem 3

The Magnitude Limit of Stars Observed by a Telescope

It is known that the maximum limit of the apparent magnitude value of stars that can be seen with the naked eye is $m_{\text{max}} = m_{\text{limit of visibility}} = 6^{\text{m}}$. Stars whose apparent magnitudes are $m > m_{\text{limit of visibility}}$ can no longer be observed with the naked eye. However, they can be observed with a telescope.

Determine the maximum limit of the apparent magnitude value of stars that can be observed with the help of a telescope, $m_{\text{limit of telescope}}$, whose objective lens has a diameter $D_{\text{telescope objective}} = 300$ mm, if the diameter of the pupil of the observer's eye is $D_{\text{pupil}} = 6$ mm.

Solution

The degree of collection of light from a star, g_{c}, by a telescope, used as shown in the drawing in Figure 3.1, is defined as the ratio of the surface area of the circle whose diameter is equal to the diameter of the lens of the telescope and the area of the circle whose diameter is equal to the diameter of the pupil of the observer's eye:

$$g_{\text{c}} = \frac{\frac{\pi}{4} D^2_{\text{telescope objective lens}}}{\frac{\pi}{4} D^2_{\text{pupil}}} = \left(\frac{D_{\text{objective telescope}}}{D_{\text{pupil}}} \right)^2;$$

$$D_{\text{objective telescope}} > D_{\text{pupil}}; g_{\text{c}} > 1.$$

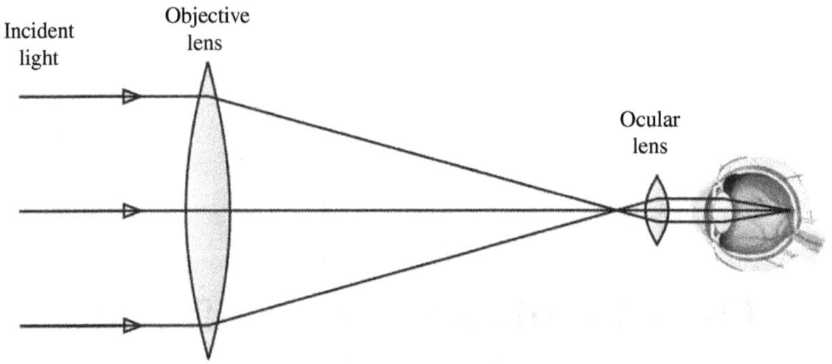

Fig. 3.1

If the illumination that a star produces for the observer's eye when the star is observed with the naked eye is E_0, and if, when the same star is observed with a telescope, the illumination that the star produces for the observer's eye is E, according to Pogson's formula, we get

$$\log \frac{E}{E_0} = -0.4 \cdot (m - m_0),$$

where m_0 is the apparent visual magnitude of the star when the star is observed with the naked eye, and m is the apparent magnitude of the star observed through the telescope.

When astronomers used a telescope to observe stars and were able to measure their brightness, they found that first-class stars (as they were called by astronomers of antiquity, i.e., stars with apparent visual magnitudes $m_{\text{I}} = 1^{\text{m}}$) have a brightness, $E_{\text{I}}, 100$ times greater than the brightness, $E_{\text{limit of visibility}}$, of stars with apparent visual magnitudes $m_{\text{limit of visibility}} = 6^{\text{m}}$ at the limit of visibility with the naked eye. Thus, according to Pogson's formula, the relationship between the brightness and the magnitudes of these stars is as follows:

$$E_{\text{I}} = 100 \cdot E_{\text{limit of visibility}};$$

$$\log \frac{E_{\text{I}}}{E_{\text{limit of visibility}}} = -0,4(m_{\text{I}} - m_{\text{limit of visibility}});$$

$$\log \frac{E_{\mathrm{I}}}{E_{\text{limit of visibility}}} = \log \frac{100 \cdot E_{\text{limit of visibility}}}{E_{\text{limit of visibility}}}$$
$$= \log 100 = \log 10^2 = 2;$$
$$-0.4(m_{\mathrm{I}} - m_{\text{limit of visibility}}) = -0.4(1-6) = -0.4 \cdot (-5) = 2.$$

Due to the use of a telescope to observe the stars, the light collected by the telescope from the stars increases, the degree of light collection being given by the expression

$$g_{\mathrm{c}} = \left(\frac{D_{\text{objective of telescope}}}{D_{\text{pupil}}}\right)^2 > 1.$$

Thus, when observed under a microscope, the brightness of the stars is higher than in the case of their direct visual observation.

Knowing that in the scale of magnitudes, the brightest stars have the smallest apparent magnitudes, and the less bright stars have the largest apparent magnitudes, it means that the apparent magnitudes of the stars observed through the telescope are lower than the apparent magnitudes of the same stars observed with the naked eye.

As a result, the same star that, observed with the naked eye, has the apparent brightness E_0 and apparent magnitude m_0, when observed through a telescope, has the brightness

$$E = g_{\mathrm{c}} E_0 = E_0 \cdot \left(\frac{D_{\text{objective of telescope}}}{D_{\text{pupil}}}\right)^2 > E_0$$

and the magnitude $m < m_0$, such that, according to Pogson's formula, the result is:

$$\log \frac{E}{E_0} = -0.4(m - m_0) = \log \frac{g_{\mathrm{c}} E_0}{E_0} = \log g_{\mathrm{c}};$$
$$m - m_0 = -2.5 \log g_{\mathrm{c}} = -2.5 \cdot \log \left(\frac{D_{\text{objective of telescope}}}{D_{\text{pupil}}}\right)^2$$
$$= -5 \cdot \log \frac{D_{\text{objective of telescope}}}{D_{\text{pupil}}};$$
$$m - m_0 = -5 \cdot \log \frac{D_{\text{objective telescope}}}{D_{\text{pupil}}};$$

$$\frac{D_{\text{objective of telescope}}}{D_{\text{pupil}}} = \frac{|D_{\text{objective of telescope (mm)}}| \cdot \text{mm}}{|D_{\text{pupil (mm)}}| \cdot \text{mm}}$$

$$= \frac{|D_{\text{objective of telescope (mm)}}|}{|D_{\text{pupil (mm)}}|};$$

$$m - m_0 = -5 \cdot \log \frac{|D_{\text{objective of telescope (mm)}}|}{|D_{\text{pupil (mm)}}|};$$

$$m - m_0 = -5 \cdot (\log |D_{\text{objective of telescope (mm)}}| - \log |D_{\text{pupil (mm)}}|);$$

$$m_0 - m = 5 \cdot \log |D_{\text{objective of telescope (mm)}}| - 5 \cdot \log |D_{\text{pupil (mm)}}|;$$

$$m_0 - m = \Delta m > 0;$$

$$\Delta m = 5 \cdot \log |D_{\text{objective of telescope (mm)}}| - 5 \cdot \log |D_{\text{pupil (mm)}}|,$$

representing the star's apparent magnitude, as a result of its observation with the telescope.

At the limit of visual observation with the naked eye, the maximum apparent visual magnitude of a star is $m_{\text{limit of visibility}} = m_{\text{max}} = 6^{\text{m}}$.

Under these conditions, at the limit of telescope observation is a star whose apparent magnitude is:

$$m_{\text{limit of telescope}} = \Delta m + m_{\text{max}} > m_{\text{max}};$$

$$m_{\text{limit of telescope}} = m_{\text{max}} + 5 \cdot \log |D_{\text{objective of telescope (mm)}}| - 5 \cdot \log |D_{\text{pupil (mm)}}|;$$

$$m_{\text{limit of telescope}} = 6 + 5 \cdot \log |D_{\text{objective of telescope (mm)}}| - 5 \cdot \log |D_{\text{pupil (mm)}}|;$$

$$D_{\text{pupil}} = 6 \text{ mm}; \quad D_{\text{objective of telescope}} = 300 \text{ mm};$$

$$m_{\text{limit of telescope}} = 6 + 5 \cdot \log |D_{\text{objective of telescope (mm)}}| - 5 \cdot \log |D_{\text{pupil (mm)}}|;$$

$$m_{\text{limit of telescope}} \approx 14.5^{\text{m}}.$$

This means that using the telescope, whose objective lens has a diameter $D_{\text{objective of telescope}} = 300$ mm, we can observe stars whose apparent magnitudes are at most $m_{\text{limit of telescope}} = 14.5^{\text{m}}$, a value beyond the limit of free visual observation, which is $m_{\text{limit of visibility}} = m_{\text{max}} = 6^{\text{m}}$.

Problem 4

Astrometric Binary System

In an astrometric binary star system, only one of its components, σ_1, is visible. Its motion is known: the period of motion, $T = 60$ years; the apparent angular radius of its orbit, $\theta = 2''$; the distance between the star and the observer, $d = 10$ pc. The second component, σ_2, has very low brightness and is not visible. However, its existence is proven by the observed movement of the visible component.

Determine the masses of the two components of the given system, M_1 and M_2, as well as the brightness of the invisible component, L_2, if the brightness of the visible component is known, $L_1 = 20 \cdot L_{\text{Sun}}$.

It is known that the mass of the visible star meets the condition $0.43 \cdot M_{\text{Sun}} < M_1 < 2 \cdot M_{\text{Sun}}$, and the mass of the invisible component meets the condition $M_2 < 0.43 \cdot M_S$. The orbits of the two stars, relative to their center of mass, are concentric circles, and the plane of their orbits is perpendicular to the plane of the sky.

Solution

According to Kepler's third law, in generalized form, when the two stars rotate in circular orbits around their center of mass, it follows that

$$T^2 = \frac{4\pi^2 a^3}{K(M_1 + M_2)},$$

where a is the distance between the centers of the two stars.

According to the notation in Figure 4.1, it results that:

$$a = \theta \cdot d;$$

$$\theta = 2''; \quad 1'' = \frac{3.14}{180 \cdot 60 \cdot 60} \text{ rad}; \quad 1 \text{ AU} = \frac{1}{206265} \text{ pc};$$

$$1 \text{ pc} = 206265 \text{ AU};$$

$$d = 10 \text{ pc};$$

$$a_{\text{AU}} = \theta_{\text{rad}} \cdot d_{\text{AU}};$$

$$a_{\text{AU}} = 2 \cdot \frac{3.14}{180 \cdot 60 \cdot 60} \text{ rad} \cdot 10 \cdot 206265 \text{ AU};$$

$$a_{\text{AU}} = 2 \cdot \frac{3.14}{180 \cdot 60 \cdot 60} \text{ rad} \cdot 10 \cdot 206265 \text{ AU} = 19.98 \text{ AU};$$

$$a \approx 20 \text{ AU};$$

$$a(\text{AU}) = \theta('') \cdot d(\text{pc}) = 2 \cdot 10 \ \cdot \text{AU} = 20 \text{ AU}.$$

Writing Kepler's third law and the motion of the Earth in a circular orbit around the Sun, it follows that:

$$T_{\text{E}}^2 = \frac{4\pi^2 a_{\text{ES}}^3}{KM_{\text{S}}}; \quad T_{\text{E}} = 1 \text{ year}; \quad a_{\text{ES}} = 1 \text{ AU};$$

$$\frac{T^2}{T_{\text{E}}^2} = \frac{a^3}{a_{\text{ES}}^3} \cdot \frac{M_{\text{S}}}{M_1 + M_2};$$

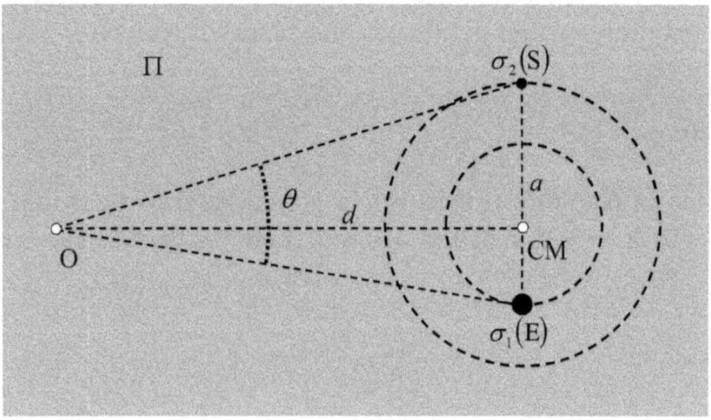

Fig. 4.1

$$\left(\frac{T}{T_{\mathrm{E}}}\right)^2 = \left(\frac{a}{a_{\mathrm{ES}}}\right)^3 \cdot \frac{M_{\mathrm{S}}}{M_1 + M_2};$$

$$\frac{M_1 + M_2}{M_{\mathrm{S}}} = \left(\frac{a}{a_{\mathrm{ES}}}\right)^3 \cdot \left(\frac{T_{\mathrm{E}}}{T}\right)^2;$$

$$a = 20\ \mathrm{AU}; \quad a_{\mathrm{ES}} = 1\ \mathrm{AU}; \quad T = 60\ \mathrm{years}; \quad T_{\mathrm{E}} = 1\ \mathrm{year};$$

$$\frac{M_1 + M_2}{M_{\mathrm{S}}} = \left(\frac{20\ \mathrm{AU}}{1\ \mathrm{AU}}\right)^3 \cdot \left(\frac{1\ \mathrm{year}}{60\ \mathrm{years}}\right)^2 = \frac{20}{9};$$

$$M_1 + M_2 = \frac{20}{9} M_{\mathrm{S}}.$$

Knowing that

$$0.43 \cdot M_{\mathrm{Sun}} < M_1 < 2 \cdot M_{\mathrm{Sun}},$$

it results that:

$$\frac{L_1}{L_{\mathrm{S}}} = \left(\frac{M_1}{M_{\mathrm{S}}}\right)^4 = 20;$$

$$M_1 = \sqrt[4]{20} \cdot M_{\mathrm{S}} \approx 2.1 \cdot M_{\mathrm{S}};$$

$$M_2 = \frac{20}{9} M_{\mathrm{S}} - M_1 \approx 0.12 \cdot M_{\mathrm{S}};$$

$$M_2 < 0.43 \cdot M_{\mathrm{S}}; \quad \frac{L_2}{L_{\mathrm{S}}} \approx 0.23 \cdot \left(\frac{M_2}{M_{\mathrm{S}}}\right)^{2,3};$$

$$\frac{L_2}{L_{\mathrm{S}}} \approx 0.23 \cdot \left(\frac{M_2}{M_{\mathrm{S}}}\right)^{2,3}; \frac{L_2}{L_{\mathrm{S}}} \approx 0.23 \cdot (0.12)^{2,3};$$

$$L_2 = 0.002 \cdot L_{\mathrm{S}}.$$

Problem 5
Distance to a Galaxy

Determine the distance Δ from Earth to a galaxy whose brightness is $L_G = 10^{11} L_S$, where L_S is the Sun's brightness, if the apparent magnitude of the galaxy is $m_G = 15.7$.

It is known that the absolute magnitude of a standard star (a star located at a standard distance), $\Delta_{st} = 10$ ps, whose brightness is equal to the brightness of the Sun, $L_{\text{standard star}} = L_{\text{Sun}}$, is:

$$m_{\text{absolute, standard star}} = m_{\text{absolute, Sun}} = 4.73;$$

$$M_{\text{standard star}} = M_{\text{Sun}} = 4.73.$$

The absolute magnitude of a star is its apparent magnitude if it were at a standard distance from the observer, $\Delta_{\text{std}} = 10$ pc. It is denoted by m_{ab} or M.

Solution

The brightness of a star or a galaxy, L, represents the energy of all radiation emitted in a unit of time through its entire surface in all directions:

$$L = \frac{W}{t}; \quad \langle L \rangle_{\text{SI}} = \frac{\text{J}}{\text{s}} = \text{W}.$$

The flux of radiation emitted by a star or a galaxy, Φ, represents the energy of all radiation emitted in a unit of time through its entire surface in all directions:

$$\Phi = \frac{W}{tS}; \quad \langle\Phi\rangle_{\text{SI}} = \text{Wm}^{-2}.$$

The relationship between the two values is

$$\Phi = \frac{W}{tS} = \frac{L}{4\pi d^2},$$

where d is the distance from the center of the star/galaxy to the observer (Earth).

The *standard flux* is defined as the radiation flux of a *standard star*, whose brightness is equal to the brightness of the Sun, for an observer at a standard distance, $\Delta_{\text{st}} = 10$ ps, from the center of the star:

$$\Phi_{\text{std}} = \frac{L_{\text{S}}}{4\pi\Delta_{\text{std}}^2}.$$

The brightness of stars or galaxies is characterized by the flow of their radiation reaching the Earth. The brightness is evaluated on a scale called the apparent magnitude (m) scale, which dates back to ancient Greece.

In the scale of apparent magnitudes, the brightest stars in the sky have apparent magnitudes of $m = 1$, and the brightest stars in the sky visible to the naked eye have apparent magnitudes of about $m = 6$.

Specifically, the difference in the apparent magnitudes of two stars, or two galaxies, relative to the same observer on Earth is defined by the relationship:

$$m_2 - m_1 = 2.5 \cdot \log\left(\frac{\Phi_1}{\Phi_2}\right) = -2.5 \cdot \log\left(\frac{\Phi_2}{\Phi_1}\right),$$

where Φ_1 and Φ_2 are the radiation flows from the two cosmic bodies, measured by the Earth observer;

$$\log\left(\frac{\Phi_1}{\Phi_2}\right) = \frac{m_2 - m_1}{2.5} = 0.4(m_2 - m_1) = (m_2 - m_1) \cdot \log\sqrt[5]{100};$$

$$\sqrt[5]{100} = 2.512; \quad \log\sqrt[5]{100} = 0.4;$$

$$\frac{\Phi_1}{\Phi_2} = 10^{0.4(m_2 - m_1)} = 10^{-0.4(m_1 - m_2)}.$$

We can also discuss the difference in the values of the apparent magnitudes for the same star in relation to two observers at different distances from the star so that, in the previous relation, m_1 and m_2 are the apparent magnitudes for the same star according to two observers at distances d_1 and d_2, respectively, and Φ_1 and Φ_2 are the light streams received by the two observers from the same star:

$$\Phi_1 = \frac{L_{\text{star}}}{4\pi d_1^2}; \quad \Phi_2 = \frac{L_{\text{star}}}{4\pi d_2^2};$$

$$m_2 - m_1 = 2.5 \cdot \log\left(\frac{\Phi_1}{\Phi_2}\right) = -2.5 \cdot \log\left(\frac{\Phi_2}{\Phi_1}\right);$$

$$m_2 - m_1 = 2.5 \cdot \log\left(\frac{d_2^2}{d_1^2}\right) = -2.5 \cdot \log\left(\frac{d_1^2}{d_2^2}\right).$$

If one of the two observers is at a standard distance from the star $d_2 = \Delta_{\text{std}} = 10$ ps, then the apparent magnitude of the star relative to this observer is called the *absolute magnitude* of the star, $m_2 = m_{\text{ab}} = M$. The apparent magnitude for the same star, $m_1 = m$, corresponds to the measurement by the observer on Earth located at the distance $d_1 = \Delta \neq \Delta_{\text{std}}$.

The relationship between the two magnitudes for the same star, according to the two observers, under the conditions specified above, is established as follows:

$$\Phi_1 = \Phi_\Delta = \frac{L_{\text{star}}}{4\pi\Delta^2}; \quad \Phi_2 = \Phi_{\Delta,\,\text{std}} = \frac{L_{\text{star}}}{4\pi\Delta_{\text{std}}^2};$$

$$m_1 = m; \quad m_2 = m_{\text{ab}} = M;$$

$$\frac{\Phi_1}{\Phi_2} = \frac{\Delta_{\text{std}}^2}{\Delta^2} = 10^{0.4(m_{\text{ab}}-m)}; \quad \frac{\Phi_1}{\Phi_2} = \frac{\Delta_{\text{std}}^2}{\Delta^2} = 10^{0.4(M-m)};$$

$$\log\left(\frac{\Delta_{\text{std}}^2}{\Delta^2}\right) = 0.4(m_{\text{ab}} - m); \quad \log\left(\frac{\Delta_{\text{std}}^2}{\Delta^2}\right) = 0.4(M - m);$$

$$m_{\text{ab}} = m + 5 \cdot \log\left(\frac{\Delta_{\text{std}}}{\Delta}\right);$$

$$m_{\text{ab}} = m + 5 \cdot \log|\Delta_{\text{std}}| - 5 \cdot \log|\Delta_{(\text{pc})}|;$$

$$M = m + 5 \cdot \log\left(\frac{\Delta_{\text{std}}}{\Delta}\right); \quad M = m + 5 \cdot \log|\Delta_{\text{std}}| - 5 \cdot \log|\Delta_{(\text{pc})}|;$$

$$\Delta_{\text{st}} = 10 \text{ ps}; \quad |\Delta_{\text{std}}| = 10; \quad \log|\Delta_{\text{std}}| = 1;$$

$$m_{\text{ab}} = m + 5 - 5 \cdot \log|\Delta_{(\text{pc})}|;$$

$$M = m + 5 - 5 \cdot \log|\Delta_{(\text{pc})}|.$$

For example, for the Sun, where $m_{\text{ab,Sun}}$ is the absolute magnitude of the Sun (the apparent magnitude of the Sun, if it were at a standard distance, $\Delta_{\text{std}} = 10$ pc, from the observer on Earth); m_{Sun} is the apparent magnitude of the Sun, estimated by the observer on Earth; $\Delta_{\text{ES(pc)}}$ is the distance between the Earth and the Sun expressed in parsecs; and $|\Delta_{\text{ES(pc)}}|$ is numerical value of the distance between the Earth and the Sun, expressed in parsecs:

$$M_{\text{Sun}} = m_{\text{Sun}} + 5 - 5 \cdot \log|\Delta_{\text{ES(pc)}}|;$$

$$\Delta_{\text{ES}} = 1 \text{ AU} = \frac{1}{206256} \text{ ps}; \quad |\Delta_{\text{ES(pc)}}| = \frac{1}{206256};$$

$$m_{\text{ab, Sun}} = -26.84 + 5 + 5 \cdot \log(206256) \approx 4.73.$$

$$M_{\text{Sun}} = -26.84 + 5 + 5 \cdot \log(206256) \approx 4.73.$$

Under these conditions, for the standard star, whose brightness is equal to the brightness of the Sun, located at the standard distance $\Delta_{\text{std}} = 10$ ps from an observer on Earth, its apparent magnitude, or absolute magnitude, will be:

$$m_{\text{ab, standard star}} = m_{\text{ab, Sun}} = 4.73;$$

$$M_{\text{standard star}} = M_{\text{Sun}} = 4.73.$$

Thus:

$$\Phi_2 = \Phi_{\text{Galaxy}} = \frac{L_{\text{Galaxy}}}{4\pi\Delta^2} = \frac{10^{11} L_{\text{Sun}}}{4\pi\Delta^2};$$

$$\Phi_1 = \Phi_{\text{standard star}} = \frac{L_{\text{standard star}}}{4\pi\Delta_{\text{std}}^2} = \frac{L_{\text{Sun}}}{4\pi\Delta_{\text{std}}^2}.$$

If m_1 is the apparent magnitude of the standard star, which, by definition, is called the absolute magnitude of the standard star and is identified with the absolute magnitude of the Sun, it results in:

$$m_1 = m_{\text{standard star}} = m_{\text{ab, standard star}} = M_{\text{standard star}}$$
$$= M_{\text{Sun}} = 4.73;$$
$$m_2 = m_{\text{Galaxy}} = 15.7;$$
$$\frac{\Phi_1}{\Phi_2} = \frac{\frac{L_S}{4\pi\Delta_{\text{std}}^2}}{\frac{10^{11} L_S}{4\pi\Delta^2}} = \frac{\Delta^2}{10^{11} \cdot \Delta_{\text{std}}^2};$$
$$m_2 - m_1 = 2.5 \cdot \log\left(\frac{\Phi_1}{\Phi_2}\right);$$
$$15.7 - 4.73 = 2.5 \cdot \log\left(\frac{\Delta^2}{10^{11} \cdot \Delta_{\text{std}}^2}\right);$$
$$\log 10^{\frac{15.7-4.73}{2.5}} = \log\left(\frac{\Delta^2}{10^{11} \cdot \Delta_{\text{std}}^2}\right);$$
$$10^{\frac{15.7-4.73}{2.5}} = \frac{\Delta^2}{10^{11} \cdot \Delta_{\text{std}}^2};$$
$$\Delta = \Delta_{\text{st}}\sqrt{10^{11} \cdot 10^{\frac{15.7-4.73}{2.5}}}; \quad \Delta = \Delta_{\text{st}} \cdot 10^{7.694};$$
$$10^{7.694} = x; \quad \log x = 7.694; \quad x \approx 5 \cdot 10^7;$$
$$\Delta_{\text{st}} = 10 \text{ ps}; \quad \Delta \approx 500 \text{ Mps}.$$

Problem 6
Forms of Kepler's Third Law

The term "visual binary star" means a system consisting of two stars gravitationally connected to each other, which can be resolved (seen separately) by a telescope, if their centers are at an angular distance of at least 1 arcsec.

The expression

$$T^2 = \frac{4\pi^2}{K(m+M)}a^3$$

represents Kepler's third law in its form given by Newton, where all the quantities involved (T – orbital period; a – distance between the two stars; m and M – the masses of the two stars; K – constant of universal attraction) are expressed in SI units.

Express Kepler's third law:

a) in "solar units";

b) depending on the parallax of the double star, p, and the angular distance, β, between the two components.

Given: the Sun's mass, $M_S = 1.989 \cdot 10^{30}$ kg; 1 year = 365 days; the constant of universal attraction, $K = 6.67 \cdot 10^{-11}$ Nm^{2}kg^{-2}; 1 AU $= 1.5 \cdot 10^{10}$ m.

Solution

a) Consider the expression

$$T^2 = \frac{4\pi^2}{K(m+M)}a^3,$$

which represents Kepler's third law in its form given by Newton.

In particular, let us clarify that: $|T_{(\text{years})}|$ is the numerical value of the period, T, when it is expressed in years; $|(m+M)_{(M_\text{S})}|$ is the numerical value of $(m+M)$ when m and M are expressed in solar masses (M_S); and $|a_{(\text{AU})}|$ is the numerical value of the distance between the two stars, a, expressed in AU. Under these conditions, Kepler's third law takes the form:

$$\frac{|T_{(\text{years})}|^2\,(1\text{ year})^2 \cdot |(m+M)_{(M_\text{S})}|\,M_\text{S}}{|a_{(\text{AU})}|^3\,(1\text{ AU})^3} = \frac{4\pi^2}{K};$$

$$\frac{|T_{(\text{years})}|^2 \cdot |(m+M)_{(M_\text{S})}|}{|a_{(\text{AU})}|^3} \cdot \frac{(1\text{ year})^2 \cdot M_\text{S}}{(1\text{ AU})^3} = \frac{4\pi^2}{K};$$

$$\frac{(1\text{ year})^2 \cdot M_\text{S}}{(1\text{ AU})^3} = \frac{(365 \cdot 24 \cdot 3600)^2\text{ s}^2 \cdot 1.989 \cdot 10^{30}\text{ kg}}{(15 \cdot 10^{10})^3\text{ m}^3}$$

$$\approx 5.86 \cdot 10^{11}\,\frac{\text{s}^2\text{kg}}{\text{m}^3};$$

$$\frac{4\pi^2}{K} = \frac{4 \cdot 3.14 \cdot 3.14}{6.67 \cdot 10^{-11}\text{ Nm}^2\text{kg}^{-2}} \approx 5.91 \cdot 10^{11}\,\frac{\text{s}^2\text{kg}}{\text{m}^3};$$

$$\frac{(1\text{ year})^2 \cdot M_\text{S}}{(1\text{AU})^3} \approx \frac{4\pi^2}{K};$$

$$\frac{|T_{(\text{years})}|^2 \cdot |(m+M)_{(M_\text{S})}|}{|a_{(\text{AU})}|^3} = 1;$$

$$[T_{(\text{years})}]^2 = \frac{1}{|(m+M)_{(M_\text{S})}|}|a_{(\text{AU})}|^3,$$

representing Kepler's third law in "solar units".

b) According to Kepler's third law, using the drawing in Figure 6.1, where p is the parallax of the star and β is the angular distance between the two components, we find that:

$$\tan\beta = \frac{a_{\text{AU}}}{d_{\text{AU}}} \approx \beta_{(\text{rad})} = \frac{|a_{(\text{AU})}| \times 1\ \text{AU}}{|d_{(\text{AU})}| \times 1\ \text{AU}}\text{rad} = \frac{|a_{(\text{AU})}|}{|d_{(\text{AU})}|}\text{rad};$$

$$\beta_{(\text{rad})} = \frac{|a_{(\text{AU})}|}{|d_{(\text{AU})}|}\text{rad} = |\beta_{(\text{rad})}| \times 1\ \text{rad} = |\beta_{(\text{rad})}| \times \frac{180 \cdot 60 \cdot 60}{3.14}\ \text{arcsec};$$

$$\beta_{(\text{rad})} = |\beta_{(\text{rad})}\frac{180 \cdot 60 \cdot 60}{3.14}| \times 1\ \text{arcsec};$$

$$\beta_{(\text{rad})} = |\beta_{(\text{arcsec})}| \times 1\ \text{arcsec} = |\beta''| \times 1\ \text{arcsec};$$

$$\beta_{(\text{rad})} = \frac{|a_{(\text{AU})}|}{|d_{(\text{AU})}|}\text{rad} = |\beta''| \times 1\ \text{arcsec};$$

$$|d_{(\text{AU})}| = \frac{|a_{(\text{AU})}|}{|\beta''|}\frac{\text{rad}}{\text{arcsec}};$$

$$\tan p = \frac{1\ \text{AU}}{d_{(\text{AU})}} = p_{(\text{rad})} = \frac{1}{|d_{(\text{AU})}|}\text{rad};$$

$$\tan p = \frac{1\ \text{AU}}{d_{(\text{AU})}} \approx p_{(\text{rad})} = \frac{1\ \text{AU}}{|d_{(\text{AU})}| \times 1\ \text{AU}}\text{rad} = \frac{1}{|d_{(\text{AU})}|}\text{rad};$$

$$p_{(\text{rad})} = \frac{1}{|d_{(\text{AU})}|}\text{rad} = |p_{(\text{rad})}| \times 1\ \text{rad} = |p_{(\text{rad})}| \times \frac{180 \cdot 60 \cdot 60}{3.14}\ \text{arcsec};$$

$$p_{(\text{rad})} = \left|p_{(\text{rad})}\frac{180 \cdot 60 \cdot 60}{3.14}\right| \times 1\ \text{arcsec};$$

$$p_{(\text{rad})} = |p_{(\text{arcsec})}| \times\ 1\ \text{arcsec} = |p''| \times 1\ \text{arcsec};$$

$$p_{(\text{rad})} = \frac{1}{|d_{(\text{AU})}|}\text{rad} = |p''| \times 1\ \text{arcsec};$$

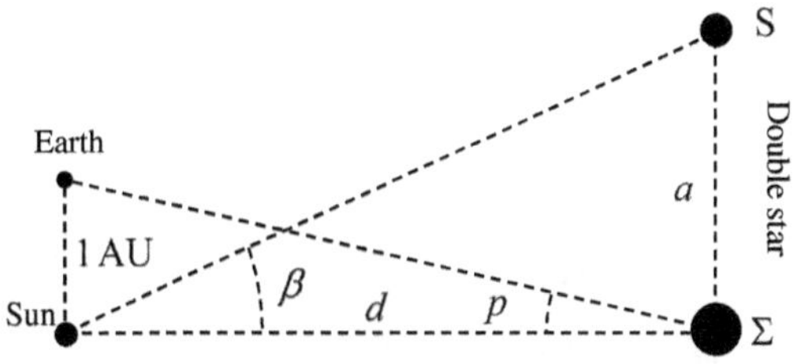

Fig. 6.1

$$|d_{(\mathrm{AU})}| = \frac{1}{|p''|}\frac{\mathrm{rad}}{\mathrm{arcsec}};$$

$$\frac{|a_{(\mathrm{AU})}|}{|\beta''|}\frac{\mathrm{rad}}{\mathrm{arcsec}} = \frac{1}{|p''|}\frac{\mathrm{rad}}{\mathrm{arcsec}};$$

$$\frac{|a_{(\mathrm{AU})}|}{|\beta''|} = \frac{1}{|p''|};$$

$$|a_{(\mathrm{AU})}| = \frac{|\beta''|}{|p''|};$$

$$|a_{(\mathrm{AU})}| = \frac{|\beta''|}{|p''|};$$

$$\frac{|T_{(\mathrm{years})}|^2 \cdot |(m+M)_{(M_\mathrm{S})}|}{\left(\frac{|\beta''|}{|p''|}\right)^3} = 1;$$

$$|(m+M)_{(M_\mathrm{S})}| = \frac{\left(\frac{|\beta''|}{|p''|}\right)^3}{|T_{(\mathrm{years})}|^2};$$

$$|(m+M)_{(M_\mathrm{S})}| = \left(\frac{|\beta''|}{|p''|}\right)^3 \frac{1}{|T_{(\mathrm{years})}|^2};$$

$$|(m+M)_{(M_\mathrm{S})}| \cdot |T_{(\mathrm{years})}|^2 = \left(\frac{|\beta''|}{|p''|}\right)^3.$$

This represents Kepler's third law according to the parallax of the double star:

$$\frac{m+M}{M_\mathrm{S}} \cdot \frac{T^2}{(1\ \mathrm{year})^2} = \left(\frac{|\beta''|}{|p''|}\right)^3 = \left(\frac{\beta_{(\mathrm{rad})}}{p_{(\mathrm{rad})}}\right)^3;$$

$$\left(\frac{|\beta''|}{|p''|}\right)^3 = \left(\frac{\beta_{(\mathrm{rad})}}{p_{(\mathrm{rad})}}\right)^3 = \left(\frac{\frac{a_{(\mathrm{AU})}}{d_{(\mathrm{AU})}}}{\frac{1\ \mathrm{AU}}{d_{(\mathrm{AU})}}}\right)^3$$

$$= \left(\frac{a_{(\mathrm{AU})}}{1\ \mathrm{AU}}\right)^3 = |a_{(\mathrm{AU})}|^3 = \frac{a^3_{(\mathrm{AU})}}{(1\ \mathrm{AU})^3};$$

$$\frac{m+M}{M_\mathrm{S}} \cdot \frac{T^2}{(1\ \mathrm{year})^2} = \left(\frac{|\beta''|}{|p''|}\right)^3 = \left(\frac{\beta_{(\mathrm{rad})}}{p_{(\mathrm{rad})}}\right)^3;$$

$$(m+M)T^2 = \left(\frac{\beta''}{p''}\right)^3 M_\mathrm{S} \times (1\ \text{year})^2 = \frac{a^3}{(1\ \text{AU})^3} M_\mathrm{S} \times (1\ \text{year})^2$$

$$= \frac{M_\mathrm{S} \times (1\ \text{year})^2}{(1\ \text{AU})^3} a^3;$$

$$\frac{M_\mathrm{S} \times (1\ \text{year})^2}{(1\ \text{AU})^3} = 5.91 \cdot 10^{11} \frac{\mathrm{s^2 kg}}{\mathrm{m^3}} = \frac{4\pi^2}{K};$$

$$(m+M)T^2 = \frac{4\pi^2}{K} a^3.$$

The quantities a, m, M, T and K are expressed in SI units.

Problem 7

A Planet with Satellites

The two tables below (Table 7.1 and Table 7.2) present data for several satellites of the same planet, P, in our solar system. They indicate the dependence of the rotation period of each satellite around the planet, $T(s)$, expressed in seconds, on the average distance between the planet and the satellite, $a_{\text{average}}(\text{m})$, expressed in meters. This dependence is expressed as $T = f(a)$. The trajectories of the satellites around planet P are elliptical.

a) *Identify* the table in which the data show the correct dependency, $T = f(a_{\text{average}})$.

b) The drawing in Figure 7.1 locates the positions of several satellites of the planet according to observations made at the same time of night over several days.

Table 7.1

$a_{\text{average}}(\text{m})$	$51 \cdot 10^7$	$64 \cdot 10^7$	$80 \cdot 10^7$	$100 \cdot 10^7$	$126 \cdot 10^7$	$162 \cdot 10^7$
$T(s)$	$2 \cdot 10^5$	$2.85 \cdot 10^5$	$4 \cdot 10^5$	$5.7 \cdot 10^5$	$8 \cdot 10^5$	$11.4 \cdot 10^5$

Table 7.2

$a_{\text{average}}(\text{m})$	$40 \cdot 10^7$	$53.1 \cdot 10^7$	$70.8 \cdot 10^7$	$106 \cdot 10^7$	$141.5 \cdot 10^7$
$T(s)$	$1.6 \cdot 10^5$	$2.55 \cdot 10^5$	$4 \cdot 10^5$	$8 \cdot 10^5$	$12.6 \cdot 10^5$

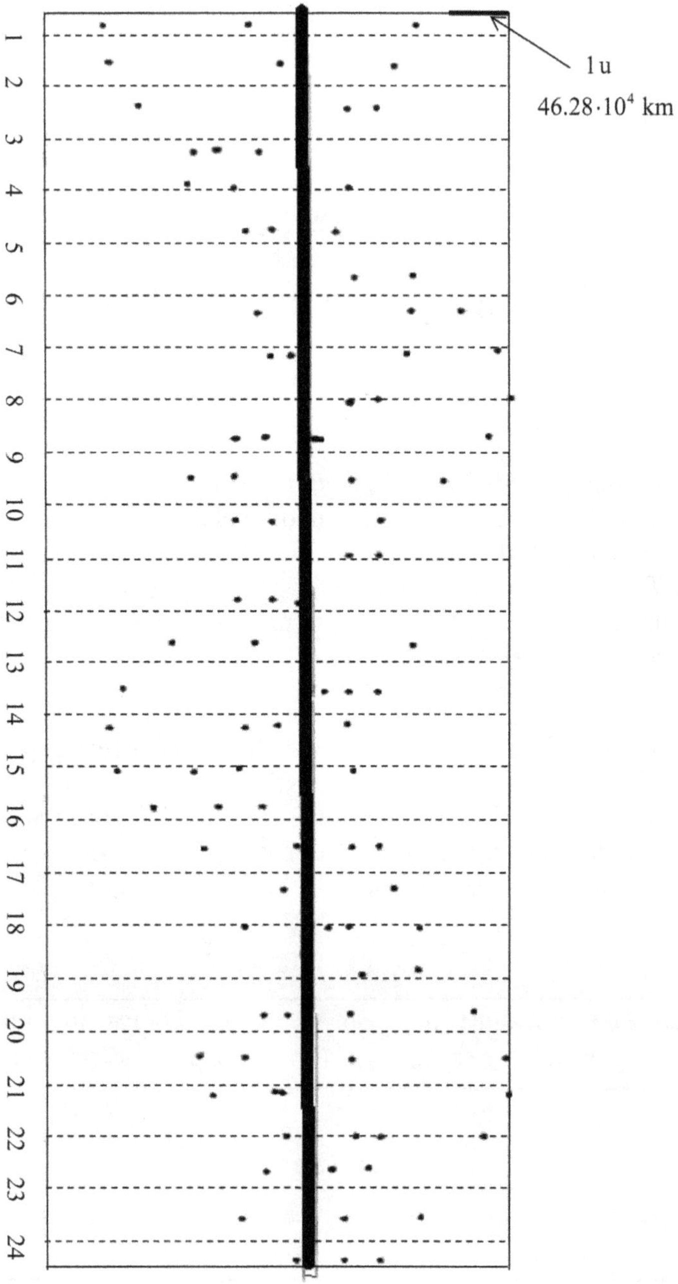
1u
46.28·10⁴ km
1
2
3
4
5
6
7
8
9
10
11
12
13
14
15
16
17
18
19
20
21
22
23
24

Fig. 7.1

Of the observed satellite trajectories, only one is circular, and the Earth observer is right in the plane of the satellite's revolutionary motion.

Identify the position of this satellite in the table that contains the correct observational data.

c) *Identify* the planet P, knowing the value of the gravitational constant, $G = 6.674 \cdot 10^{-11}\,\mathrm{Nm^2kg^{-2}}$, and $\pi = 3.14159265$. It is known that the mass of planet P is much larger than the mass of any of its satellites.

Solution

a) The average distance between each satellite and planet P is the major half-axis of its elliptical orbit, $a_{\text{average}} = a$.

Kepler's third law, written for planet P and any of its satellites,

$$T^2 = \frac{4\pi^2}{G(M+m)} \cdot a^3,$$

results in:

$$M \gg m;$$

$$T^2 = \frac{4\pi^2}{GM} \cdot a^3.$$

From this, by logarithmic calculation, we obtain:

$$2\log T = \log\frac{4\pi^2}{G} - \log M + 3\log a;$$

$$\log T = \frac{1}{2}\cdot\log\frac{4\pi^2}{G} - \frac{1}{2}\cdot\log M + \frac{3}{2}\cdot\log a;$$

$$T = |T|\cdot 1\,\mathrm{s}; \quad G = |G|\cdot 1\,\mathrm{Nm^2kg^{-2}};$$

$$M = |M|\cdot 1\,\mathrm{kg}; \quad a = |a|\cdot 1\,\mathrm{m};$$

$$\log|T|\cdot \mathrm{s} = \frac{1}{2}\cdot\log\frac{4\pi^2}{|G|\cdot \mathrm{Nm^2kg^{-2}}} - \frac{1}{2}\cdot\log|M|\cdot\mathrm{kg} + \frac{3}{2}\cdot\log|a|\cdot\mathrm{m};$$

$$\log|T| + \log(\mathrm{s}) = \frac{1}{2}\cdot\log\frac{4\pi^2}{|G|} + \frac{1}{2}\cdot\log\frac{1}{(\mathrm{Nm^2kg^{-2}})} - \frac{1}{2}\cdot\log|M| - \frac{1}{2}\cdot\log(\mathrm{kg}) + \frac{3}{2}\cdot\log|a| + \frac{3}{2}\cdot\log(\mathrm{m});$$

$$\log|T| = \frac{3}{2}\cdot\log|a| + \frac{1}{2}\cdot\log\frac{4\pi^2}{|G|} - \frac{1}{2}\cdot\log|M| + \frac{1}{2}\cdot\log\frac{1}{(\mathrm{Nm^2kg^{-2}})} - \frac{1}{2}\cdot\log(\mathrm{kg}) + \frac{3}{2}\cdot\log(\mathrm{m}) - \log(\mathrm{s});$$

$$\begin{aligned}
&\frac{1}{2}\cdot\log\frac{1}{(\mathrm{Nm^2kg^{-2}})} - \frac{1}{2}\cdot\log(\mathrm{kg}) + \frac{3}{2}\cdot\log(\mathrm{m}) - \log(\mathrm{s}) \\
&= \frac{1}{2}\left(\log\frac{1}{\mathrm{Nm^2kg^{-2}}} - \log(\mathrm{kg}) + \log(\mathrm{m^3}) - \log(s^2)\right) \\
&= \frac{1}{2}(-\log(\mathrm{Nm^2kg^{-2}}) - \log(\mathrm{kg}) + \log(\mathrm{m^3}) - \log(s^2)) \\
&= \frac{1}{2}\cdot\log\frac{\mathrm{m^3}}{\mathrm{Nm^2kg^{-2}}kg\cdot\mathrm{s^2}} = \frac{1}{2}\cdot\log\frac{\mathrm{m^3}}{\mathrm{kg}\cdot\frac{m}{s^2}\cdot\mathrm{m^2}\cdot\mathrm{kg^{-2}}\cdot\mathrm{kg}\cdot\mathrm{s^2}} \\
&= \frac{1}{2}\cdot\log(1) = 0;
\end{aligned}$$

$$\log|T| = \frac{3}{2}\cdot\log|a| + \frac{1}{2}\cdot\log\frac{4\pi^2}{|G|} - \frac{1}{2}\cdot\log|M|.$$

This proves that

$$\log|T| = f(\log|a|)$$

is a linear dependency of the form

$$y = mx + b,$$

where:

$$y = \log|T|; \quad x = \log|a|;$$

$$m = \frac{3}{2} = 1.5,$$

representing the slope of the line

$$b = \frac{1}{2} \cdot \log\frac{4\pi^2}{|G|} - \frac{1}{2} \cdot \log|M|.$$

In these conditions, to identify the table in the statement of the problem in which the dependency $T = f(a)$ is correct, using the data entered in Tables 7.3 and 7.4, we draw the graphs of the two dependencies, $\log|T| = f(\log|a|)$, where: $|T|$ – the numerical value of the period of each satellite expressed in seconds; $|a|$ – the numerical value of the average distance between the planet and any of its satellites, expressed in meters.

The graphs of the two dependencies are represented in the drawings in Figures 7.2 and 7.3.

Table 7.3

$\|a\|$	$51 \cdot 10^7$	$64 \cdot 10^7$	$80 \cdot 10^7$	$100 \cdot 10^7$	$126 \cdot 10^7$	$162 \cdot 10^7$
$\log\|a\|$	8.70	8.80	8.90	9.00	9.10	9.20
$\|T\|$	$2 \cdot 10^5$	$2.82 \cdot 10^5$	$4 \cdot 10^5$	$5.63 \cdot 10^5$	$8 \cdot 10^5$	$10.2 \cdot 10^5$
$\log\|T\|$	5.30	5.45	5.60	5.75	5.90	6.00

Table 7.4

$\|a\|$	$40 \cdot 10^7$	$53.1 \cdot 10^7$	$70.8 \cdot 10^7$	$106 \cdot 10^7$	$141.5 \cdot 10^7$
$\log\|a\|$	8.6	8.725	8.85	9.025	9.15
$\|T\|$	$1.6 \cdot 10^5$	$2.55 \cdot 10^5$	$4 \cdot 10^5$	$8 \cdot 10^5$	$12.6 \cdot 10^5$
$\log\|T\|$	5.2	5.4	5.6	5.9	6.1

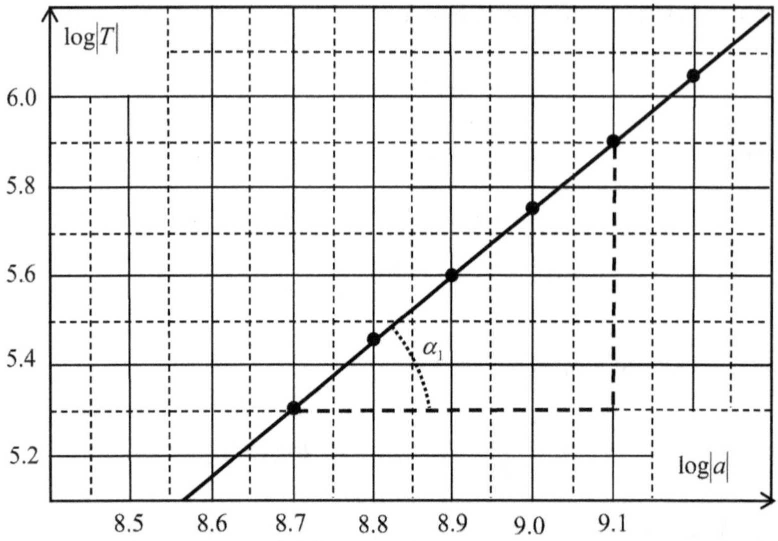
log|T|
6.0
5.8
5.6
5.4
5.2
α_1
log|a|
8.5
8.6
8.7
8.8
8.9
9.0
9.1

Fig. 7.2

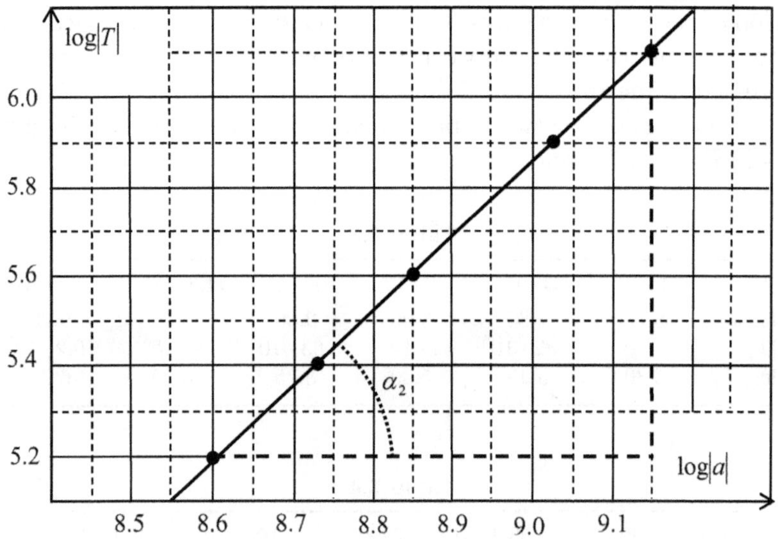
log|T|
6.0
5.8
5.6
5.4
5.2
α_2
log|a|
8.5
8.6
8.7
8.8
8.9
9.0
9.1

Fig. 7.3

From the line graphs of the two dependencies, represented in the drawings in Figures 7.2 and 7.3, calculating the slope of each line, it results that:

$$\tan \alpha_1 = m_1 = \frac{5.9 - 5.3}{9.1 - 8.7} = \frac{0.6}{0.4} = \frac{3}{2} = 1.5;$$

$$\tan \alpha_2 = m_2 = \frac{6.1 - 5.2}{9.15 - 8.6} = \frac{0.9}{0.55} = \frac{90}{55} = 1.63.$$

As $m_1 \equiv m$ and $m_2 \neq m$, it turns out that the data listed in Table 7.1 of the problem statement correctly show the dependence $T = f(a)$.

b) The images of the satellites shown in the drawing in Figure 7.1 constitute the projections of the satellites in the plane of the sky, a plane perpendicular to the direction of view of the observer. This direction is in the plane of the only circular orbit.

The trajectories of the satellites around planet P are ellipses located in different planes. As a result, the movements of their projections in the plane of the sky are not sinusoidal harmonic oscillating movements.

There is only one exception – the satellite whose orbit is a circle. Since the Earth observer is right in the plane of the revolution motion of this satellite, the projection of its motion in the plane of the sky is a sinusoidal harmonic oscillating motion whose amplitude is equal to the circle's radius.

As a result, under these conditions, we can easily recognize what the drawing in Figure 7.4 shows.

Namely, the observed satellite, whose trajectory is circular, is the satellite whose data are shown in the rightmost column in Table 7.1:

$$T = 13.2 \text{ days} = 13.2 \cdot 24 \cdot 3600\,\text{s} = 1140480\,\text{s} \approx 11.4 \cdot 10^5\ \text{s};$$

$$a = 3.5\,\text{u} \cdot \frac{46.28 \cdot 10^7\ \text{m}}{1\,\text{u}} = 3.5 \cdot 1\,\text{cm} \cdot \frac{46.28 \cdot 10^7}{1\,\text{cm}} = 161.98 \cdot 10^7\ \text{m}.$$

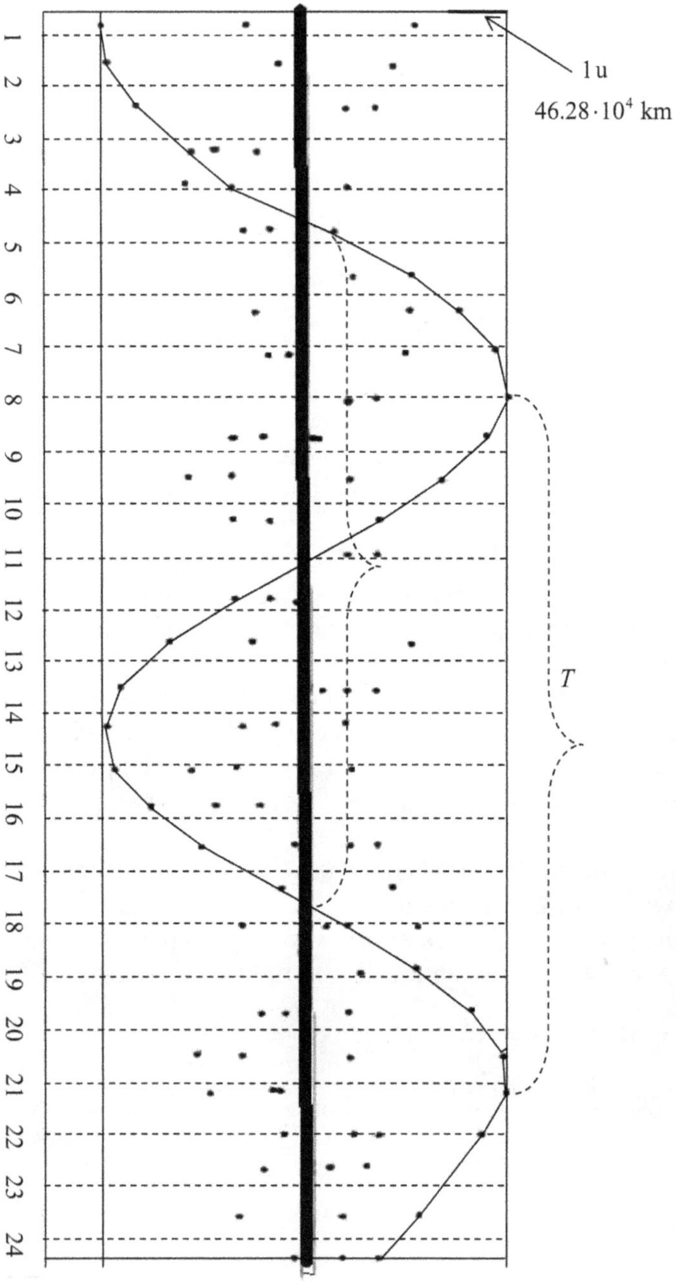
1 u
46.28·10⁴ km
T
1
2
3
4
5
6
7
8
9
10
11
12
13
14
15
16
17
18
19
20
21
22
23
24

Fig. 7.4

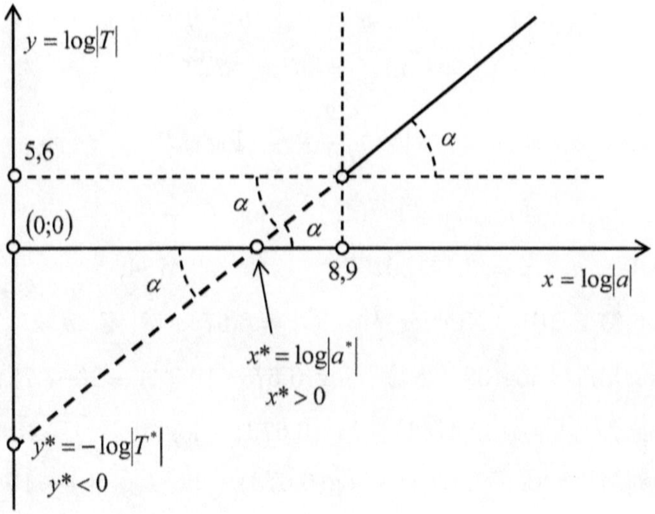

Fig. 7.5

c) With the notation and numerical values inscribed on the system of axes represented in the drawing in Figure 7.5, it results that:

$$\tan\alpha = \frac{5.6}{8.9 - x^*} = m = 1.5;$$

$$\frac{5.6}{8.9 - x^*} = 1.5; \quad x^* = 5.166 = \log(a^*);$$

$$\tan\alpha = \frac{5.6 + |y^*|}{8.9} = m = 1.5;$$

$$\frac{5.6 + |y^*|}{8.9} = 1.5; \quad |y^*| = 7.75; \quad y^* = -7.75 = -\log|T^*|;$$

$$\tan\alpha = \frac{|y^*|}{x^*} = m = 1.5;$$

$$\frac{|y^*|}{x^*} = 1.5; \quad x^* = 5.166;$$

$$y = mx + b;$$

$$y = 0; \quad x = x^*; \quad 0 = mx^* + b;$$

$$b = -mx^* = -(1.5) \cdot (5.166) = -7.75;$$

$$x = 0; \quad y = y^*; \quad b = y^*;$$
$$b = -mx^* = y^* = -7.75;$$
$$b = \frac{1}{2} \cdot \log \frac{4\pi^2}{|G|} - \frac{1}{2} \cdot \log |M|;$$
$$\log |M| = \log(4\pi^2) - \log |G| - 2b;$$
$$\pi = 3.14159265; \quad 4\pi^2 = 39.4784;$$
$$G = 6.674 \cdot 10^{-11}\,\mathrm{Nm^2kg^{-2}}; \quad |G| = 6.674 \cdot 10^{11}; \quad b = -7.75;$$
$$\log |M| = \log(39.4784) - \log(6.673 \cdot 10^{-11}) - 2(-7.75);$$
$$\log |M| = \log(39.4784) - \log(6.673) - \log(10^{-11}) + 15.5;$$
$$\log |M| = \log(39.4784) - \log(6.673) + 11 \cdot \log(10) + 15.5;$$
$$\log |M| = 1.5963 - 0.8243 + 11 + 15.5 = 27.272;$$
$$\log |M| = 27.272 = \log(10^{27.27}) = \log(10^{27} \cdot 10^{0.27});$$
$$|M| = 10^{0.27} \cdot 10^{27};$$
$$10^{0.27} = x; \quad \log x = 0.27; \quad x = 1.9;$$
$$|M| = 1.9 \cdot 10^{27};$$
$$M = 1.9 \cdot 10^{27}\,\mathrm{kg}.$$

We recognize this value as:

$$M_{\mathrm{Jupiter}} = 1.9 \cdot 10^{27}\,\mathrm{kg}.$$

Problem 8

Changing the Orbit of a Satellite

A satellite S evolves in an elliptical orbit, with the Earth in one of its foci. *Determine* the necessary corrections to the speed of the satellite at the apogee or at the perigee of the elliptical orbit so that the subsequent evolutions of the satellite can occur in circular orbits around the Earth. *Compare* the two corrections.

The following are known: M – the mass of the Earth; K – the gravitational attraction constant; a – the large semi-axis of the elliptical orbit; e – the eccentricity of the elliptical orbit.

Solution

At **apogee**, the satellite's speed must be increased, and at perigee, the speed of the satellite must be reduced, the orientation of the speed vector remaining unchanged each time.

1) The gravitational transfer of the special satellite S from the elliptical orbit when the satellite is at its **apogee**, at a distance $r_{\text{max}} = a(1+e)$ from the center of the Earth, is in a circular orbit with radius $r_{\text{ext}} = r_{\text{max}}$, as shown in the drawing in Figure 8.1. The speed of the satellite $\vec{v}_{\text{ext}} // \vec{v}_{\text{min}}$, having the modulus $v_{\text{ext}} > \text{v}_{\text{apg}}$, is

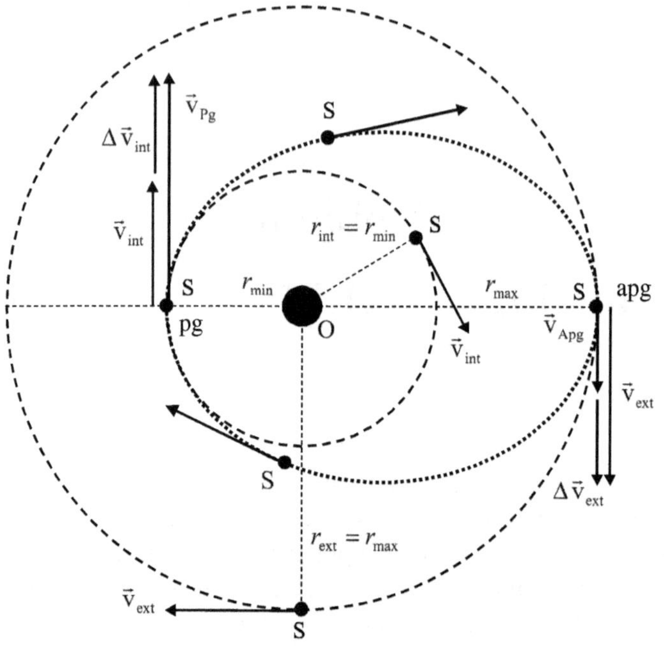

Fig. 8.1

necessary for the evolution in a circle with radius $r_{\max}$, meaning:

$$\mathrm{v_{ext}} = \sqrt{K\frac{M}{r_{\text{ext}}}} = \sqrt{K\frac{M}{r_{\max}}} = \sqrt{K\frac{M}{a(1+e)}} > \mathrm{v_{apg}},$$

where $\mathrm{v_{apg}}$ is the speed of the satellite at the **apogee** of the elliptical orbit;

$$\mathrm{v_{apg}} = \mathrm{v_{min}} = \sqrt{KM\frac{1-e}{a(1+e)}}.$$

Thus, the speed correction required for the transfer of the satellite from the elliptical orbit to its **apogee** in the outer circular orbit is:

$$\Delta\vec{\mathrm{v}}_{\text{ext}} = \vec{\mathrm{v}}_{\text{ext}} - \vec{\mathrm{v}}_{\text{apg}};$$

$$\Delta\mathrm{v_{ext}} = \sqrt{\mathrm{v_{ext}^2} + \mathrm{v_{apg}^2} - 2\mathrm{v_{ext}}\mathrm{v_{apg}}\cos 0^\circ};$$

$$\Delta\mathrm{v_{ext}} = \mathrm{v_{ext}} - \mathrm{v_{apg}} = \mathrm{v_{ext}} - \mathrm{v_{min}};$$

$$\Delta v_{ext} = \sqrt{K\frac{M}{a(1+e)}} - \sqrt{KM\frac{1-e}{a(1+e)}};$$

$$\Delta v_{ext} = \sqrt{K\frac{M}{a(1+e)}}(1-\sqrt{1-e}).$$

2) The gravitational transfer of the special satellite S from the elliptical orbit when the satellite is in its **perigee**, at a distance $r_{min} = a(1-e)$ from the center of the Earth, is in a circular orbit with radius $r_{int} = r_{min}$, as shown in the drawing in Figure 8.1. The speed of the satellite $\vec{v}_{int}//\vec{v}_{max}$, having the modulus $v_{int} < v_{pg}$, is necessary for the evolution in a circle with radius r_{min}. Hence:

$$v_{int} = \sqrt{K\frac{M}{r_{int}}} = \sqrt{K\frac{M}{r_{min}}} = \sqrt{K\frac{M}{a(1-e)}} < v_{pg},$$

where v_{pg} is the speed of the satellite at the **perigee** of the elliptical orbit;

$$v_{pg} = v_{max} = \sqrt{KM\frac{1+e}{a(1-e)}}.$$

Thus, the speed correction required for the transfer of the satellite from the elliptical orbit to its **apogee** in the inner circular orbit is:

$$\Delta\vec{v}_{int} = \vec{v}_{pg} - \vec{v}_{int};$$

$$\Delta v_{int} = \sqrt{v_{pg}^2 + v_{int}^2 - 2v_{pg}v_{int}\cos 0°};$$

$$\Delta v_{int} = v_{pg} - v_{int} = v_{max} - v_{int};$$

$$\Delta v_{int} = \sqrt{KM\frac{1+e}{a(1-e)}} - \sqrt{K\frac{M}{a(1-e)}};$$

$$\Delta v_{int} = \sqrt{K\frac{M}{a(1-e)}}(\sqrt{1+e}-1).$$

3) Assuming that

$$\Delta v_{\text{int}} > \Delta v_{\text{ext}},$$

the result is:

$$\sqrt{K\frac{M}{a(1-e)}}(\sqrt{1+e}-1) > \sqrt{K\frac{M}{a(1+e)}}(1-\sqrt{1-e});$$

$$\sqrt{\frac{1}{(1-e)}}(\sqrt{1+e}-1) > \sqrt{\frac{1}{(1+e)}}(1-\sqrt{1-e});$$

$$\frac{\sqrt{1+e}-1}{\sqrt{1-e}} > \frac{1-\sqrt{1-e}}{\sqrt{1+e}};$$

$$1+e-\sqrt{1+e} > \sqrt{1-e}-(1-e);$$

$$1+e-\sqrt{1+e} > \sqrt{1-e}-1+e;$$

$$1-\sqrt{1+e} > \sqrt{1-e}-1;$$

$$2 > \sqrt{1+e}+\sqrt{1-e};$$

$$4 > 1+e+2\sqrt{1-e^2}+1-e;$$

$$4 > 2+2\sqrt{1-e^2};$$

$$2 > 2\sqrt{1-e^2};$$

$$1 > \sqrt{1-e^2};$$

$$e < 1; \quad 1-e^2 < 1; \quad \sqrt{1-e^2} < 1.$$

Conclusion:

$$\Delta v_{\text{int}} > \Delta v_{\text{ext}}.$$

Problem 9

The Third Cosmic Speed

Determine the approximate minimum value of the escape velocity that must be imprinted on a body launched from the Earth, in relation to the Earth, so that it leaves the Solar System forever (the third cosmic velocity).

Given: $V_0 \approx 30\,\frac{\text{km}}{\text{s}}$, the speed of the Earth in its circular orbit around the Sun; $\text{v}_0 \approx 7.9\,\frac{\text{km}}{\text{s}}$, the speed of a terrestrial satellite orbiting the Earth in a very low circular orbit (the first cosmic speed).

It is known that: $\frac{M_{\text{T}}}{R_{\text{T}}} \ll \frac{M_{\text{S}}}{R_{\text{TS}}}$. The variation of the body's kinetic energy in relation to the Sun is neglected during the body's evolution from the surface of the Earth to the limit of the Earth's gravitational attraction.

Solution

1) Let $\vec{\text{v}}_{\text{B}}$ represent the speed of body B at the time of its launch from Earth, in relation to the Sun, so that the body reaches the limit of the Sun's gravitational attraction, and so that it is at rest in relation to the Sun. Using the details in Figure 9.1, following the

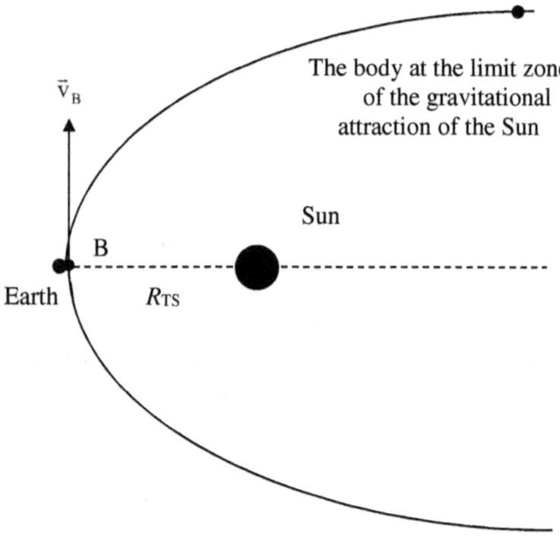

Fig. 9.1

law of conservation of mechanical energy, it follows that:

$$R_T \ll R_{TS};$$

$$\frac{m v_B^2}{2} - K\frac{mM_T}{R_T} - K\frac{mM_S}{R_{TS}} = 0;$$

$$\frac{M_T}{R_T} \ll \frac{M_S}{R_{TS}};$$

$$\frac{m v_B^2}{2} - K\frac{mM_S}{R_{TS}} = 0;$$

$$v_B = \sqrt{2}\cdot\sqrt{K\frac{M_S}{R_{TS}}};$$

$$v = \sqrt{2}\sqrt{K\frac{M_S}{R_{TS}}};$$

$$\sqrt{K\frac{M_S}{R_{TS}}} = v_{TS} = V_{\text{orbital}} = V_0,$$

representing the orbital speed of the Earth on the circular trajectory around the Sun;

$$v_p = \sqrt{2}\,V_0,$$

representing the second cosmic speed in relation to the Sun (parabolic speed);

$$R_{\mathrm{TS}} \approx 1.5 \cdot 10^8\,\mathrm{km}; \quad T_{\mathrm{TS}} = 1\,\mathrm{year};$$

$$V_0 = \frac{2\pi R_{\mathrm{TS}}}{T_{\mathrm{TS}}} \approx 30\,\frac{\mathrm{km}}{\mathrm{s}}; \quad \mathrm{v_p} \approx 42.42\,\frac{\mathrm{km}}{\mathrm{s}}.$$

Conclusion: The body launched from the Earth reaches the limit of the Sun's gravitational attraction on a parabolic trajectory in relation to the Sun, with the Sun in its focus.

2) Let $\vec{\mathrm{v}}$ represent the speed of body C at the time of its launch from Earth, in relation to the Earth, so that the body reaches the limit of the Sun's gravitational attraction, and so that it is at rest in relation to the Sun.

This escape velocity, $\vec{\mathrm{v}}$, will have the minimum value when its orientation is the same as the orientation of the vector representing the speed of the Earth in relation to the Sun, $\vec{V}_0$.

As a result, at the limit of the Earth's gravitational pull, before reaching the Sun's gravitational pull, the speed of the body relative to the Earth, $\vec{\mathrm{v}}_\infty$, will not be null, $\mathrm{v}_\infty \neq 0$. Under these conditions, in relation to the Earth, following the law of conservation of mechanical energy, it results that:

$$\frac{m\mathrm{v}^2}{2} - K\frac{mM_{\mathrm{T}}}{R_{\mathrm{T}}} = \frac{m\mathrm{v}_\infty^2}{2};$$

$$\mathrm{v}_\infty^2 = \mathrm{v}^2 - 2K\frac{M_{\mathrm{T}}}{R_{\mathrm{T}}};$$

$$\sqrt{K\frac{M_{\mathrm{T}}}{R_{\mathrm{T}}}} = \mathrm{v}_{\mathrm{orbital}} = \mathrm{v}_0 = 7.9\,\frac{\mathrm{km}}{\mathrm{s}},$$

representing the orbital speed of the body if it were to evolve in a circle around the Earth at a very low altitude (the first cosmic speed);

$$\mathrm{v}_\infty^2 = \mathrm{v}^2 - 2\mathrm{v}_0^2;$$

$$\mathrm{v}_\infty = \sqrt{\mathrm{v}^2 - 2\mathrm{v}_0^2},$$

representing the speed of the body in relation to the Earth at the limit of the Earth's gravitational attraction.

Conclusion: The body launched from the Earth reaches the limit of the zone of terrestrial gravitational attraction on a hyperbolic trajectory in relation to the Earth, with the center of the Earth in its focus.

3) Let $\vec{v}_{CS\infty}$ be the speed of body C in relation to the Sun at the limit of the gravitational attraction of the Earth:

$$\vec{v}_{CS\infty} = \vec{v}_\infty + \vec{v}_{TS} = \vec{v}_\infty + \vec{V}_0,$$

where the orientations of the vectors $\vec{v}_\infty$ and $\vec{V}_0$ must be identical;

$$v_{CS\infty} = v_\infty + V_0 = \sqrt{v^2 - 2v_0^2} + V_0.$$

4) The variation of the kinetic energy of the body in relation to the Sun during the evolution of the body from the Earth's surface to the limit of the Earth's gravitational attraction is negligible, according to the law of conservation of mechanical energy, resulting in:

$$v_{CS\infty} \approx v_p = \sqrt{2}\, V_0;$$

$$\sqrt{2}\, V_0 = \sqrt{v^2 - 2v_0^2} + V_0;$$

$$v = \sqrt{(\sqrt{2} - 1)^2 V_0^2 + 2v_0^2} \approx 16.7\,\frac{\text{km}}{\text{s}},$$

representing the speed of body C at the time of its launch from Earth, in relation to the Earth, so that the body reaches the limit of the gravitational attraction of the Sun, and so that it is at rest in relation to the Sun (the third cosmic speed).

Problem 10

A Terrestrial Satellite

A small satellite (a luminous sphere) that orbits the Earth in an equatorial circular orbit at altitude $h = 100\,\text{km}$ is photographed with a camera on the ground, the lens of which is a converging lens with a focal length $f = 50\,\text{cm}$. The photograph is taken when the satellite aligns with the camera and the center of the Earth.

Determine the length of the image on the photo after the development of the photographic plate, placed in the focal plane of the lens, as shown in the drawing in Figure 10.1, if the exposure time of the camera was $\Delta t = 1\,\text{s}$, knowing that the radius of the Earth is $R = 6400\,\text{km}$.

Given: The mass of the Earth, $M = 6 \cdot 10^{24}\,\text{kg}$; the constant of gravitational attraction, $K = 6.67 \cdot 10^{-11}\,\text{m}^3 s^{-2}\text{kg}^{-1}$. During the photographic exposure, the proper rotation of the Earth is neglected.

Solution

Sector AB, in the circular orbit of the satellite, represented in the drawing in Figure 10.2, which is traveled by the satellite during the

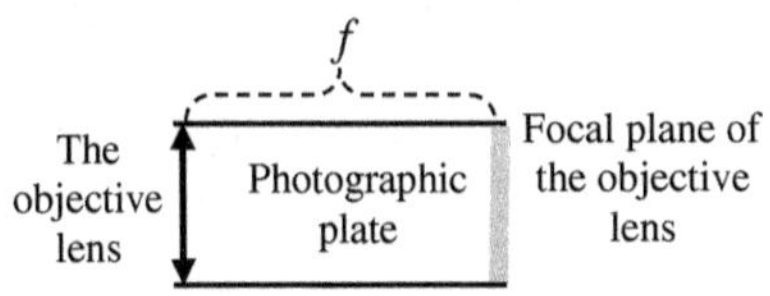

Fig. 10.1

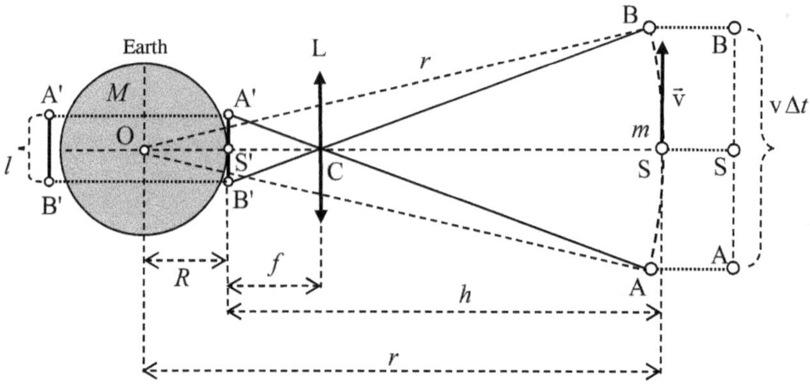

Fig. 10.2

time Δt and is equal to the exposure time of the camera, has the length

$$AB = \mathrm{v}_{\mathrm{rel}} \cdot \Delta t,$$

where v is the speed of the satellite in its circular orbit relative to the center of the Earth;

$$K\frac{mM}{r^2} = \frac{m\mathrm{v}^2}{r};$$

$$\mathrm{v} = \sqrt{K\frac{M}{r}} = \sqrt{K\frac{M}{R+h}};$$

$$\mathrm{v} = \sqrt{6.67 \cdot 10^{-11} \cdot \frac{\mathrm{m}^3}{s^2\mathrm{kg}} \cdot \frac{6 \cdot 10^{24}\,\mathrm{kg}}{65 \cdot 10^5\,\mathrm{m}}} = 7846.6\,\frac{\mathrm{m}}{\mathrm{s}}.$$

Because the distance between the satellite and the camera lens is large, the image will be formed in the focal plane of the camera lens.

Since the circle sector AB is short and has a large radius ($r = R + h$), it can be assimilated with the line segment AB, so that:

$$\Delta(\mathrm{ACS}) \sim \Delta(\mathrm{A'CS'});$$

$$\frac{\mathrm{A'S'}}{\mathrm{AS}} = \frac{\mathrm{CS'}}{\mathrm{CS}}; \quad \frac{\frac{l}{2}}{\frac{\mathrm{v}\cdot\Delta t}{2}} = \frac{f}{h-f},$$

where l is the length of the image on the photo;

$$l = \frac{f}{h-f} \cdot \mathrm{v} \cdot \Delta t;$$

$$l = \frac{f}{h-f} \cdot \sqrt{K\frac{M}{R+h}} \cdot \Delta t;$$

$$f = 50\,\mathrm{cm}; \quad h = 100\,\mathrm{km}; \quad K = 6.67 \cdot 10^{-11}\,\mathrm{m^3 s^{-2} kg^{-1}};$$

$$M = 6 \cdot 10^{24}\,\mathrm{kg};$$

$$R = 6400\,\mathrm{km}; \quad \Delta t = 1\,\mathrm{s}; \quad h = 100\,\mathrm{km};$$

$$f \ll h; \quad l = \frac{f}{h} \cdot \sqrt{K\frac{M}{R+h}} \cdot \Delta t;$$

$$l = \frac{50}{10^7} \cdot \sqrt{6.67 \cdot 10^{-11} \cdot \frac{\mathrm{m}^3}{s^2\mathrm{kg}} \cdot \frac{6 \cdot 10^{24}\,\mathrm{kg}}{65 \cdot 10^5\,\mathrm{m}}} \cdot 1\,\mathrm{s};$$

$$l = \frac{50}{10^7} \cdot \sqrt{6.67 \cdot 10^{-11} \cdot \frac{6 \cdot 10^{24}}{65 \cdot 10^5}}\,\mathrm{m};$$

$$l = 3.9 \cdot 10^{-2}\,\mathrm{m} = 3.9\,\mathrm{cm}.$$

Problem 11

The Expansion of a Nebula

The planetary nebula ESO 465-67 (located in the constellation Sagittarius) is expanding, represented at two different times in the images in Figure 11.1. With respect to the Earth, it is at a distance of $\Delta = 6.100\,\text{pc}$.

Knowing the field of view (FOV) of the converging lens of the optical instrument with which the two images were created, $\text{FOV} = (202'' \times 202'')$, *determine*:

a) the speed, v, at which the ESO 465-67 nebula expands in space, considering that this expansion is uniform;
b) the diameters D_1 and D_2, corresponding to the two moments.

Given: $1\,\text{parsec} = 1\,\text{pc} = 3 \cdot 10^{13}\,\text{km}$; $\text{tg}(101'') = 0.000489661$.

The field of view of an optical instrument is the angular extension, $\angle\text{FOV} = (\alpha; \alpha)$, of the image formed by that optical instrument in its focal plane, as shown in the drawing in Figure 11.2.

Solution

a) The field of view of an optical instrument can be measured horizontally, vertically, or diagonally.

For a converging lens, which projects a linear image into its focal plane as shown in the drawing in Figure 11.3, the angle α, representing the angular extension of the image in its focal plane, results from

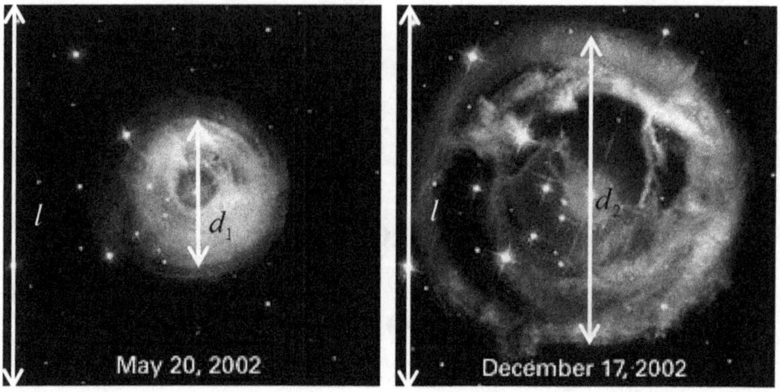

Fig. 11.1

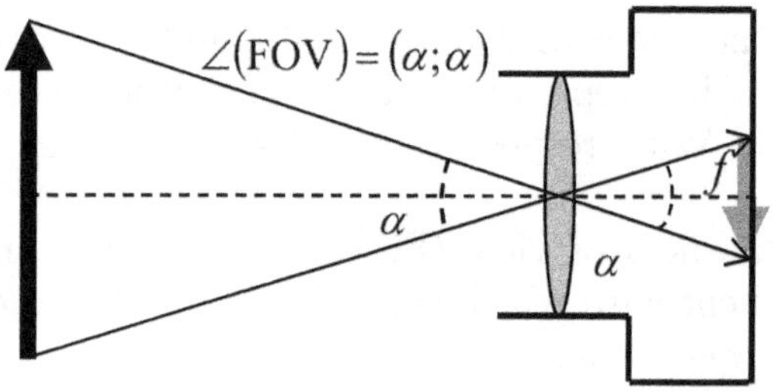

Fig. 11.2

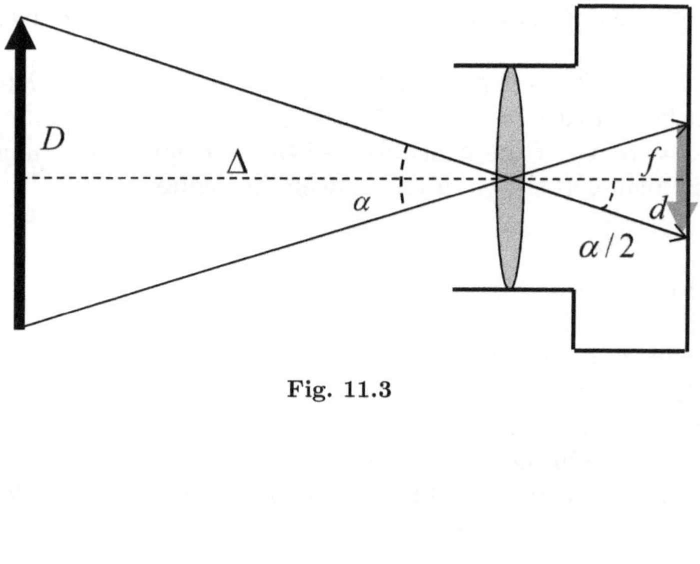

Fig. 11.3

the expression

$$\tan\frac{\alpha}{2} = \frac{d/2}{f} = \frac{d}{2f},$$

where d is the image height and f is the focal length of the camera lens;

$$\tan\frac{\alpha}{2} = \frac{D/2}{\Delta} = \frac{D}{2\Delta},$$

where D is the height of the object and Δ is the distance from the object to the lens.

Measuring with a ruler the lengths (d_1, d_2, l) in the images in Figure 11.1, it results that:

$$\tan\frac{\alpha_1}{2} = \frac{d_1/2}{f} = \frac{d_1}{2f}; \quad \tan\frac{\alpha_1}{2} = \frac{D_1/2}{\Delta} = \frac{D_1}{2\Delta}; \quad \frac{D_1}{2\Delta} = \frac{d_1}{2f};$$

$$D_1 = 2\Delta\frac{d_1}{2f},$$

representing the diameter of the nebula at the time t_1;

$$\tan\frac{\alpha_2}{2} = \frac{d_2/2}{f} = \frac{d_2}{2f}; \quad \tan\frac{\alpha_2}{2} = \frac{D_2/2}{\Delta} = \frac{D_2}{2\Delta}; \quad \frac{D_2}{2\Delta} = \frac{d_2}{2f};$$

$$D_2 = 2\Delta\frac{d_2}{2f},$$

representing the diameter of the nebula at the time t_2;

$$D_2 - D_1 = 2\Delta\frac{d_2}{2f} - 2\Delta\frac{d_1}{2f} = \frac{2\Delta}{2f}(d_2 - d_1),$$

with the maximum angular aperture of the two images being the same;

$$\tan\frac{\alpha_{\max}}{2} = \frac{l/2}{f} = \frac{l}{2f}; \quad 2f = \frac{l}{\tan\frac{\alpha_{\max}}{2}}.$$

So, we get:

$$D_2 - D_1 = \frac{2\Delta}{2f}(d_2 - d_1) = \frac{2\Delta}{\frac{l}{\tan\frac{\alpha_{\max}}{2}}}(d_2 - d_1);$$

$$D_2 - D_1 = 2\Delta\frac{d_2 - d_1}{l}\tan\frac{\alpha_{\max}}{2};$$

$$\alpha_{\max} = 202''; \quad \tan\frac{\alpha_{\max}}{2} = \tan(101'') = 0.000489661;$$

$$\Delta = 6.100\,\text{pc}; \quad 1\,\text{pc} = 3 \cdot 10^{13}\,\text{km};$$

$$t_1 = 20 \text{ May } 2002; \quad t_2 = 17 \text{ December } 2002;$$

$$t_2 - t_1 = 212 \text{ days}; \quad l = 50\,\text{mm}; \quad d_1 = 20\,\text{mm}; \quad d_2 = 40\,\text{mm};$$

$$\text{v} = \frac{D_2 - D_1}{t_2 - t_1};$$

$$\text{v} = \frac{2\Delta}{t_2 - t_1}\frac{d_2 - d_2}{l}\tan\frac{\alpha_{\max}}{2};$$

$$\text{v} = \frac{2 \cdot 6.1 \cdot 3 \cdot 10^{13}\,\text{km}}{212 \cdot 24 \cdot 3600\,\text{s}} \cdot \frac{20\,\text{mm}}{50\,\text{mm}} \cdot 0.000489661;$$

$$\text{v} = \frac{6.1 \cdot 10^{11}}{212 \cdot 6 \cdot 12 \cdot 5} \cdot 0.000489661\,\frac{\text{km}}{\text{s}}; \quad \text{v} = 3913.6951\,\frac{\text{km}}{\text{s}};$$

b)

$$D_1 = 2\Delta\frac{d_1}{2f}; \quad 2f = \frac{l}{\tan\frac{\alpha_{\max}}{2}}; \quad D_1 = 2\Delta\frac{d_1}{l} \cdot \tan\frac{\alpha_{\max}}{2};$$

$$\tan\frac{\alpha_{\max}}{2} = \tan(101'') = 0.000489661;$$

$$\Delta = 6.1\,\text{pc}; \quad 1pc = 3 \cdot 10^{13}\,\text{km};$$

$$l = 50\,\text{mm}; \quad d_1 = 20\,\text{mm};$$

$$D_1 = 0.00238\,\text{pc} = 69 \cdot 10^9\,\text{km};$$

$$D_1 = 2\Delta\frac{d_2}{2f}; \quad 2f = \frac{l}{\tan\frac{\alpha_{\max}}{2}}; \quad D_2 = 2\Delta\frac{d_2}{l} \cdot \tan\frac{\alpha_{\max}}{2};$$

$$\tan\frac{\alpha_{\text{max}}}{2} = \tan(101'') = 0.000489661;$$

$$\Delta = 6,1\,\text{pc}; \quad 1\,\text{pc} = 3 \cdot 10^{13}\,\text{km};$$

$$l = 50\,\text{mm}; \quad d_2 = 40\,\text{mm};$$

$$D_2 = 0.00489\,\text{pc} = 138 \cdot 10^{9}\,\text{km}.$$

Problem 12

Transit of Airplanes and Helicopters in Front of the Moon

I. Photographic images 1.1 and 1.2 in Figure 12.1 show two identical airplanes, A_1 and A_2. They are shown in horizontal, rectilinear, and uniform flights, in opposite directions, with identical speeds and at different times during the transit of each aircraft in front of the Moon's disk. In each case, the direction of the Moon's motion is the same (as shown in the two images). When capturing the images, it was perpendicular to the image plane.

Determine the duration of the full transit of each airplane, A_1 and A_2, in front of the Moon's disk, in each of the two variants, t_1 and t_2, represented in photographic images 1.1 and 1.2, respectively, in Figure 12.1. *Compare* the two durations, t_1 and t_2.

The duration of each transit is considered from the beginning of the overlap of the aircraft image with the Moon image until the moment of complete separation of the aircraft image from the Moon image.

Consider:

a) During t_1, the transit of the airplane A_1, the Moon is in uniform circular motion, and the airplane A_1 is in rectilinear and uniform

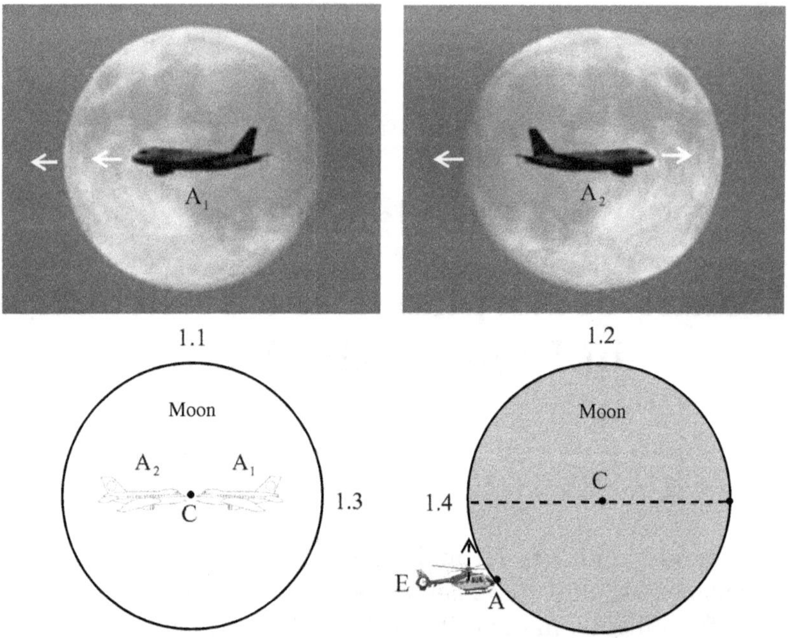

Fig. 12.1

motion, with the variants: **1)** the airplane A_1 comes behind, and its image gradually overlaps with the image of the Moon; **2)** the Moon comes last, and its image gradually enters behind the image of the airplane A_1;

b) During t_1, the transit of the airplane A_1, the Moon is in approximately rectilinear and uniform motion, and the airplane A_1 is in rectilinear and uniform motion, with the variants: **1)** the airplane A_1 comes after, and its image gradually overlaps with the image of the Moon; **2)** the Moon comes last, and its image gradually enters behind the image of the airplane A_1;

c) During t_2, the transit of the airplane A_2, the Moon is in uniform circular motion, and the airplane A_2, in rectilinear and uniform motion, meets the Moon so that the image of the airplane A_2 gradually overlaps with the image of the Moon;

d) During t_2, the transit of the airplane A_2, the Moon is in approximately rectilinear and uniform motion, and the airplane A_2, in rectilinear and uniform motion, meets the Moon so that the image of the airplane A_2 gradually overlaps with the image of the Moon.

Given: the minimum distance between each airplane and the observer, $\Delta_{A,min}$; the distance between the observer and the Moon, $\Delta_L \gg \Delta_{A,min}$; the radius of the Moon, R_L; the speed of the Moon's rotation around the Earth, v_L; the speeds of the two airplane, A_1 and A_2, in relation to the observer, $v_{A,1} = v_{A,2} = v_A < v_L$; the length of each airplane, L_A.

During the entire duration of each transit, the rotation of the Earth around the Sun and around its own axis are neglected.

II. Let us now consider that the flights of the two identical airplanes, A_1 and A_2, occur at different speeds, $v_{A,1} \neq v_{A,2}$, under the conditions specified above, and that they are simultaneous, on rectilinear trajectories, parallel, very close, and in opposite directions, so that the observer on Earth appreciates that airplane A_1 and A_2 meet at some point during their transit across the Moon's disk. The projection of this point on the Moon's disk is the center C of the Moon's disk, as shown in image 1.3 in Figure 12.1.

Determine the time required for each aircraft to complete the transit across the Moon's disk, τ_1 and τ_2, considering the time at which the aircraft meet, until each aircraft is entirely out of the Moon's disk, depending on the speed of each aircraft, $v_{A,1}$ and v_{A2}, respectively.

Compare the two durations, τ_1 and τ_2. In this particular case: $v_{A,1} = v_{A,2} = v_A$.

Consider:

a) During τ_1, the transit of the airplane A_1, the Moon is in uniform circular motion, and the airplane A_1, in rectilinear and uniform motion, came last, gradually overlapping with the image of the Moon, flying over it and overtaking it.

b) During τ_1, the transit of the airplane A_1, the Moon is in approximately rectilinear and uniform motion, and the airplane A_1, in rectilinear and uniform motion, came after, gradually overlapping with the image of the Moon, flying over it, and overtaking it.

c) During τ_2, the transit of the Airplanes A_2, the Moon is in uniform circular motion, and the Airplanes A_2, in rectilinear and uniform motion, comes in front of the Moon, gradually overlapping the image of the Moon, flying over it, and overtaking it.

d) During τ_2, the transit of the Airplanes A_2, the Moon is in approximately rectilinear and uniform motion, and the Airplanes A_2, in rectilinear and uniform motion, comes in front of the Moon, gradually overlaps with the image of the Moon, flies over it, and overtakes it.

III. Image 1.4 in Figure 12.1 shows a helicopter, E, with a straight and uniform flight length L_E, ascending in a vertical direction in front of the Earth observer at the beginning of the transit in front of the Moon's disk at point A, whose position coordinates in relation to the center C of the Moon's disk are $(x_0; y_0)$.

Determine:

a) The speed of the vertical ascent of the helicopter, v_E, so that the complete exit of the helicopter from its transit in front of the Moon's disk occurs at the rightmost point of the horizontal diameter of the Moon's disk.

b) The surface area of the Moon's disk traversed by the helicopter during the entire transit ΔS.

It will be considered that the angles from which the observer sees different stages of the whole process, from the moment of the beginning of the transit until the moment of the end of the transit, are small, so that for a small angle φ: $\sin\varphi \approx \varphi$; $\cos\varphi \approx 1$; $\text{tg}\,\varphi \approx \varphi$.

Solution

I.

I.a.1) *Transit of the airplane* A_1 *in variant I.a.1*

According to the variant represented in photographic image 1.1 of Figure 12.1, the airplane transits the Moon's disk as indicated in the drawing in Figure 12.2, when the directions of motion of the Moon and the airplane A_1 in relation to the observer O on Earth are identical.

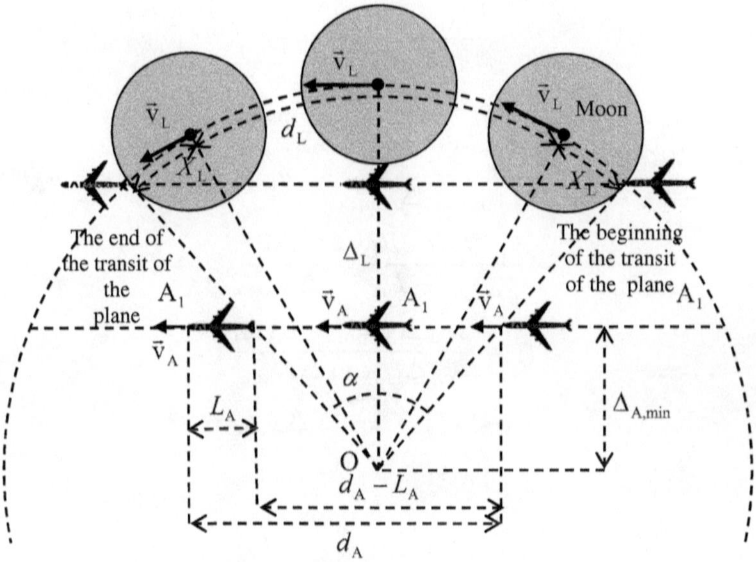

Fig. 12.2

The plane of the drawing in Figure 12.2 is the plane of the airplane's flight, the same as the plane of the Moon's orbit, and the same as the plane of the observation line.

Under these conditions, if $\vec{v}_L$ is the speed of the Moon in its circular orbit around the Earth, d_L is the distance traveled by the center of the Moon during the transit of the airplane A_1, and d_A is the distance traveled by the airplane A_1 during its transit, it results that:

$$d_L + 2X_L = \alpha \cdot \Delta_L; \quad X_L = \frac{\delta_O}{2}\Delta_L;$$

$$\delta_O = 2\frac{R_L}{\Delta_L}; \quad X_L = \frac{R_L}{\Delta_L}\Delta_L = R_L;$$

$$d_L + 2R_L = \alpha \cdot \Delta_L;$$

$$\tan\frac{\alpha}{2} = \frac{(d_A - L_A)/2}{\Delta_{A,\min}}; \quad d_A - L_A \ll \Delta_{A,\min}; \quad \tan\frac{\alpha}{2} \approx \frac{\alpha}{2};$$

$$\frac{(d_A - L_A)/2}{\Delta_{A,\min}} = \frac{\alpha}{2};$$

$$d_A - L_A = \alpha \cdot \Delta_{A,\min};$$

$$\frac{d_L + 2R_L}{d_A - L_A} = \frac{\alpha \cdot \Delta_L}{\alpha \cdot \Delta_{A,\min}} = \frac{\Delta_L}{\Delta_{A,\min}}; \quad d_L = v_L t_1; \quad d_A = v_A t_1;$$

$$\frac{v_L \cdot t_1 + 2R_L}{v_A \cdot t_1 - L_A} = \frac{\Delta_L}{\Delta_{A,\min}};$$

$$t_1 = \frac{2R_L \Delta_{A,\min} + L_A \Delta_L}{v_A \Delta_L - v_L \Delta_{A,\min}};$$

$$v_A < v_L; \quad \Delta_L \gg \Delta_{A,\min};$$

$$v_A \Delta_L > v_L \Delta_{A,\min};$$

$$t_1 > 0.$$

I.a.2) *Transit of the airplane* A_1 *in variant I.a.2*

Corresponding to the same variant, represented in the photographic image 1.1 in Figure 12.1, let us consider that the evolution of the airplane transit A_1 is represented in the drawing in Figure 12.3, when the Moon follows the airplane A_1, and its disk enters behind the airplane A_1:

$$d_L - 2R_L = \alpha \cdot \Delta_L;$$

$$\tan\frac{\alpha}{2} = \frac{(d_A + L_A)/2}{\Delta_{A,\min}}; \quad d_A + L_A \ll \Delta_{A,\min}; \quad \tan\frac{\alpha}{2} \approx \frac{\alpha}{2};$$

$$\frac{(d_A + L_A)/2}{\Delta_{A,\min}} = \frac{\alpha}{2};$$

$$d_A + L_A = \alpha \cdot \Delta_{A,\min};$$

$$\frac{d_L - 2R_L}{d_A + L_A} = \frac{\alpha \cdot \Delta_L}{\alpha \cdot \Delta_{A,\min}} = \frac{\Delta_L}{\Delta_{A,\min}}; \quad d_L = v_L t_1; \quad d_A = v_A t_1;$$

$$\frac{v_L \cdot t_1 - 2R_L}{v_A \cdot t_1 + L_A} = \frac{\Delta_L}{\Delta_{A,\min}};$$

$$t_1 = \frac{2R_L \Delta_{A,\min} + L_A \Delta_L}{v_L \Delta_{A,\min} - v_A \Delta_L};$$

$$\mathrm{v_A} < \mathrm{v_L}; \quad \Delta_\mathrm{L} \gg \Delta_\mathrm{A,min};$$
$$\mathrm{v_L}\Delta_\mathrm{A,min} < \mathrm{v_A}\Delta_\mathrm{L};$$
$$t < 0.$$

This means that the evolution of aircraft $\mathrm{A_1}$'s transit cannot be what is proposed in the drawing in Figure 12.3.

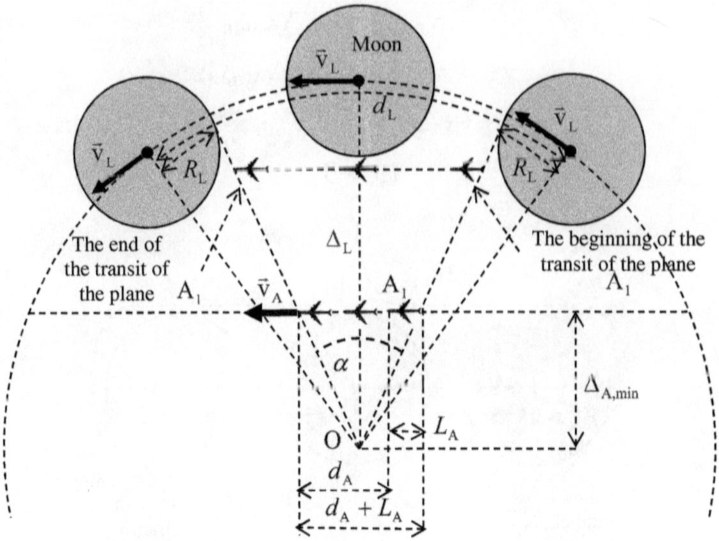

Fig. 12.3

I.b.1) ***Transit of the airplane*** $\mathbf{A_1}$ ***in variant I.b.1***

If, during the very short transit time of the aircraft $\mathrm{A_1}$, the motion of the center of the Moon's disk can be considered to be rectilinear and uniform, as shown in Figure 12.4, corresponding to the same variant shown in photo 1.1 of Figure 12.1, considering that the aircraft $\mathrm{A_1}$ passes over the disk of the Moon from behind, it results that:

$$\tan\frac{\alpha}{2} = \frac{\frac{d_\mathrm{L}+2R_\mathrm{L}}{2}}{\Delta_\mathrm{L}} \approx \frac{\alpha}{2}; \quad d_\mathrm{L} + 2R_\mathrm{L} = \alpha \cdot \Delta_\mathrm{L};$$

$$\tan\frac{\alpha}{2} = \frac{\frac{d_\mathrm{A}-L_\mathrm{A}}{2}}{\Delta_\mathrm{A,min}} \approx \frac{\alpha}{2}; \quad d_\mathrm{A} - L_\mathrm{A} = \alpha \cdot \Delta_\mathrm{A,min};$$

$$\frac{d_L + 2R_L}{d_A - L_A} = \frac{\Delta_L}{\Delta_{A,\min}};$$

$$d_L = v_L t_1; \quad d_A = v_A t_1;$$

$$\frac{v_L \cdot t_1 + 2R_L}{v_A \cdot t_1 - L_A} = \frac{\Delta_L}{\Delta_{A,\min}};$$

$$t_1 = \frac{2R_L\Delta_{A,\min} + L_A\Delta_L}{v_A\Delta_L - v_L\Delta_{A,\min}};$$

$$v_A < v_L; \quad \Delta_L \gg \Delta_{A,\min};$$

$$v_A\Delta_L > v_L\Delta_{A,\min};$$

$$t_1 > 0.$$

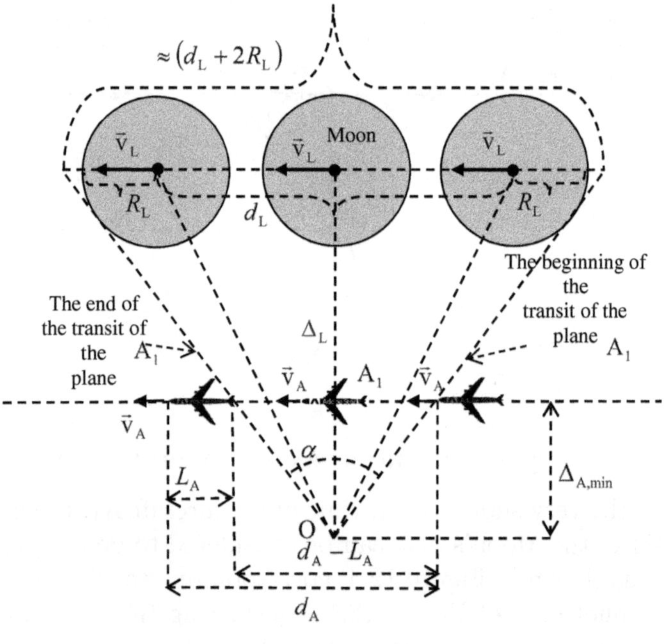

Fig. 12.4

I.b.2) *Transit of the airplane* A_1 *in variant I.b.2*

Corresponding to the same variant, represented in the photographic image 1.1 from Figure 12.1, let us consider that the evolution of

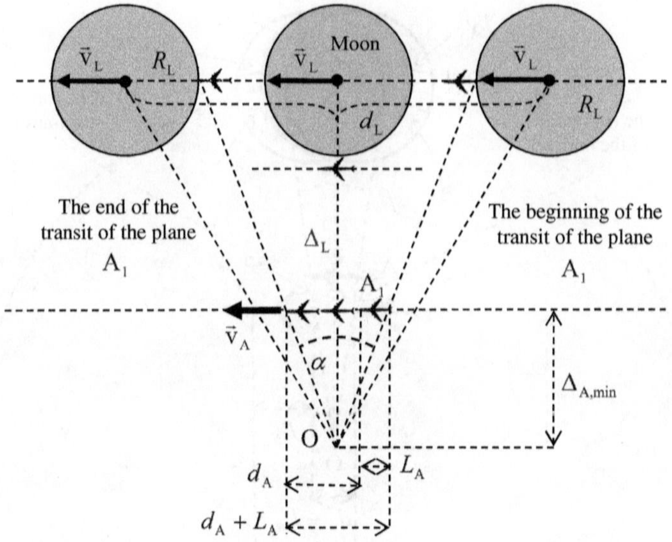

Fig. 12.5

the transit of the airplane A_1 is that represented in the drawing in Figure 12.5, when the Moon follows the airplane A_1 and its disk enters behind the airplane A_1:

$$\tan\frac{\alpha}{2} = \frac{(d_L - 2R_L)/2}{\Delta_L} = \frac{(d_A + L_A)/2}{\Delta_{A,min}};$$

$$\frac{(d_L - 2R_L)}{\Delta_L} = \frac{(d_A + L_A)}{\Delta_{A,min}};$$

$$d_L = v_L t_1; \quad d_A = v_A t_1;$$

$$\frac{v_L t_1 - 2R_L}{\Delta_L} = \frac{v_A t_1 + L_A}{\Delta_{A,min}};$$

$$t_1 = \frac{2R_L\Delta_{A,min} + L_A\Delta_L}{v_L\Delta_{A,min} - v_A\Delta_L};$$

$$v_A < v_L; \quad \Delta_L \gg \Delta_{A,min};$$

$$v_L\Delta_{A,min} < v_A\Delta_L;$$

$$t < 0.$$

This means that the evolution of the aircraft A_1 cannot be the one proposed in the drawing in Figure 12.5.

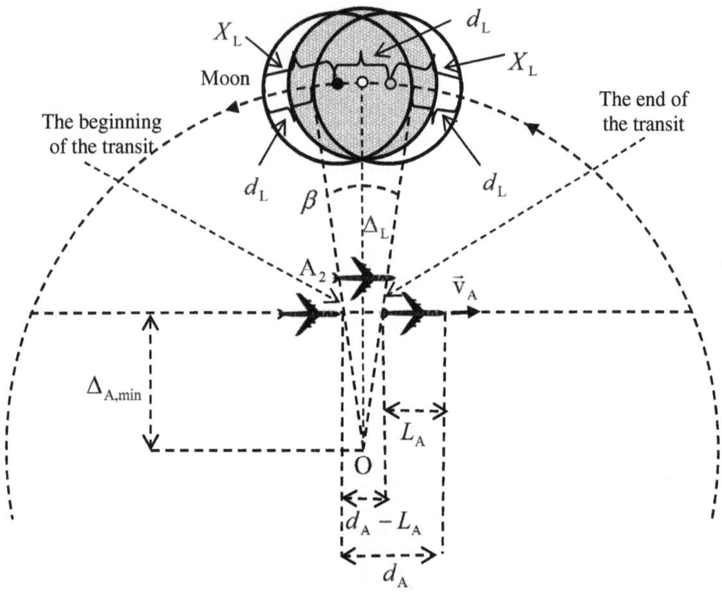

Fig. 12.6

I.c) *Transit of the airplane* A_2 *in variant I.c.*

According to the variant represented in the photographic image 1.2 in Figure 12.1, the transit of the airplane A_2 over the disk of the Moon occurs, as indicated in the drawing in Figure 12.6, when the directions of motion of the Moon and the airplane A_2 in relation to the observer O on Earth are opposite.

The plane of the drawing in Figure 12.6 is the plane of airplane A_2's flight, the same as the plane of the Moon's orbit and the plane of the observation line.

Under these conditions, it results that:

$$X_L - d_L + d_L + X_L - d_L = \beta \cdot \Delta_L;$$

$$2X_L - d_L = \beta \cdot \Delta_L;$$

$$X_L = \frac{\delta_O}{2}\Delta_L; \quad \delta_O = 2\frac{R_L}{\Delta_L}; \quad X_L = \frac{R_L}{\Delta_L}\Delta_L = R_L;$$

$$2R_L - d_L = \beta \cdot \Delta_L;$$

$$\tan\frac{\beta}{2} = \frac{(d_{\mathrm{A}} - L_{\mathrm{A}})/2}{\Delta_{\mathrm{A,min}}} \approx \frac{\beta}{2}; \quad d_{\mathrm{A}} - L_{\mathrm{A}} = \beta \cdot \Delta_{\mathrm{A,min}};$$

$$\frac{2R_{\mathrm{L}} - d_{\mathrm{L}}}{d_{\mathrm{A}} - L_{\mathrm{A}}} = \frac{\beta \cdot \Delta_{\mathrm{L}}}{\beta \cdot \Delta_{\mathrm{A,min}}} = \frac{\Delta_{\mathrm{L}}}{\Delta_{\mathrm{A,min}}}; \quad d_{\mathrm{L}} = \mathrm{v_L} t_2; \quad d_{\mathrm{A}} = \mathrm{v_A} t_2;$$

$$\frac{2R_{\mathrm{L}} - \mathrm{v_L} t_2}{\mathrm{v_A} t_2 - L_{\mathrm{A}}} = \frac{\Delta_{\mathrm{L}}}{\Delta_{\mathrm{A,min}}};$$

$$t_2 = \frac{2R_{\mathrm{L}}\Delta_{\mathrm{A,min}} + L_{\mathrm{A}}\Delta_{\mathrm{L}}}{\mathrm{v_A}\Delta_{\mathrm{L}} + \mathrm{v_L}\Delta_{\mathrm{A,min}}} > 0.$$

I.d) *Transit of the airplane* $\mathbf{A_2}$ *in variant I.d*

Corresponding to the same variant, represented in the photographic image 1.2 in Figure 12.1, if, during the short duration of airplane A_2's transit the motion of the center of the Moon's disk can be considered rectilinear and uniform, as shown in the drawing in Figure 12.7, then:

$$\tan\frac{\beta}{2} = \frac{\frac{2R_{\mathrm{L}} - d_{\mathrm{L}}}{2}}{\Delta_{\mathrm{L}}} \approx \frac{\beta}{2}; \quad 2R_{\mathrm{L}} - d_{\mathrm{L}} = \beta \cdot \Delta_{\mathrm{L}};$$

$$\tan\frac{\beta}{2} = \frac{\frac{d_{\mathrm{A}} - L_{\mathrm{A}}}{2}}{\Delta_{\mathrm{A,min}}} \approx \frac{\beta}{2}; \quad d_{\mathrm{A}} - L_{\mathrm{A}} = \beta \cdot \Delta_{\mathrm{A,min}};$$

$$\frac{2R_{\mathrm{L}} - d_{\mathrm{L}}}{d_{\mathrm{A}} - L_{\mathrm{A}}} = \frac{\Delta_{\mathrm{L}}}{\Delta_{\mathrm{A,min}}};$$

$$d_{\mathrm{L}} = \mathrm{v_L} t_2; \quad d_{\mathrm{A}} = \mathrm{v_A} t_2;$$

$$\frac{2R_{\mathrm{L}} - \mathrm{v_L} t_2}{\mathrm{v_A} \cdot t_2 - L_{\mathrm{A}}} = \frac{\Delta_{\mathrm{L}}}{\Delta_{\mathrm{A,min}}};$$

$$\frac{2R_{\mathrm{L}} - \mathrm{v_L} t_2}{\mathrm{v_A} t_2 - L_{\mathrm{A}}} = \frac{\Delta_{\mathrm{L}}}{\Delta_{\mathrm{A,min}}};$$

$$t_2 = \frac{2R_{\mathrm{L}}\Delta_{\mathrm{A,min}} + L_{\mathrm{A}}\Delta_{\mathrm{L}}}{\mathrm{v_A}\Delta_{\mathrm{L}} + \mathrm{v_L}\Delta_{\mathrm{A,min}}}.$$

Conclusion: $t_2 < t_1$.

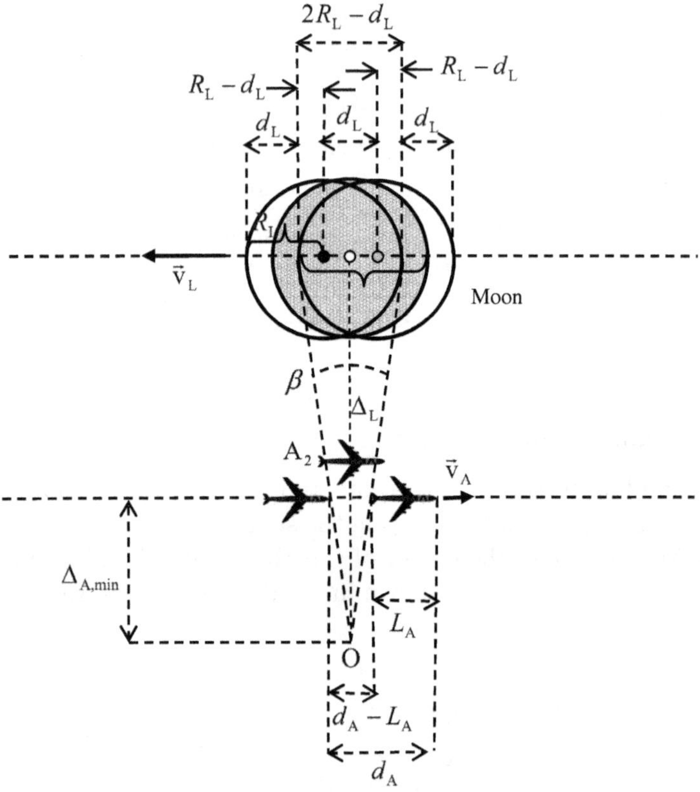

Fig. 12.7

II. The drawing in Figure 12.8 shows the two airplanes, A_1 and A_2, at the time of their meeting when the projection of their meeting point on the plane of the Moon's disk is the center of the Moon's disk. The plane of the drawing is the plane of flight of the two airplanes A_1 and A_2, where the plane of the Moon's orbit is the same as the plane of the line of observation.

The first airplane to complete its transit in front of the Moon's disk is the airplane whose direction of motion is the opposite of the direction of the Moon's motion, meaning the airplane A_2.

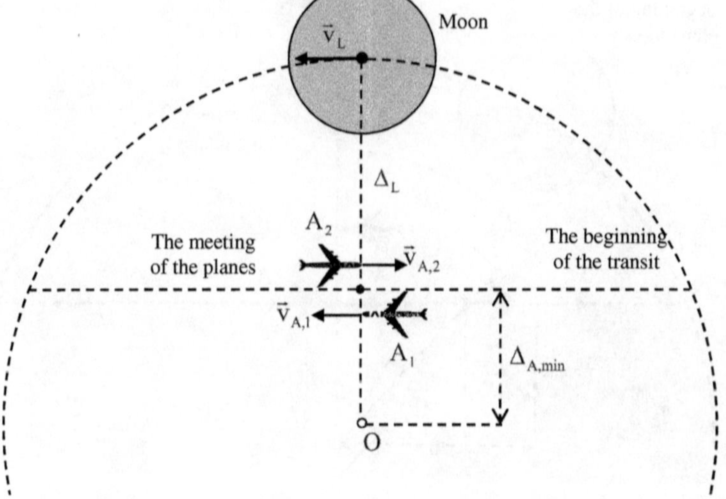

Fig. 12.8

II.a) *Transit of the airplane* A_1 *in variant II.a*

The drawing in Figure 12.9 shows the evolution of the airplane A_1 from the moment it meets the airplane A_2 to the moment it ends its transit in front of the disk of the Moon.

Knowing the meaning of the notation in Figure 12.9, it follows that:

$$d_{A,1} = v_{A,1}\tau_1;$$

$$d_L = \beta \cdot \Delta_L = v_L\tau_1;$$

$$\beta = \frac{v_L\tau_1}{\Delta_L};$$

$$\mathrm{tg}\left(90^\circ - \left(\beta + \frac{\delta_O}{2}\right)\right) = \frac{\Delta_{A,\min}}{d_{A,1} - L_A} = \frac{\Delta_{A,\min}}{v_{A,1}\tau_1 - L_A};$$

$$\mathrm{tg}\left(90^\circ - \left(\beta + \frac{\delta_O}{2}\right)\right) = \mathrm{ctg}\left(\beta + \frac{\delta_O}{2}\right) = \frac{\cos\left(\beta + \frac{\delta_O}{2}\right)}{\sin\left(\beta + \frac{\delta_O}{2}\right)} \approx \frac{1}{\beta + \frac{\delta_O}{2}};$$

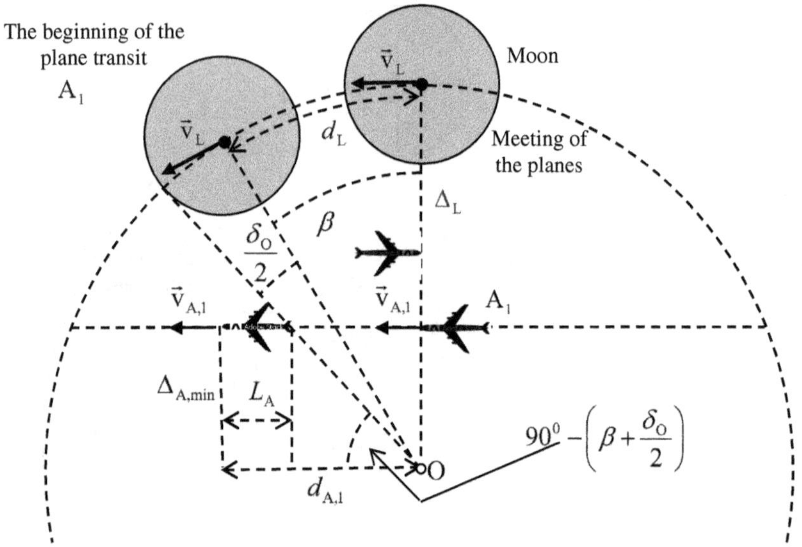

Fig. 12.9

$$\frac{\Delta_{A,min}}{v_{A,1}\tau_1 - L_A} = \frac{1}{\beta + \frac{\delta_O}{2}}; \quad \beta = \frac{v_L \tau_1}{\Delta_L};$$

$$\frac{\Delta_{A,min}}{v_{A,1}\tau_1 - L_A} = \frac{1}{\frac{v_L \tau_1}{\Delta_L} + \frac{\delta_O}{2}}; \quad \delta_O = 2\frac{R_L}{\Delta_L};$$

$$\frac{\Delta_{A,min}}{v_{A,1}\tau_1 - L_A} = \frac{1}{\frac{v_L \tau_1}{\Delta_L} + \frac{R_L}{\Delta_L}};$$

$$v_L \frac{\Delta_{A,min}}{\Delta_L}\tau_1 + \frac{R_L \Delta_{A,min}}{\Delta_L} = v_{A,1}\tau_1 - L_A;$$

$$L_A + \frac{R_L \Delta_{A,min}}{\Delta_L} = v_{A,1}\tau_1 - v_L \frac{\Delta_{A,min}}{\Delta_L}\tau_1;$$

$$\tau_1 = \frac{L_A \Delta_L + R_L \Delta_{A,min}}{v_{A,1}\Delta_L - v_L \Delta_{A,min}};$$

$$v_{A,1} = v_A;$$

$$\tau_1 = \frac{L_A \Delta_L + R_L \Delta_{A,min}}{v_A \Delta_L - v_L \Delta_{A,min}},$$

representing the time after which the aircraft A_1 completes its transit in front of the Moon's disk;

$$\tau_2 = \frac{L_A \Delta_L + R_L \Delta_{A,\min}}{v_L \Delta_{A,\min} + v_A \Delta_L},$$

representing the time after which the aircraft A_2 completes its transit in front of the Moon's disk;

$$\tau_1 > \tau_2.$$

II.b) *Transit of the airplane* A_1 *in variant II.b*

If, during the short transit time, the motion of the center of the Moon's disk can be considered to be rectilinear and uniform, as shown in the drawing in Figure 12.10, it results that:

$$d_{A,1} = v_{A,1}\tau_1;$$

$$\tan\beta = \frac{d_L}{\Delta_L} \approx \beta; \quad d_L = \beta \cdot \Delta_L;$$

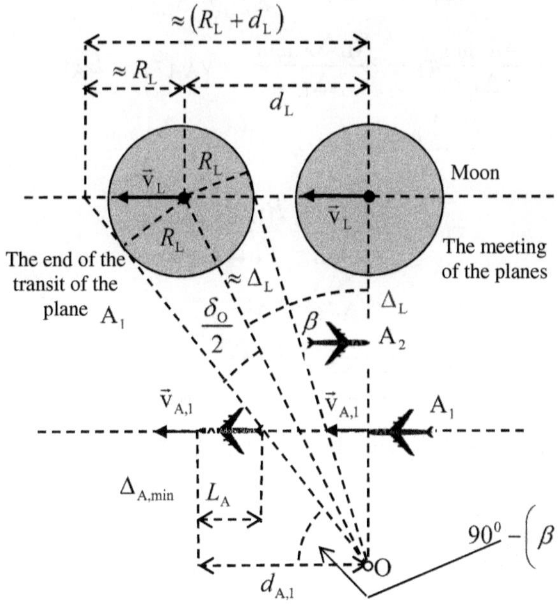

Fig. 12.10

$$d_{\mathrm{L}} = \beta \cdot \Delta_{\mathrm{L}} = \mathrm{v}_{\mathrm{L}} \tau_1;$$

$$\beta = \frac{\mathrm{v}_{\mathrm{L}} \tau_1}{\Delta_{\mathrm{L}}};$$

$$\mathrm{tg}\left(90^\circ - \left(\beta + \frac{\delta_{\mathrm{O}}}{2}\right)\right) = \frac{\Delta_{\mathrm{A,min}}}{d_{\mathrm{A},1} - L_{\mathrm{A}}} = \frac{\Delta_{\mathrm{A,min}}}{\mathrm{v}_{\mathrm{A},1}\tau_1 - L_{\mathrm{A}}};$$

$$\mathrm{tg}\left(90^\circ - \left(\beta + \frac{\delta_{\mathrm{O}}}{2}\right)\right) = \mathrm{ctg}\left(\beta + \frac{\delta_{\mathrm{O}}}{2}\right) = \frac{\cos\left(\beta + \frac{\delta_{\mathrm{O}}}{2}\right)}{\sin\left(\beta + \frac{\delta_{\mathrm{O}}}{2}\right)} \approx \frac{1}{\beta + \frac{\delta_{\mathrm{O}}}{2}};$$

$$\frac{\Delta_{\mathrm{A,min}}}{\mathrm{v}_{\mathrm{A},1}\tau_1 - L_{\mathrm{A}}} = \frac{1}{\beta + \frac{\delta_{\mathrm{O}}}{2}}; \quad \beta = \frac{\mathrm{v}_{\mathrm{L}}\tau_1}{\Delta_{\mathrm{L}}};$$

$$\frac{\Delta_{\mathrm{A,min}}}{\mathrm{v}_{\mathrm{A},1}\tau_1 - L_{\mathrm{A}}} = \frac{1}{\frac{\mathrm{v}_{\mathrm{L}}\tau_1}{\Delta_{\mathrm{L}}} + \frac{\delta_{\mathrm{O}}}{2}};$$

$$\sin\frac{\delta_{\mathrm{O}}}{2} \approx \frac{R_{\mathrm{L}}}{\Delta_{\mathrm{L}}} \approx \frac{\delta_{\mathrm{O}}}{2}; \quad \delta_{\mathrm{O}} = 2\frac{R_{\mathrm{L}}}{\Delta_{\mathrm{L}}};$$

$$\frac{\Delta_{\mathrm{A,min}}}{\mathrm{v}_{\mathrm{A},1}\tau_1 - L_{\mathrm{A}}} = \frac{1}{\frac{\mathrm{v}_{\mathrm{L}}\tau_1}{\Delta_{\mathrm{L}}} + \frac{R_{\mathrm{L}}}{\Delta_{\mathrm{L}}}};$$

$$\mathrm{v}_{\mathrm{L}}\frac{\Delta_{\mathrm{A,min}}}{\Delta_{\mathrm{L}}}\tau_1 + \frac{R_{\mathrm{L}}\Delta_{\mathrm{A,min}}}{\Delta_{\mathrm{L}}} = \mathrm{v}_{\mathrm{A},1}\tau_1 - L_{\mathrm{A}};$$

$$L_{\mathrm{A}} + \frac{R_{\mathrm{L}}\Delta_{\mathrm{A,min}}}{\Delta_{\mathrm{L}}} = \mathrm{v}_{\mathrm{A},1}\tau_1 - \mathrm{v}_{\mathrm{L}}\frac{\Delta_{\mathrm{A,min}}}{\Delta_{\mathrm{L}}}\tau_1;$$

$$\tau_1 = \frac{L_{\mathrm{A}}\Delta_{\mathrm{L}} + R_{\mathrm{L}}\Delta_{\mathrm{A,min}}}{\mathrm{v}_{\mathrm{A},1}\Delta_{\mathrm{L}} - \mathrm{v}_{\mathrm{L}}\Delta_{\mathrm{A,min}}};$$

$$\mathrm{v}_{\mathrm{A},1} = \mathrm{v}_{\mathrm{A}};$$

$$\tau_1 = \frac{L_{\mathrm{A}}\Delta_{\mathrm{L}} + R_{\mathrm{L}}\Delta_{\mathrm{A,min}}}{\mathrm{v}_{\mathrm{A}}\Delta_{\mathrm{L}} - \mathrm{v}_{\mathrm{L}}\Delta_{\mathrm{A,min}}},$$

representing the time after which the aircraft A_1 completes its transit in front of the Moon's disk;

$$\tau_2 = \frac{L_{\mathrm{A}}\Delta_{\mathrm{L}} + R_{\mathrm{L}}\Delta_{\mathrm{A,min}}}{\mathrm{v}_{\mathrm{L}}\Delta_{\mathrm{A,min}} + \mathrm{v}_{\mathrm{A}}\Delta_{\mathrm{L}}},$$

representing the time after which the aircraft A_2 completes its transit in front of the Moon's disk;

$$\tau_1 > \tau_2.$$

II.c) *Transit of the airplane* A_2 *in variant II.c*

The drawing in Figure 12.11 shows the evolution of the airplane A_2 from the moment it meets the airplane A_1 until the moment when it ends its transit in front of the disk of the Moon.

Knowing the meaning of the notation in Figure 12.11, it follows that:

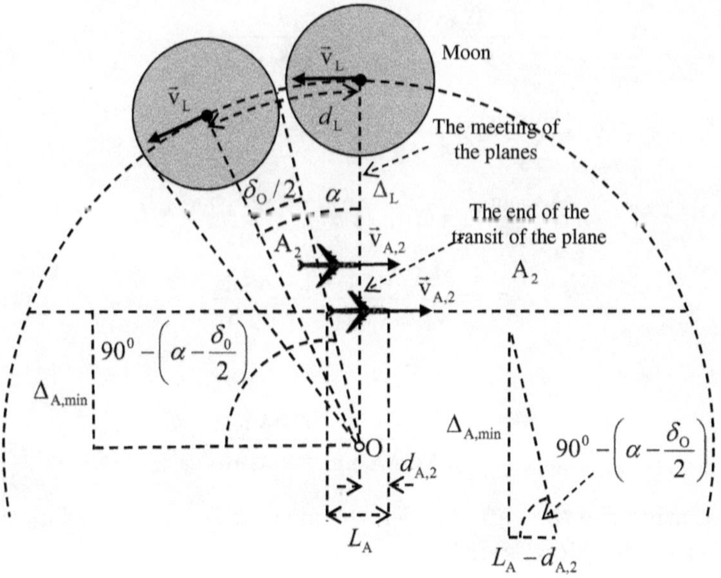

Fig. 12.11

$$d_{A,2} = v_{A,2}\tau_2;$$

$$d_L = v_L\tau_2 = \alpha \cdot \Delta_L; \quad \alpha = \frac{v_L\tau_2}{\Delta_L};$$

$$\text{tg}\left(90^\circ - \left(\alpha - \frac{\delta_O}{2}\right)\right) = \frac{\Delta_{A,\min}}{L_A - d_{A,2}} = \frac{\Delta_{A,\min}}{L_A - v_{A,2}\tau_2};$$

$$\operatorname{tg}\left(90^\circ - \left(\alpha - \frac{\delta_O}{2}\right)\right)$$

$$= \operatorname{ctg}\left(\alpha - \frac{\delta_O}{2}\right) = \frac{\cos\left(\alpha - \frac{\delta_O}{2}\right)}{\sin\left(\alpha - \frac{\delta_O}{2}\right)} \approx \frac{1}{\alpha - \frac{\delta_O}{2}};$$

$$\frac{\Delta_{A,\min}}{L_A - v_{A,2}\tau_2} = \frac{1}{\alpha - \frac{\delta_O}{2}}; \quad \alpha = \frac{v_L \tau_2}{\Delta_L};$$

$$\frac{\Delta_{A,\min}}{L_A - v_{A,2}\tau_2} = \frac{1}{\frac{v_L \tau_2}{\Delta_L} - \frac{\delta_O}{2}}; \quad \delta_O = 2\frac{R_L}{\Delta_L};$$

$$\frac{\Delta_{A,\min}}{L_A - v_{A,2}\tau_2} = \frac{1}{\frac{v_L \tau_2}{\Delta_L} - \frac{R_L}{\Delta_L}};$$

$$v_L \frac{\Delta_{A,\min}}{\Delta_L}\tau_2 - \frac{R_L \Delta_{A,\min}}{\Delta_L} = L_A - v_{A,2}\tau_2;$$

$$v_L \frac{\Delta_{A,\min}}{\Delta_L}\tau_2 + v_{A,2}\tau_2 = \frac{R_L \Delta_{A,\min}}{\Delta_L} + L_A;$$

$$\tau_2 = \frac{L_A \Delta_L + R_L \Delta_{A,\min}}{v_L \Delta_{A,\min} + v_{A,2}\Delta_L};$$

$$v_{A,2} = v_A;$$

$$\tau_2 = \frac{L_A \Delta_L + R_L \Delta_{A,\min}}{v_L \Delta_{A,\min} + v_A \Delta_L},$$

representing the time after which the aircraft A_2 completes its transit in front of the Moon's disk.

II.d) *Transit of the airplane* A_2 *in variant II.d*

If, during the very short transit time, the motion of the center of the Moon's disk can be considered to be rectilinear and uniform, as shown in the drawing in Figure 12.12, the result is:

$$\tan\alpha = \frac{d_L}{\Delta_L} \approx \alpha; \quad d_L = \alpha \cdot \Delta_L;$$

$$v_L \tau_2 = \alpha \cdot \Delta_L; \quad \alpha = \frac{v_L \tau_2}{\Delta_L};$$

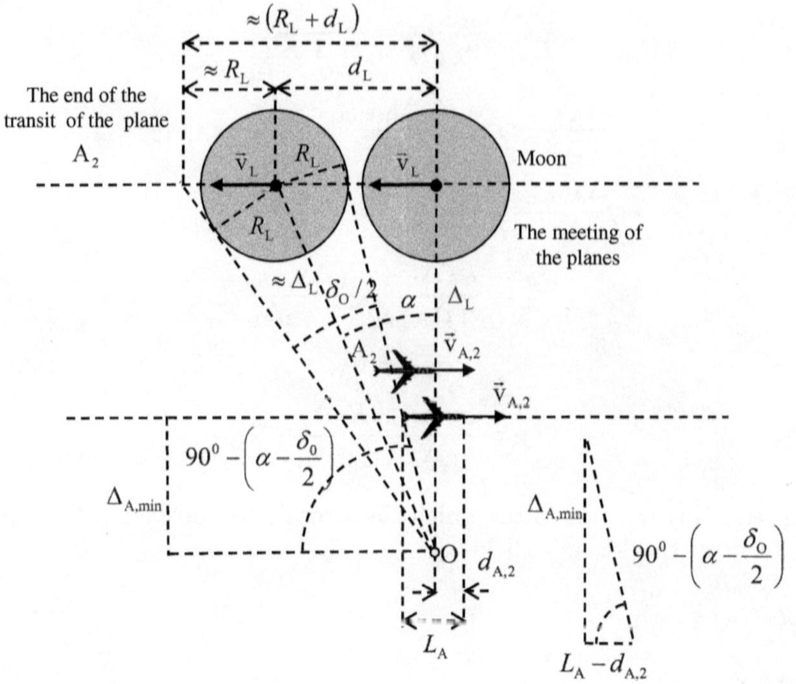

Fig. 12.12

$$d_{A,2} = v_{A,2}\tau_2;$$

$$\operatorname{tg}\left(90^\circ - \left(\alpha - \frac{\delta_O}{2}\right)\right) = \frac{\Delta_{A,\min}}{L_A - d_{A,2}} = \frac{\Delta_{A,\min}}{L_A - v_{A,2}\tau_2};$$

$$\operatorname{tg}\left(90^\circ - \left(\left(\alpha - \frac{\delta_O}{2}\right)\right)\right)$$

$$= \operatorname{ctg}\left(\alpha - \frac{\delta_O}{2}\right) = \frac{\cos\left(\alpha - \frac{\delta_O}{2}\right)}{\sin\left(\alpha - \frac{\delta_O}{2}\right)} \approx \frac{1}{\alpha - \frac{\delta_O}{2}};$$

$$\frac{\Delta_{A,\min}}{L_A - v_{A,2}\tau_2} = \frac{1}{\alpha - \frac{\delta_O}{2}}; \quad \alpha = \frac{v_L\tau_2}{\Delta_L};$$

$$\frac{\Delta_{A,\min}}{L_A - v_{A,2}\tau_2} = \frac{1}{\frac{v_L\tau_2}{\Delta_L} - \frac{\delta_O}{2}}; \quad \delta_O = 2\frac{R_L}{\Delta_L};$$

$$\frac{\Delta_{\mathrm{A,min}}}{L_{\mathrm{A}} - \mathrm{v}_{\mathrm{A},2}\tau_2} = \frac{1}{\frac{\mathrm{v_L}\tau_2}{\Delta_{\mathrm{L}}} - \frac{R_{\mathrm{L}}}{\Delta_{\mathrm{L}}}};$$

$$\mathrm{v_L}\frac{\Delta_{\mathrm{A,min}}}{\Delta_{\mathrm{L}}}\tau_2 - \frac{R_{\mathrm{L}}\Delta_{\mathrm{A,min}}}{\Delta_{\mathrm{L}}} = L_{\mathrm{A}} - \mathrm{v}_{\mathrm{A},2}\tau_2;$$

$$\mathrm{v_L}\frac{\Delta_{\mathrm{A,min}}}{\Delta_{\mathrm{L}}}\tau_2 + \mathrm{v}_{\mathrm{A},2}\tau_2 = \frac{R_{\mathrm{L}}\Delta_{\mathrm{A,min}}}{\Delta_{\mathrm{L}}} + L_{\mathrm{A}};$$

$$\tau_2 = \frac{L_{\mathrm{A}}\Delta_{\mathrm{L}} + R_{\mathrm{L}}\Delta_{\mathrm{A,min}}}{\mathrm{v_L}\Delta_{\mathrm{A,min}} + \mathrm{v}_{\mathrm{A},2}\Delta_{\mathrm{L}}};$$

$$\mathrm{v}_{\mathrm{A},2} = \mathrm{v_A};$$

$$\tau_2 = \frac{L_{\mathrm{A}}\Delta_{\mathrm{L}} + R_{\mathrm{L}}\Delta_{\mathrm{A,min}}}{\mathrm{v_L}\Delta_{\mathrm{A,min}} + \mathrm{v_A}\Delta_{\mathrm{L}}},$$

representing the time after which the aircraft A_2 completes its transit in front of the Moon's disk.

III.

III.a)

Method 1. If the helicopter's entry into its transit in front of the Moon's disk occurs with the cockpit at point A, as shown in the drawing in Figure 12.13, the helicopter's exit from the transit in front of the Moon's disk will occur with the helicopter's tail at point B.

In the sequences in Figure 12.13, as the helicopter climbs vertically in front of the observer, the Moon's disk moves to the left, behind the helicopter, so that:

$$y_0 = \mathrm{v_E}t; \quad d_{\mathrm{L}} = \mathrm{v_L}t; \quad d_{\mathrm{L}} = R_{\mathrm{L}} + L_{\mathrm{E}} + x_0;$$

$$R_{\mathrm{L}} + L_{\mathrm{E}} + x_0 = \mathrm{v_L}\frac{y_0}{\mathrm{v_E}};$$

$$\mathrm{v_E} = \frac{\mathrm{v_L}y_0}{R_{\mathrm{L}} + L_{\mathrm{E}} + x_0}.$$

Method 2. The condition of this transit is that the direction of the relative velocity vector of the point at the helicopter's tail, relative to the Moon's disk, $\vec{\mathrm{v}}_{\mathrm{E,L}}$, passes through the point at the end of the

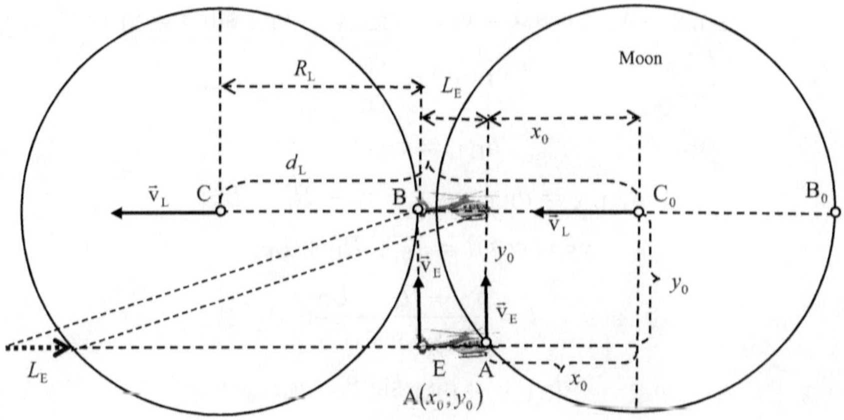

Fig. 12.13

horizontal diameter of the Moon's disk, as shown in Figure 12.14:

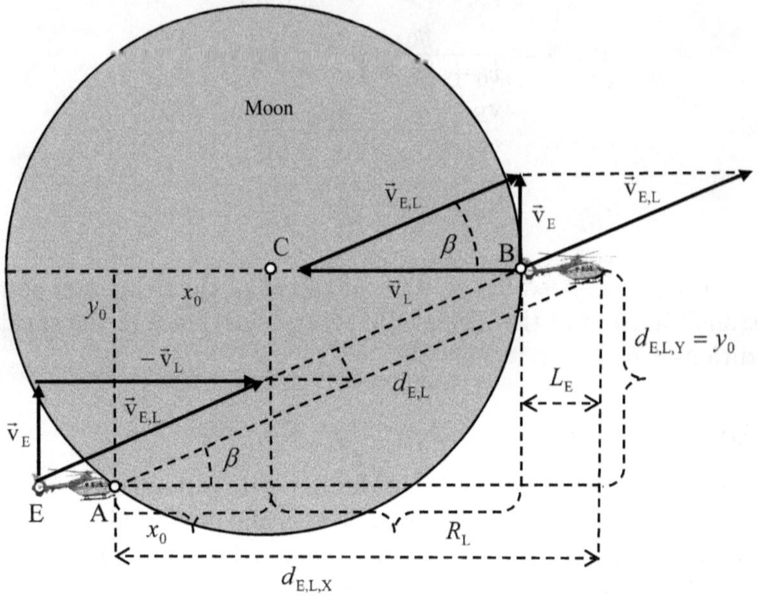

Fig. 12.14

$$\vec{v}_{E,L} = \vec{v}_E - \vec{v}_L = \vec{v}_E + (-\vec{v}_L);$$

$$v_{E,L} = \sqrt{v_E^2 + v_L^2};$$

$$\mathrm{v_{E,L,X}} = \mathrm{v_{E,L}} \cos\beta = \mathrm{v_L}; \quad \mathrm{v_{E,L,Y}} = \mathrm{v_{E,L}} \sin\beta = \mathrm{v_E};$$

$$\tan\beta = \frac{\mathrm{v_E}}{\mathrm{v_L}};$$

$$d_{\mathrm{E,L}} = \mathrm{v_{E,L}} t;$$

$$d_{\mathrm{E,L,X}} = d_{\mathrm{E,L}} \cos\beta = x_0 + R_{\mathrm{L}} + L_{\mathrm{E}};$$

$$\mathrm{v_{E,L}} t \cos\beta = x_0 + R_{\mathrm{L}} + L_{\mathrm{E}};$$

$$t = \frac{x_0 + R_{\mathrm{L}} + L_{\mathrm{E}}}{\mathrm{v_{E,L}} \cos\beta};$$

$$d_{\mathrm{E,L,Y}} = d_{\mathrm{E,L}} \sin\beta = y_0;$$

$$\mathrm{v_{E,L}} t \sin\beta = y_0; \quad t = \frac{y_0}{\mathrm{v_{E,L}} \sin\beta};$$

$$\frac{x_0 + R_{\mathrm{L}} + L_{\mathrm{E}}}{\mathrm{v_{E,L}} \cos\beta} = \frac{y_0}{\mathrm{v_{E,L}} \sin\beta};$$

$$\tan\beta = \frac{y_0}{x_0 + R_{\mathrm{L}} + L_{\mathrm{E}}}; \quad \tan\beta = \frac{\mathrm{v_E}}{\mathrm{v_L}};$$

$$\frac{\mathrm{v_E}}{\mathrm{v_L}} = \frac{y_0}{x_0 + R_{\mathrm{L}} + L_{\mathrm{E}}};$$

$$\mathrm{v_E} = \frac{y_0 \mathrm{v_L}}{x_0 + R_{\mathrm{L}} + L_{\mathrm{E}}},$$

representing the speed of the vertical ascent of the helicopter so that its transit in front of the disk of the Moon takes place in the specified conditions.

III.b)

The sector on the surface of the Moon's disk traversed by the helicopter during the entire transit can be identified, approximately, with the isosceles trapezoid ANBM, represented in the drawing in Figure 12.15, whose area is

$$\Delta S = \frac{(B + b)h}{2};$$

$$\mathrm{v_{EL}} = \sqrt{\mathrm{v_E^2} + \mathrm{v_L^2}};$$

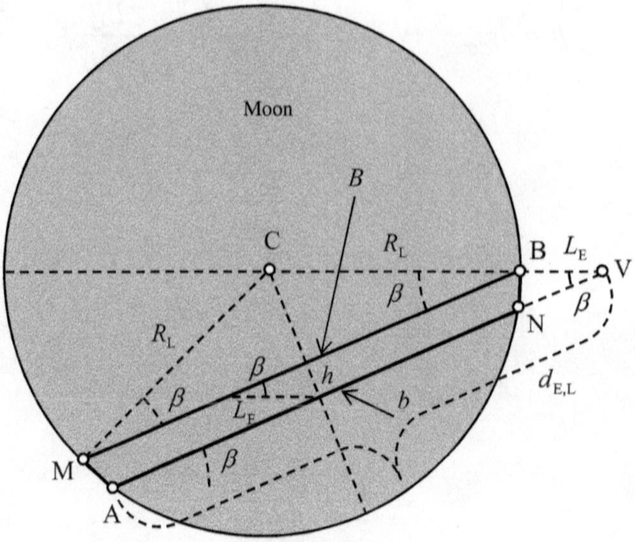

Fig. 12.15

$$v_{E,L,Y} = v_{E,L} \sin\beta = v_E; \quad \sin\beta = \frac{v_E}{v_{EL}} = \frac{v_E}{\sqrt{v_E^2 + v_L^2}};$$

$$\cos\beta = \sqrt{1 - \sin^2\beta} = \frac{v_L}{\sqrt{v_E^2 + v_L^2}};$$

$$B = 2R_L \cos\beta;$$

$$B = \frac{2R_L v_L}{\sqrt{v_E^2 + v_L^2}};$$

$$d_{E,L} = v_{E,L} t;$$

$$v_{EL} = \sqrt{v_E^2 + v_L^2}; \quad t = \frac{x_0 + R_L + L_E}{v_{E,L} \cos\beta};$$

$$d_{E,L} = \sqrt{v_E^2 + v_L^2} \cdot \frac{x_0 + R_L + L_E}{v_{E,L} \cos\beta};$$

$$d_{E,L} = \frac{x_0 + R_L + L_E}{\cos\beta};$$

$$NV = \frac{L_E}{\cos\beta};$$

$$b = d_{\mathrm{EL}} - \mathrm{NV} = \frac{x_0 + R_{\mathrm{L}} + L_{\mathrm{E}}}{\cos\beta} - \frac{L_{\mathrm{E}}}{\cos\beta};$$

$$b = \frac{x_0 + R_{\mathrm{L}}}{\cos\beta};$$

$$\cos\beta = \frac{\mathrm{v_L}}{\sqrt{\mathrm{v_E^2} + \mathrm{v_L^2}}};$$

$$b = \frac{x_0 + R_{\mathrm{L}}}{\mathrm{v_L}} \cdot \sqrt{\mathrm{v_E^2} + \mathrm{v_L^2}};$$

$$h = \frac{L_{\mathrm{E}}}{\sin\beta}; \quad \sin\beta = \frac{\mathrm{v_E}}{\sqrt{\mathrm{v_E^2} + \mathrm{v_L^2}}};$$

$$h = \frac{L_{\mathrm{E}}\sqrt{\mathrm{v_E^2} + \mathrm{v_L^2}}}{\mathrm{v_E}};$$

$$\Delta S = \frac{(B + b)h}{2};$$

$$\Delta S = \frac{(x_0 + R_{\mathrm{L}})\mathrm{v_E^2} + (x_0 + 3R_{\mathrm{L}})\mathrm{v_L^2}}{2\mathrm{v_L}\mathrm{v_E}} L_{\mathrm{E}},$$

where

$$\mathrm{v_E} = \frac{y_0 \mathrm{v_L}}{x_0 + R_{\mathrm{L}} + L_{\mathrm{E}}}.$$

Problem 13

Andromeda Constellation Rotation

A telescope whose mirror has a diameter of D = 6 m can help us observe our galaxy and the constellation Andromeda.

The photographs obtained will be used to *determine* the observation time required to discern the rotational motions of our galaxy and the constellation Andromeda around their common center of mass.

Given: where R_0 is the distance between Earth and Sun, the mass of our galaxy is $M_G = 2.5 \cdot 10^{11} M_0$, where M_0 is the mass of the Sun; the mass of the constellation Andromeda is $M_A = 3.6 \cdot 10^{11} M_0$.

Knowing that:

1) the photos were taken using visible light, with a wavelength $\lambda = 5 \cdot 10^{-7}$ m;

2) the angular distance between two objects for which they can be observed separately is $\varphi_0 = 1.22 \cdot \frac{\lambda}{D}$.

Solution

The study of light diffraction has revealed that the angular distance between two objects for which they can be observed separately is approximately λ/D. This means that the change in the position of the constellation Andromeda can be observed if its angular displacement is

$$\varphi_0 = 1.22 \cdot \frac{\lambda}{D} = 1.22 \cdot \frac{5 \cdot 10^{-7}}{6} = 10.16 \cdot 10^{-8} \text{ radians.}$$

If T is the period of rotation of our galaxy and the Andromeda constellation around their common center of mass, then the time required for the angular displacement φ_0 of the Andromeda constellation is

$$\tau = \frac{\varphi_0}{\omega} = \frac{\varphi_0}{2\pi} T.$$

For two binary systems, using the generalized form of Kepler's third law, we can write that

$$\left(\frac{T_1}{T_2}\right)^2 \left(\frac{M_1 + m_1}{M_2 + m_2}\right) = \left(\frac{a_1}{a_2}\right)^3.$$

For the galaxy–Andromeda and Sun–Earth systems, this results in:

$$\left(\frac{T}{T_0}\right)^2 \left(\frac{M_\mathrm{G} + M_\mathrm{A}}{M_0 + M_\mathrm{E}}\right) = \left(\frac{R}{R_0}\right)^3; \quad M_\mathrm{E} \ll M_0;$$

$$T_0 = 1 \text{ terrestrial year;}$$

$$\left(\frac{T}{T_0}\right)^2 \left(\frac{M_\mathrm{G} + M_\mathrm{A}}{M_0}\right) = \left(\frac{R}{R_0}\right)^3;$$

$$T^2 = T_0^2 \cdot \left(\frac{R}{R_0}\right)^3 \cdot \left(\frac{M_0}{M_G + M_A}\right); \quad \tau = \frac{\varphi_0}{2\pi} T; \quad \varphi_0 = 1.22 \cdot \frac{\lambda}{D};$$

$$\tau = 1.22 \cdot \frac{\lambda}{2\pi D} T_0 \left(\frac{R}{R_0}\right)^{3/2} \left(\frac{M_0}{M_\mathrm{G} + M_\mathrm{A}}\right)^{1/2} \approx 10^3 \text{ years.}$$

Problem 14

Spectroscopic Binary Star System

Determine the masses of the components of a spectroscopic binary star system, M_1 and M_2, respectively; the maximum radial velocities of the two components relative to the observer on Earth, $v_{rad, max, 1}$ and $v_{rad, max, 2}$; the angle between the sky and the relative orbits of the two components, i; the period of rotation of the two components of the system in relation to the center of mass, T; and the constant of gravitational attraction, G.

It is known that the system's center of mass is at rest relative to the observer on Earth, and the two components move in concentric circles around their center of mass.

Solution

If the observation of the two stars occurs as shown in the drawing in Figure 14.1 when the plane of the relative orbits, Π, forming an angle $i < 90°$ with the plane perpendicular to the direction of observation (the plane of the sky), Π_c, and the CM of the system, is at rest relative to the observer on Earth, $\vec{v}_{CM, Ob} = 0$, and the radial velocities of the two stars are maximum in value, $v_{rad, 1, max}$ and $v_{rad, 2, max}$, respectively, but in the opposite directions, in the positions Σ_1 and Σ_2, it results in:

$$\vec{H}_{CM, Ob} = (M_1 + M_2) \cdot \vec{v}_{CM, Ob} = M_1 \vec{v}_{1, CM} + M_2 \vec{v}_{2, CM} = 0;$$

$$\vec{v}_{CM, Ob} = \frac{M_1 \vec{v}_{1, CM} + M_2 \vec{v}_{2, CM}}{M_1 + M_2} = 0;$$

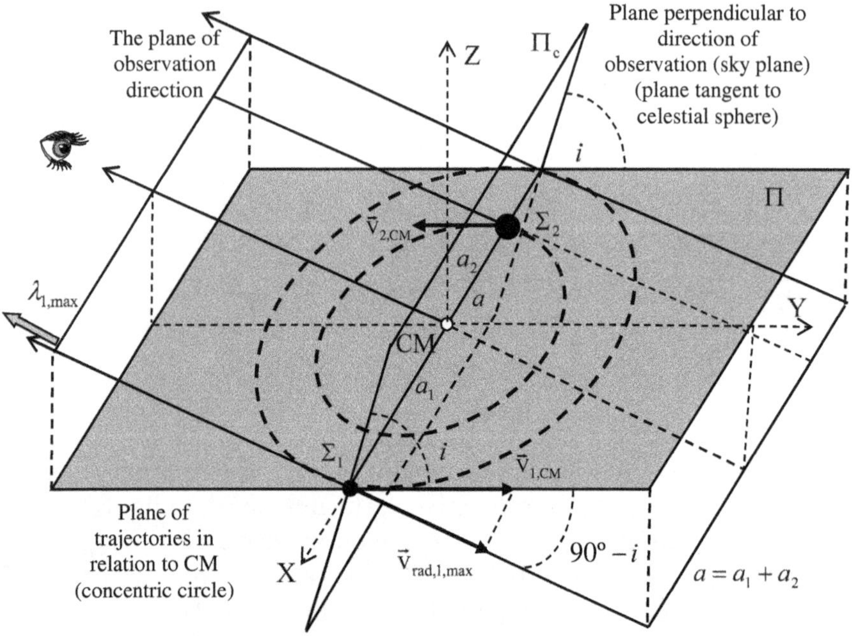

Fig. 14.1

$$M_1\vec{v}_{1,\,\mathrm{CM}} + M_2\vec{v}_{2,\,\mathrm{CM}} = 0; \quad M_1 v_{1,\,\mathrm{CM}} = M_2 v_{2,\,\mathrm{CM}}; \quad \frac{M_2}{M_1} = \frac{v_{1,\,\mathrm{CM}}}{v_{2,\,\mathrm{CM}}};$$

$$v_{\mathrm{rad},\,1,\,\mathrm{max}} = v_{1,\,\mathrm{CM}}\cos(90^\circ - i) = v_{1,\,\mathrm{CM}}\sin i;$$

$$v_{1,\,\mathrm{CM}} = \frac{v_{\mathrm{rad},\,1,\,\mathrm{max}}}{\sin i};$$

$$v_{\mathrm{rad},\,2,\,\mathrm{max}} = v_{2,\,\mathrm{CM}}\cos(90^\circ - i) = v_{2,\,\mathrm{CM}}\sin i;$$

$$v_{2,\,\mathrm{CM}} = \frac{v_{\mathrm{rad},\,2,\,\mathrm{max}}}{\sin i};$$

$$\frac{M_2}{M_1} = \frac{v_{1,\,\mathrm{CM}}}{v_{2,\,\mathrm{CM}}} = \frac{v_{\mathrm{rad},\,1,\,\mathrm{max}}}{v_{\mathrm{rad},\,2,\,\mathrm{max}}};$$

$$v_{1,\,\mathrm{CM}} = \omega a_1 = \frac{2\pi}{T}a_1; \quad v_{2,\,\mathrm{CM}} = \omega a_2 = \frac{2\pi}{T}a_2;$$

$$a_1 = \frac{v_{1,\,\mathrm{CM}}T}{2\pi}; \quad a_2 = \frac{v_{2,\,\mathrm{CM}}T}{2\pi};$$

$$a_1 + a_2 = \frac{T}{2\pi}(\mathrm{v}_{1,\,\mathrm{CM}} + \mathrm{v}_{2,\,\mathrm{CM}}) = a;$$

$$M_1 a_1 = M_2 a_2; \quad a_1 + a_2 = a;$$

$$a_1 = a\frac{M_2}{M_1 + M_2}; \quad a_2 = a\frac{M_1}{M_1 + M_2};$$

$$F = K\frac{M_1 M_2}{a^2} = M_1 \frac{4\pi^2}{T^2} a_1 = M_2 \frac{4\pi^2}{T^2} a_2;$$

$$K\frac{M_1 M_2}{a^2} = M_1 \frac{4\pi^2}{T^2} a \frac{M_2}{M_1 + M_2};$$

$$K\frac{1}{a^2} = \frac{4\pi^2}{T^2} a \frac{1}{M_1 + M_2};$$

$$\frac{T^2}{a^3} = \frac{4\pi^2}{K(M_1 + M_2)}; \quad M_1 + M_2 = \frac{4\pi^2 a^3}{KT^2};$$

$$\frac{T^2}{(a_1 + a_2)^3} = \frac{4\pi^2}{K(M_1 + M_2)};$$

$$\frac{T^2}{\frac{T^3}{(2\pi)^3}(\mathrm{v}_{1,\,\mathrm{CM}} + \mathrm{v}_{2,\,\mathrm{CM}})^3} = \frac{4\pi^2}{K(M_1 + M_2)};$$

$$M_1 + M_2 = \frac{T}{2\pi K}(\mathrm{v}_{1,\,\mathrm{CM}} + \mathrm{v}_{2,\,\mathrm{CM}})^3;$$

$$\mathrm{v}_{1,\,\mathrm{CM}} = \frac{\mathrm{v}_{\mathrm{rad},\,1,\,\mathrm{max}}}{\sin i}; \quad \mathrm{v}_{2,\,\mathrm{CM}} = \frac{\mathrm{v}_{\mathrm{rad},\,2,\,\mathrm{max}}}{\sin i};$$

$$M_1 + M_2 = \frac{T}{2\pi K}\left(\frac{\mathrm{v}_{\mathrm{rad},\,1,\,\mathrm{max}}}{\sin i} + \frac{\mathrm{v}_{\mathrm{rad},\,2,\,\mathrm{max}}}{\sin i}\right)^3;$$

$$M_1 + M_2 = \frac{T}{2\pi K \sin^3 i} \cdot (\mathrm{v}_{\mathrm{rad},\,1,\,\mathrm{max}} + \mathrm{v}_{\mathrm{rad},\,2,\,\mathrm{max}})^3;$$

$$\frac{M_2}{M_1} = \frac{\mathrm{v}_{\mathrm{rad},\,1,\,\mathrm{max}}}{\mathrm{v}_{\mathrm{rad},\,2,\,\mathrm{max}}};$$

$$M_2 = M_1 \cdot \frac{\mathrm{v}_{\mathrm{rad},\,1,\,\mathrm{max}}}{\mathrm{v}_{\mathrm{rad},\,2,\,\mathrm{max}}};$$

$$M_1 + M_1 \cdot \frac{\mathrm{v}_{\mathrm{rad},\,1,\,\mathrm{max}}}{\mathrm{v}_{\mathrm{rad},\,2,\,\mathrm{max}}} = \frac{T}{2\pi K \sin^3 i} \cdot (\mathrm{v}_{\mathrm{rad},\,1,\,\mathrm{max}} + \mathrm{v}_{\mathrm{rad},\,2,\,\mathrm{max}})^3;$$

$$M_1 \cdot \frac{\mathrm{v}_{\mathrm{rad},\,1,\,\mathrm{max}} + \mathrm{v}_{\mathrm{rad},\,2,\,\mathrm{max}}}{\mathrm{v}_{\mathrm{rad},\,2,\,\mathrm{max}}} = \frac{T}{2\pi K \sin^3 i} \cdot \left(\mathrm{v}_{\mathrm{rad},\,1,\,\mathrm{max}} + \mathrm{v}_{\mathrm{rad},\,2,\,\mathrm{max}}\right)^3;$$

$$M_1 = \frac{T \cdot \mathrm{v}_{\mathrm{rad},\,2,\,\mathrm{max}}}{2\pi K \sin^3 i} \cdot \left(\mathrm{v}_{\mathrm{rad},\,1,\,\mathrm{max}} + \mathrm{v}_{\mathrm{rad},\,2,\,\mathrm{max}}\right)^2;$$

$$M_2 = \frac{T \cdot \mathrm{v}_{\mathrm{rad},\,1,\,\mathrm{max}}}{2\pi K \sin^3 i} \cdot \left(\mathrm{v}_{\mathrm{rad},\,1,\,\mathrm{max}} + \mathrm{v}_{\mathrm{rad},\,2,\,\mathrm{max}}\right)^2.$$

Problem 15
Altair Star

For the star Altair in the constellation Aquila, the absolute magnitude is $M = 2.3$.

Determine its mass, μ, expressing it in solar masses, μ_S, if the brightness–mass dependence for a star is given by the law $L = k \cdot \mu$, where $k =$ constant. It is known that the absolute magnitude of the Sun is $M_S = 4.8$.

Solution

Knowing the relationship between absolute magnitude and apparent magnitude, written for the star Altair and the Sun, as well as Pogson's formula, the result is:

$$M = m + 5 \cdot \log \frac{\Delta_{\text{std}}}{\Delta}; \quad M_S = m_S + 5 \cdot \log \frac{\Delta_{\text{std}}}{\Delta_S};$$

$$m = M - 5 \cdot \log \frac{\Delta_{\text{std}}}{\Delta}; \quad m_S = M_S - 5 \cdot \log \frac{\Delta_{\text{std}}}{\Delta_S};$$

$$\log \frac{E}{E_S} = -0.4 \cdot (m - m_S) = \log \frac{\frac{L}{4\pi\Delta^2}}{\frac{L_S}{4\pi\Delta_S^2}} = \log \frac{L}{L_S} + 2 \cdot \log \frac{\Delta_S}{\Delta};$$

$$\log \frac{L}{L_S} + 2 \cdot \log \frac{\Delta_S}{\Delta} = -0.4 \cdot (m - m_S);$$

$$\log \frac{L}{L_S} + 2 \cdot \log \frac{\Delta_S}{\Delta} = -0.4 \cdot \left(M - 5 \cdot \log \frac{\Delta_{\text{std}}}{\Delta} - M_S + 5 \cdot \log \frac{\Delta_{\text{std}}}{\Delta_S}\right);$$

$$\log\frac{L}{L_S}+2\cdot\log\frac{\Delta_S}{\Delta}=-0.4\cdot(M-5\cdot\log\Delta_{\mathrm{std}}$$
$$+5\cdot\log\Delta-M_S+5\cdot\log\Delta_{\mathrm{std}}-5\cdot\log\Delta_S);$$
$$\log\frac{L}{L_S}+2\cdot\log\frac{\Delta_S}{\Delta}=-0.4\cdot(M+5\cdot\log\Delta-M_S-5\cdot\log\Delta_S);$$
$$\log\frac{L}{L_S}+2\cdot\log\frac{\Delta_S}{\Delta}=-0.4\cdot(M-M_S)-0.4\cdot5\cdot(\log\Delta-\log\Delta_S);$$
$$\log\frac{L}{L_S}+2\cdot\log\frac{\Delta_S}{\Delta}=-0.4\cdot(M-M_S)+2\cdot(\log\Delta_S-\log\Delta);$$
$$\log\frac{L}{L_S}+2\cdot\log\frac{\Delta_S}{\Delta}=-0.4\cdot(M-M_S)+2\cdot\log\frac{\Delta_S}{\Delta};$$
$$\log\frac{L}{L_S}=-0.4\cdot(M-M_S);$$
$$\frac{L}{L_S}=10^{-0.4\cdot(M-M_S)};$$
$$L=L_S\cdot10^{0.4\cdot(M_S-M)};$$
$$L=k\cdot\mu^{3.5};\quad L_S=k\cdot\mu_S^{3.5};$$
$$k\cdot M^{3.5}=k\cdot M_S^{3.5}\cdot10^{0.4\cdot(M_S-M)};$$
$$\mu^{3.5}=\mu_S^{3.5}\cdot10^{0.4\cdot(M_S-M)};$$
$$\mu=\mu_S\cdot\sqrt[3.5]{10^{0.4\cdot(M_S-M)}},$$

representing the relation for the mass of the star Altair;

$$M=2.3;\quad M_S=4.8;$$
$$\mu=\mu_S\cdot\sqrt[3.5]{10^{0.4\cdot(4.8-2.3)}};\quad \mu=\mu_S\cdot\sqrt[3.5]{10^{0.4\cdot2.5}};$$
$$\mu=\mu_S\cdot\sqrt[3.5]{10};$$
$$\mu=1.94\cdot\mu_S.$$

Problem 16

Kepler's Laws and the Planets of the Solar System

Figure 16.1 depicts the first four planets in our solar system (1 – Mercury; 2 – Venus; 3 – Earth; 4 – Mars), captured at a certain moment during their evolution around the Sun in circular orbits.

a) Using the data in the adjacent table referring to the four planets, *determine* the time intervals after which each planet will align for the first time with the Earth and the Sun. *Specify* the order of the three alignments.

Planet	Revolution period (sidereal)
Mercury	87.97 days
Venus	224.7 days
Earth	365.3 days
Mars	687.0 days

b) According to the moments of alignment of the Earth with each planet and with the Sun, *establish* the positions of the other two planets in relation to the line of alignment between the Earth, each planet, and the Sun.

c) *Determine* the time intervals after which the same three alignments will be repeated: Mercury–Earth–Sun; Venus–Earth–Sun; Mars–Earth–Sun. *Specify* the order of the three realignments.

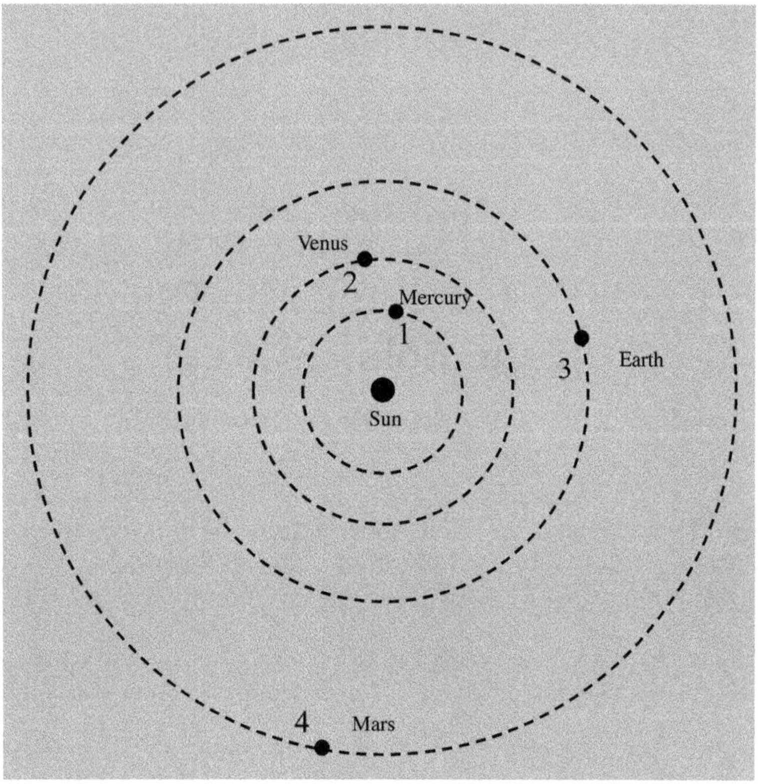

Fig. 16.1

Solution

a) From the measurements made on the drawing in Figure 16.2, where the initial positions of the planets are indicated, we find that:

$$\Delta\alpha_{31} = \Delta\alpha_{\mathrm{EMe}} = \Delta\alpha_{\mathrm{E,\,Mercury}} = 66^\circ = \frac{66}{180}\pi;$$

$$\Delta\alpha_{32} = \Delta\alpha_{\mathrm{EV}} = \Delta\alpha_{\mathrm{E,\,Venus}} = 84^\circ = \frac{84}{180}\pi;$$

$$\Delta\alpha_{34} = \Delta\alpha_{\mathrm{EMa}} = \Delta\alpha_{\mathrm{E,\,Mars}} = 104^\circ = \frac{104}{180}\pi.$$

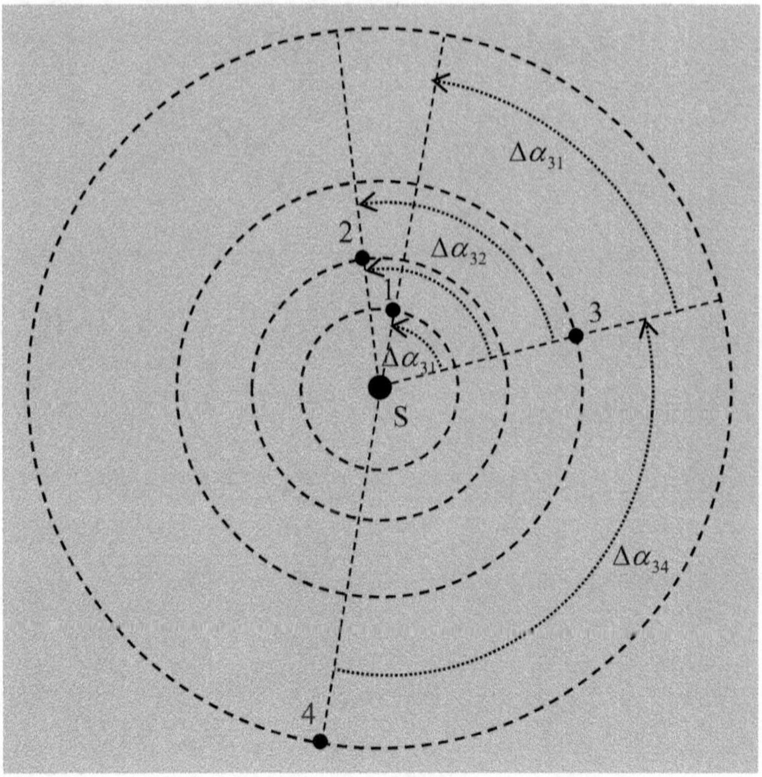

Fig. 16.2

1)

The alignment of Mercury–Earth–Sun is highlighted in the drawing in Figure 16.3. Due to the tangential velocity of Mercury, which is greater than the tangential velocity of the Earth, the angles at the center described by the vector rays of the planets Earth and Mercury, from their initial moment until their first alignment with the Sun, are:

$$\alpha_{\rm E} = \alpha_{\rm Earth} = \omega_{\rm E} t_{\rm EMe} = \frac{2\pi}{T_{\rm E}} t_{\rm EMe};$$

$$\alpha_{\rm M} = \alpha_{\rm Mercury} = \omega_{\rm Me} t_{\rm EMe} = \frac{2\pi}{T_{\rm Me}} t_{\rm EMe}.$$

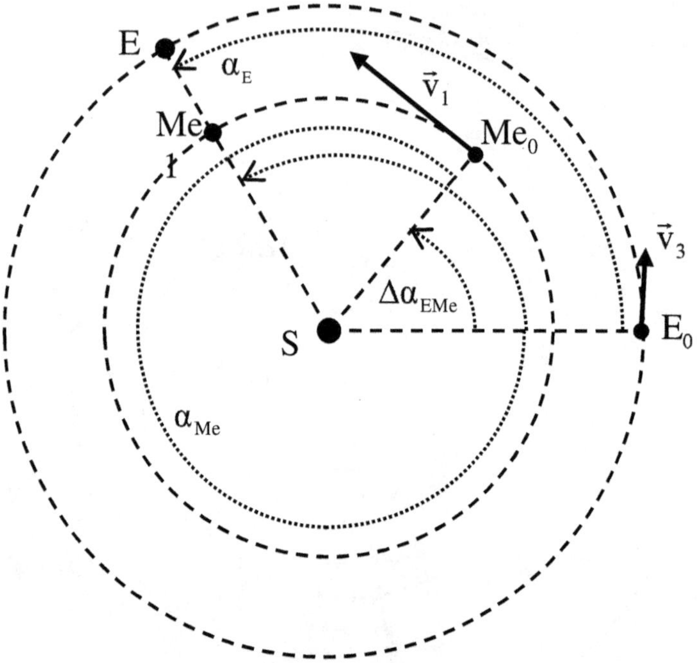

Fig. 16.3

Thus:

$$\alpha_{\text{Me}} = 2\pi + (\alpha_{\text{E}} - \Delta\alpha_{\text{EMe}});$$

$$\alpha_{\text{Me}} - \alpha_{\text{E}} = 2\pi - \Delta\alpha_{\text{EMe}};$$

$$(\omega_{\text{Me}} - \omega_{\text{E}}) \cdot t_{\text{EMe}} = 2\pi - \Delta\alpha_{\text{EMe}};$$

$$2\pi\left(\frac{1}{T_{\text{Me}}} - \frac{1}{T_{\text{E}}}\right) \cdot t_{\text{EMe}} = 2\pi - \Delta\alpha_{\text{EMe}};$$

$$2\pi \cdot \frac{T_{\text{E}} - T_{\text{Me}}}{T_{\text{E}}T_{\text{Me}}} \cdot t_{\text{EMe}} = 2\pi - \Delta\alpha_{\text{EMe}};$$

$$t_{\text{EMe}} = \frac{T_{\text{E}}T_{\text{Me}}}{T_{\text{E}} - T_{\text{Me}}}\left(1 - \frac{\Delta\alpha_{\text{EMe}}}{2\pi}\right);$$

$T_{\text{E}} = 365.3$ days; $\quad T_{\text{Mercury}} = T_{\text{Me}} = 87.97$ days; $\quad \Delta\alpha_{\text{EMe}} = \dfrac{66}{180}\pi;$

$$t_{\text{EMe}} = \frac{365.3 \cdot 87.97}{365.3 - 87.97}\left(1 - \frac{\frac{66}{180}\pi}{2\pi}\right) \text{ days;}$$

$$t_{\text{EMe}} = \frac{365.3 \cdot 87.97}{365.3 - 87.97}\left(1 - \frac{33}{180}\right) \text{ days;}$$

$$t_{\text{EMe}} = \frac{365.3 \cdot 87.97}{277.33} \cdot \frac{147}{180} \text{ days;}$$

$$t_{\text{Earth, Mercury}} = 94.63 \text{ days.}$$

2)

The alignment of Venus–Earth–Sun is highlighted in the drawing in Figure 16.4.

Because the tangential velocity of Venus is greater than the tangential velocity of the Earth, the center angles described by the vector rays of the planets Earth and Venus, from the initial moment to the moment of their first alignment with the Sun, are:

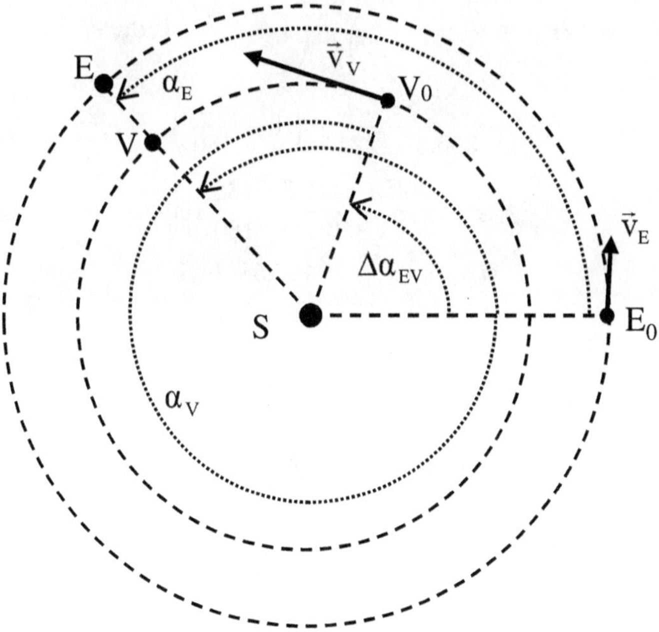

Fig. 16.4

$$\alpha_{\rm E} = \alpha_{\rm Earth} = \omega_{\rm E} t_{\rm EV} = \frac{2\pi}{T_{\rm E}} t_{\rm EV}; \quad \alpha_{\rm V} = \alpha_{\rm Venus} = \omega_{\rm V} t_{\rm EV} = \frac{2\pi}{T_{\rm V}} t_{\rm EV}.$$

Thus:

$$\alpha_{\rm V} = 2\pi + (\alpha_{\rm E} - \Delta\alpha_{\rm EV});$$

$$\alpha_{\rm M} - \alpha_{\rm E} = 2\pi - \Delta\alpha_{\rm EV};$$

$$(\omega_{\rm M} - \omega_{\rm E}) \cdot t_{\rm EV} = 2\pi - \Delta\alpha_{\rm EV};$$

$$2\pi \left(\frac{1}{T_{\rm V}} - \frac{1}{T_{\rm E}} \right) \cdot t_{\rm EV} = 2\pi - \Delta\alpha_{\rm EV};$$

$$2\pi \cdot \frac{T_{\rm E} - T_{\rm V}}{T_{\rm E} T_{\rm V}} \cdot t_{\rm EV} = 2\pi - \Delta\alpha_{\rm EV};$$

$$t_{\rm EV} = \frac{T_{\rm E} T_{\rm V}}{T_{\rm E} - T_{\rm V}} \left(1 - \frac{\Delta\alpha_{\rm EV}}{2\pi} \right);$$

$$T_{\rm E} = 365.3 \text{ days}; \quad T_{\rm Venus} = T_{\rm V} = 224.7 \text{ days}; \quad \Delta\alpha_{\rm EV} = \frac{84}{180}\pi;$$

$$t_{\rm EV} = \frac{365.3 \cdot 224.7}{365.3 - 224.7} \left(1 - \frac{\frac{84}{180}\pi}{2\pi} \right) \text{ days};$$

$$t_{\rm EV} = \frac{365.3 \cdot 224.7}{365.3 - 224.7} \left(1 - \frac{42}{180} \right) \text{ days};$$

$$t_{\rm EV} = \frac{365.3 \cdot 224.7}{140.6} \cdot \frac{138}{180} \text{ days};$$

$$t_{\rm Earth,\, Venus} = 447.58 \text{ days}.$$

3)

The alignment of Mars–Earth–Sun is highlighted in the drawing in Figure 16.5.

Because the tangential velocity of the Earth is greater than the tangential velocity of Mars, the center angles described by the vector rays of the planets Earth and Mars from the initial moment to the moment of their first alignment with the Sun are:

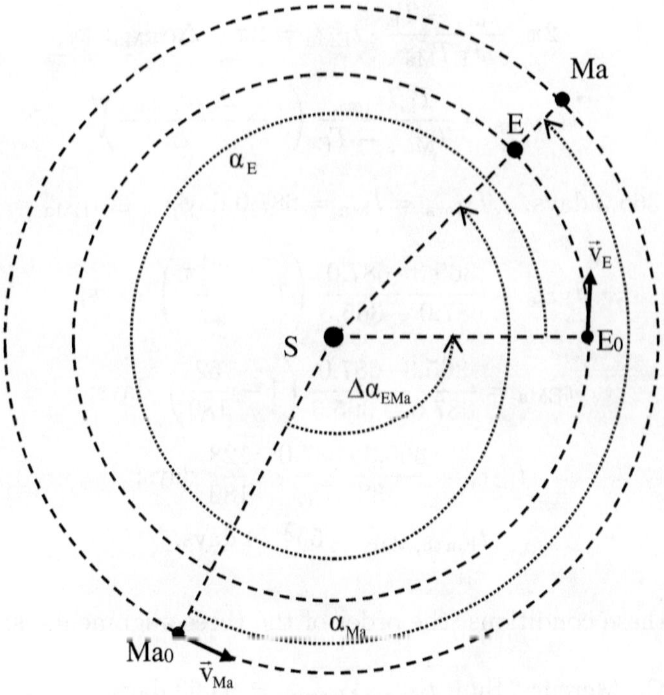

Fig. 16.5

$$\alpha_{\mathrm{E}} = \alpha_{\mathrm{Earth}} = \omega_{\mathrm{E}} t_{\mathrm{EMa}} = \frac{2\pi}{T_{\mathrm{E}}} t_{\mathrm{EMa}};$$

$$\alpha_{\mathrm{Ma}} = \alpha_{\mathrm{Mars}} = \omega_{\mathrm{Ma}} t_{\mathrm{EMa}} = \frac{2\pi}{T_{\mathrm{Ma}}} t_{\mathrm{EMa}}.$$

Thus:

$$\alpha_{\mathrm{E}} = 2\pi + (\alpha_{\mathrm{Ma}} - \Delta\alpha_{\mathrm{EMa}});$$

$$\alpha_{\mathrm{E}} - \alpha_{\mathrm{Ma}} = 2\pi - \Delta\alpha_{\mathrm{EMa}};$$

$$(\omega_{\mathrm{E}} - \omega_{\mathrm{Ma}}) \cdot t_{\mathrm{EMa}} = 2\pi - \Delta\alpha_{\mathrm{EMa}};$$

$$2\pi \left(\frac{1}{T_{\mathrm{E}}} - \frac{1}{T_{\mathrm{Ma}}} \right) \cdot t_{\mathrm{EMa}} = 2\pi - \Delta\alpha_{\mathrm{EMa}};$$

$$2\pi \cdot \frac{T_{\text{Ma}} - T_{\text{E}}}{T_{\text{E}}T_{\text{Ma}}} \cdot t_{\text{EMa}} = 2\pi - \Delta\alpha_{\text{EMa}};$$

$$t_{\text{E, Mars}} = \frac{T_{\text{E}}T_{\text{Mars}}}{T_{\text{Mars}} - T_{\text{E}}}\left(1 - \frac{\Delta\alpha_{\text{E, Mars}}}{2\pi}\right);$$

$$T_{\text{E}} = 365.3 \text{ days}; \quad T_{\text{Mars}} = T_{\text{Ma}} = 687.0 \text{ days}; \quad \Delta\alpha_{\text{EMa}} = \frac{104}{180}\pi;$$

$$t_{\text{EMa}} = \frac{365.3 \cdot 687.0}{687.0 - 365.3}\left(1 - \frac{\frac{104}{180}\pi}{2\pi}\right) \text{ days};$$

$$t_{\text{EMa}} = \frac{365.3 \cdot 687.0}{687.0 - 365.3}\left(1 - \frac{52}{180}\right) \text{ days};$$

$$t_{\text{EMa}} = \frac{365.3 \cdot 687.0}{321.7} \cdot \frac{128}{180} \text{ days};$$

$$t_{\text{Earth, Mars}} = 554.74 \text{ days}.$$

Under these conditions, the order of the three alignments is:

1) Earth, Mercury, Sun; $t_{\text{Earth, Mercury}} = 94.63$ days;
2) Earth, Venus, Sun; $t_{\text{Earth, Venus}} = 447.58$ days;
3) Earth, Mars, Sun; $t_{\text{Earth, Mars}} = 554.74$ days.

b)

First alignment: Earth, Mercury, Sun.

The angles at the center described by the vector rays of the planets Mercury, Venus, Earth, and Mars until the first alignment, Earth–Mercury–Sun, in the calculated time interval, $t_{\text{Earth, Mercury}} = 94.63$ days, are:

1)

$$\alpha_{\text{Mercury}} = \omega_{\text{Me}}t_{\text{EMe}} = \frac{2\pi}{T_{\text{Me}}}t_{\text{EMe}};$$

$$T_{\text{Mercury}} = 87.97 \text{ days}; \quad t_{\text{EMe}} = 94.63 \text{ days};$$

$$\alpha_{\text{Mercury}} = 2.15 \cdot \pi = 387.25^\circ = 360^\circ + 27.25^\circ.$$

2)

$$\alpha_{\text{Venus}} = \omega_{\text{V}} t_{\text{EMe}} = \frac{2\pi}{T_{\text{V}}} t_{\text{EMe}};$$

$$T_{\text{Venus}} = 224.7 \text{ days}; \quad t_{\text{EMe}} = 94.63 \text{ days};$$

$$\alpha_{\text{Venus}} = 0.84 \cdot \pi = 151.61^\circ.$$

3)

$$\alpha_{\text{Earth}} = \omega_{\text{E}} t_{\text{EMe}} = \frac{2\pi}{T_{\text{E}}} t_{\text{EMe}};$$

$$T_{\text{Earth}} = 365.3 \text{ days}; \quad t_{\text{EMe}} = 94.63 \text{ days};$$

$$\alpha_{\text{Earth}} = 0.51 \cdot \pi = 93.25^\circ.$$

4)

$$\alpha_{\text{Mars}} = \omega_{\text{Me}} t_{\text{EMe}} = \frac{2\pi}{T_{\text{Me}}} t_{\text{EMe}};$$

$$T_{\text{Mars}} = 687.0 \text{ days}; \quad t_{\text{EMe}} = 94.63 \text{ days};$$

$$\alpha_{\text{Mars}} = 0.27 \cdot \pi = 49.58^\circ.$$

The values of the four angles calculated above are compared to the four planets' initial positions.

The relative positions of the four planets at the time of the first alignment, Earth–Mercury–Sun, are shown in the drawing in Figure 16.6.

Second alignment: Earth, Venus, Sun.

The angles at the center described by the vector rays of the planets Mercury, Venus, Earth, and Mars, up to the moment of the second alignment, Earth–Venus–Sun, in the calculated time interval, $t_{\text{Earth, Venus}} = 447.58$ days, are:

1)

$$\alpha_{\text{Mercury}} = \omega_{\text{Me}} t_{\text{EV}} = \frac{2\pi}{T_{\text{M}}} t_{\text{EV}};$$

$$T_{\text{Mercury}} = 87.97 \text{ days}; \quad t_{\text{EV}} = 447.58 \text{ days};$$

$$\alpha_{\text{Mercury}} = 10.17 \cdot \pi = 1831.63^\circ = 5 \cdot 360^\circ + 31.63^\circ.$$

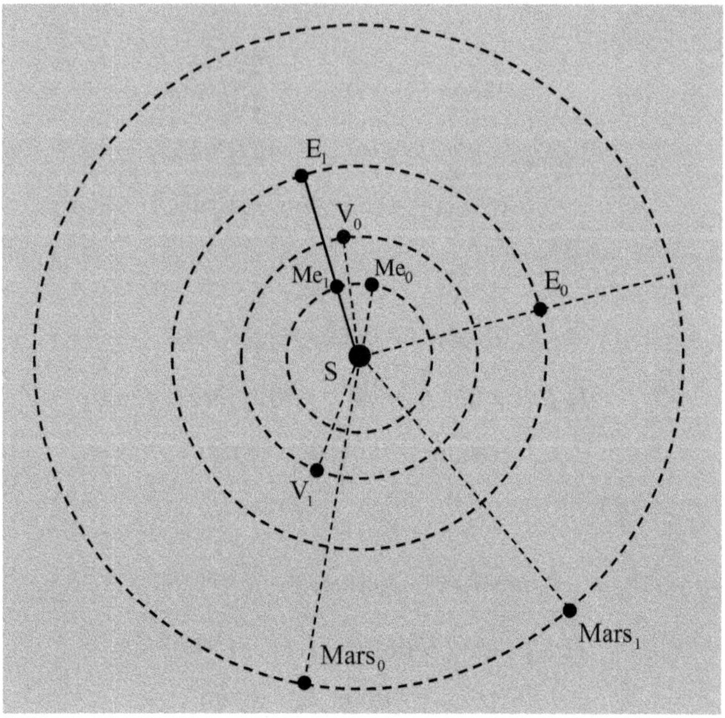

Fig. 16.6

2)

$$\alpha_{\text{Venus}} = \omega_{\text{V}} t_{\text{EV}} = \frac{2\pi}{T_{\text{V}}} t_{\text{EV}};$$

$$T_{\text{Venus}} = 224.7 \text{ days}; \quad t_{\text{EV}} = 447.58 \text{ days};$$

$$\alpha_{\text{Venus}} = 3.98 \cdot \pi = 717.08^\circ = 360^\circ + 357.08^\circ.$$

3)

$$\alpha_{\text{Earth}} = \omega_{\text{E}} t_{\text{EV}} = \frac{2\pi}{T_{\text{E}}} t_{\text{EV}};$$

$$T_{\text{Earth}} = 365.3 \text{ days}; \quad t_{\text{EV}} = 447.58 \text{ days};$$

$$\alpha_{\text{Earth}} = 2.45 \cdot \pi = 441.08^\circ = 360^\circ + 81.08^\circ.$$

4)

$$\alpha_{\text{Mars}} = \omega_{\text{Ma}} t_{\text{EV}} = \frac{2\pi}{T_{\text{Ma}}} t_{\text{EV}};$$

$$T_{\text{Mars}} = 687.0 \text{ days}; \quad t_{\text{EV}} = 445.58 \text{ days};$$

$$\alpha_{\text{Mars}} = 1.29 \cdot \pi = 233.49^{\circ}.$$

The values of the four angles calculated above are compared to the four planets' initial positions.

The relative positions of the four planets at the time of the second alignment, Earth–Venus–Sun, are shown in the drawing in Figure 16.7.

Third alignment: Earth, Mars, Sun.

The angles at the center described by the vector rays of the planets Mercury, Venus, Earth, and Mars, up to the moment of the third alignment, Earth–Mars–Sun, in the calculated time interval, $t_{\text{Earth, Mars}} = 554.74$ days, are:

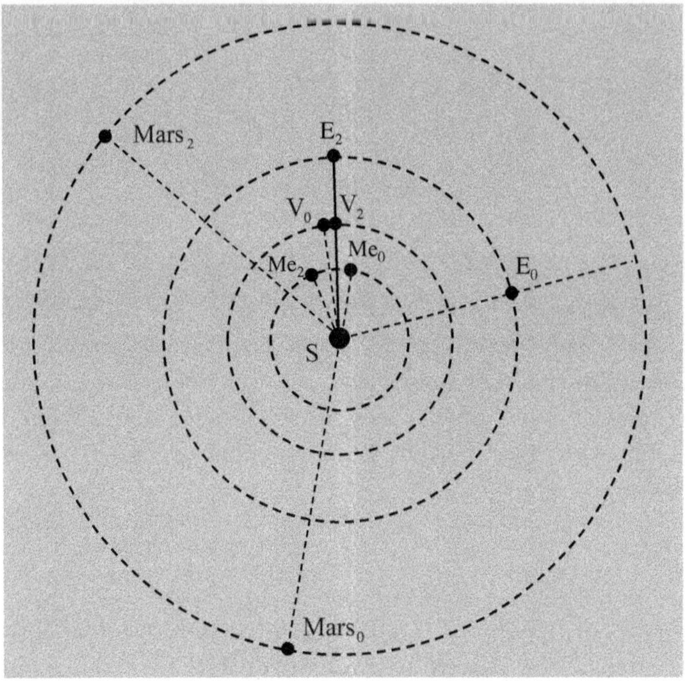

Fig. 16.7

1)

$$\alpha_{\text{Mercury}} = \omega_{\text{Ma}} t_{\text{EMa}} = \frac{2\pi}{T_{\text{M}}} t_{\text{EMa}};$$

$$T_{\text{Mercury}} = 87.97 \text{ days}; \quad t_{\text{EMa}} = 554.74 \text{ days};$$

$$\alpha_{\text{Mercury}} = 12.61 \cdot \pi = 2270.16^\circ = 6 \cdot 360^\circ + 90^\circ + 20.16^\circ.$$

2)

$$\alpha_{\text{Venus}} = \omega_{\text{V}} t_{\text{EMa}} = \frac{2\pi}{T_{\text{V}}} t_{\text{EMa}};$$

$$T_{\text{Venus}} = 224.7 \text{ days}; \quad t_{\text{EMa}} = 554.74 \text{ days};$$

$$\alpha_{\text{Venus}} = 3.98 \cdot \pi = 717.08^\circ = 360^\circ + 357.08^\circ.$$

3)

$$\alpha_{\text{Earth}} = \omega_{\text{P}} t_{\text{EMa}} = \frac{2\pi}{T_{\text{P}}} t_{\text{EMa}};$$

$$T_{\text{Earth}} = 365.3 \text{ days}; \quad t_{\text{EMa}} = 554.74 \text{ days};$$

$$\alpha_{\text{Earth}} = 3.03 \cdot \pi = 546.69^\circ = 360^\circ + 180^\circ + 6.69^\circ.$$

4)

$$\alpha_{\text{Mars}} = \omega_{\text{M}} t_{\text{EMa}} = \frac{2\pi}{T_{\text{M}}} t_{\text{EMa}};$$

$$T_{\text{Mars}} = 687.0 \text{ days}; \quad t_{\text{EMa}} = 554.74 \text{ days};$$

$$\alpha_{\text{Mars}} = 1.61 \cdot \pi = 290.69^\circ = 180^\circ + 90^\circ + 20.69^\circ.$$

The values of the four angles calculated above are compared to the four planets' initial positions.

The relative positions of the four planets at the time of the third alignment, Earth–Mars–Sun, are shown in the drawing in Figure 16.8.

c)

Each alignment, under the required conditions, will be repeated after a time equal to the period of synodic revolution of the planet that realigns with the Earth and the Sun.

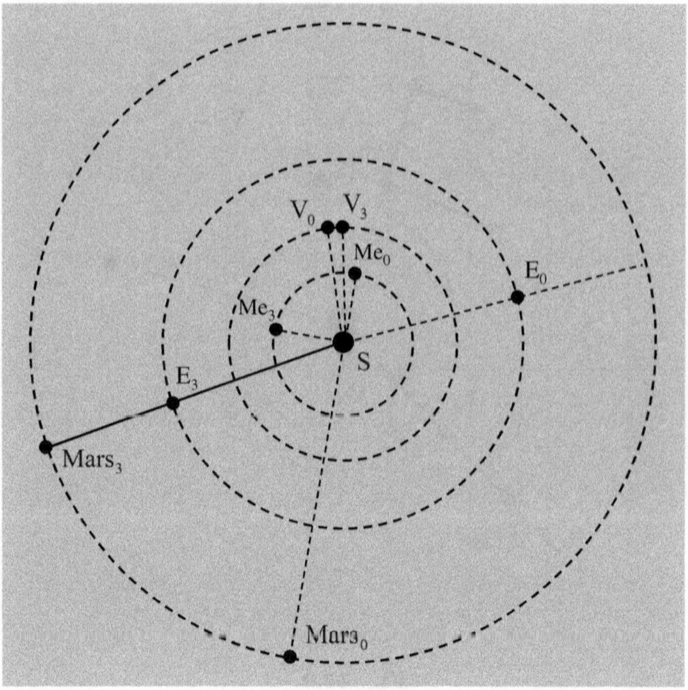

Fig. 16.8

1)

The repetition of the alignment (Earth–Mercury–Sun), shown in the drawing in Figure 16.9, will occur after the time interval representing the synodic period of the planet Mercury (inner planet), $T_{\text{synodic, Mercury}}$.

$$\alpha_{\text{Me}} = 2\pi + \alpha_{\text{E}};$$

$$\alpha_{\text{Me}} - \alpha_{\text{E}} = 2\pi; \quad \alpha_{\text{Me}} = \omega_{\text{Me}} \cdot T_{\text{synodic, Me}}; \quad \alpha_{\text{E}} = \omega_{\text{E}} \cdot T_{\text{synodic, Me}};$$

$$(\omega_{\text{Me}} - \omega_{\text{E}}) \cdot T_{\text{synodic, Me}} = 2\pi;$$

$$\left(\frac{2\pi}{T_{\text{sidereal, Me}}} - \frac{2\pi}{T_{\text{sidereal, E}}}\right) \cdot T_{\text{synodic, Me}} = 2\pi;$$

$$\left(\frac{1}{T_{\text{sidereal, Me}}} - \frac{1}{T_{\text{sidereal, E}}}\right) \cdot T_{\text{synodic, Me}} = 1;$$

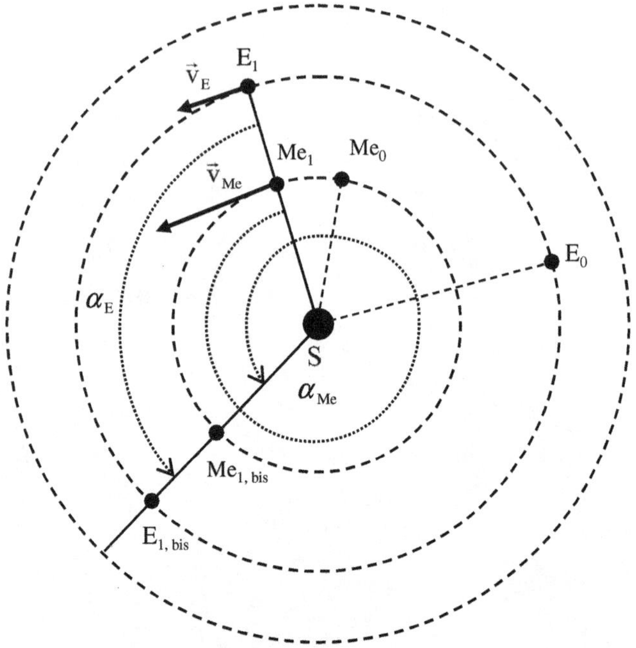

Fig. 16.9

$$T_{\text{synodic, Me}} = T_{\text{synodic, Mercury}} = \frac{T_{\text{sidereal, Earth}} \cdot T_{\text{sidereal, Mercury}}}{T_{\text{sidereal, Earth}} - T_{\text{sidereal, Mercury}}};$$

$$T_{\text{E}} = 365.3 \text{ days}; \quad T_{\text{Mercury}} = T_{\text{Me}} = 87.97 \text{ days};$$

$$T_{\text{synodic, Mercury}} = \frac{365.3 \text{ days} \cdot 87.97 \text{ days}}{365.3 \text{ days} - 87.97 \text{ days}} = 115.87 \text{ days};$$

$$\alpha_{\text{E}} = \omega_{\text{E}} \cdot T_{\text{synodic, Me}};$$

$$\alpha_{\text{E}} = \frac{2\pi}{T_{\text{sidereal, E}}} \cdot T_{\text{synodic, Me}} = \frac{2\pi}{365.3 \text{ days}} \cdot 115.87 \text{ days};$$

$$\alpha_{\text{E}} = 0.63 \cdot \pi = 113.4°;$$

$$\alpha_{\text{Me}} = \omega_{\text{Me}} \cdot T_{\text{synodic, Me}};$$

$$\alpha_{\text{Me}} = \frac{2\pi}{T_{\text{sidereal, Me}}} \cdot T_{\text{synodic, Me}} = \frac{2\pi}{87.97 \text{ days}} \cdot 115.87 \text{ days};$$

$$\alpha_{\text{Me}} = 2.63 \cdot \pi = 473.4° = 360° + 90° + 23.4°.$$

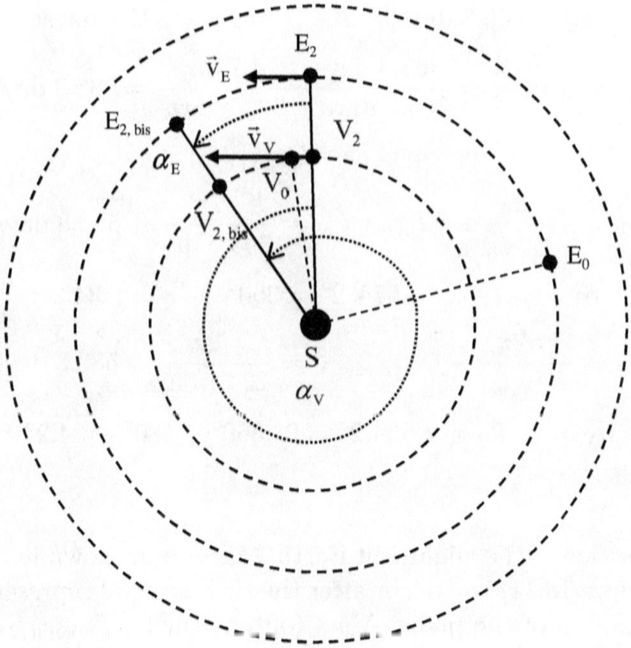

Fig. 16.10

2)

The repetition of the alignment Earth–Venus–Sun, shown in the drawing in Figure 16.10, will occur after the time interval representing the synodic period of the planet Venus (inner planet), $T_{\text{synodic, Venus}}$.

$$\alpha_V = 2\pi + \alpha_E;$$

$$\alpha_V - \alpha_E = 2\pi; \quad \alpha_V = \omega_V \cdot T_{\text{synodic, V}}; \quad \alpha_E = \omega_E \cdot T_{\text{synodic, V}};$$

$$(\omega_V - \omega_E) \cdot T_{\text{synodic, V}} = 2\pi;$$

$$\left(\frac{2\pi}{T_{\text{sidereal, V}}} - \frac{2\pi}{T_{\text{sidereal, E}}}\right) \cdot T_{\text{synodic, V}} = 2\pi;$$

$$\left(\frac{1}{T_{\text{sidereal, V}}} - \frac{1}{T_{\text{sidereal, E}}}\right) \cdot T_{\text{synodic, V}} = 1;$$

$$T_{\text{synodic, V}} = T_{\text{synodic, Venus}} = \frac{T_{\text{sidereal, Earth}} \cdot T_{\text{sidereal, Venus}}}{T_{\text{sidereal, Earth}} - T_{\text{sidereal, Venus}}};$$

$$T_{\text{E}} = 365.3 \text{ days}; \quad T_{\text{Venus}} = T_{\text{V}} = 224.7 \text{ days};$$

$$T_{\text{synodic, Venus}} = \frac{365.3 \text{ days} \cdot 224.7 \text{ days}}{365.3 \text{ days} - 224.7 \text{ days}} = 583.8 \text{ days};$$

$$\alpha_{\text{E}} = \omega_{\text{E}} \cdot T_{\text{synodic, V}};$$

$$\alpha_{\text{E}} = \frac{2\pi}{T_{\text{sidereal, E}}} \cdot T_{\text{synodic, V}} = \frac{2\pi}{365.3 \text{ days}} \cdot 583.8 \text{ days};$$

$$\alpha_{\text{E}} = 3.19 \cdot \pi = 574,2^\circ = 360^\circ + 180^\circ + 34.2^\circ;$$

$$\alpha_{\text{V}} = \frac{2\pi}{T_{\text{sidereal, V}}} \cdot T_{\text{synodic, V}} = \frac{2\pi}{224.7 \text{ days}} \cdot 583.8 \text{ days};$$

$$\alpha_{\text{V}} = 5.19 \cdot \pi = 934.2^\circ = 2 \cdot 360^\circ + 180^\circ + 34.2^\circ.$$

3)

The repetition of the alignment Earth–Mars–Sun, shown in the drawing in Figure 16.11, will occur after the time interval representing the synodic period of the planet Mars (outer planet), $T_{\text{synodic, Mars}}$.

$$\alpha_{\text{E}} = 2\pi + \alpha_{\text{Ma}};$$

$$\alpha_{\text{E}} - \alpha_{\text{Ma}} = 2\pi; \quad \alpha_{\text{Ma}} = \omega_{\text{Ma}} \cdot T_{\text{synodic, Mars}}; \quad \alpha_{\text{E}} = \omega_{\text{E}} \cdot T_{\text{synodic, Mars}};$$

$$(\omega_{\text{E}} - \omega_{\text{Ma}}) \cdot T_{\text{synodic, Mars}} = 2\pi;$$

$$\left(\frac{2\pi}{T_{\text{sidereal, E}}} - \frac{2\pi}{T_{\text{sidereal, Ma}}}\right) \cdot T_{\text{synodic, Mars}} = 2\pi;$$

$$\left(\frac{1}{T_{\text{sidereal, E}}} - \frac{1}{T_{\text{sidereal, Ma}}}\right) \cdot T_{\text{synodic, Mars}} = 1;$$

$$T_{\text{synodic, Ma}} = T_{\text{synodic, Mars}} = \frac{T_{\text{sidereal, Mars}} \cdot T_{\text{sidereal, Earth}}}{T_{\text{sidereal, Mars}} - T_{\text{sidereal, Earth}}};$$

$$T_{\text{E}} = 365.3 \text{ days}; \quad T_{\text{Mars}} = T_{\text{Ma}} = 687.0 \text{ days};$$

$$T_{\text{synodic, Mars}} = \frac{365.3 \text{ days} \cdot 687.0 \text{ days}}{687.0 \text{ days} - 365.3 \text{ days}} = 780.1 \text{ days};$$

$$\alpha_{\text{E}} = \omega_{\text{E}} \cdot T_{\text{synodic, Ma}};$$

$$\alpha_{\text{E}} = \frac{2\pi}{T_{\text{sidereal, E}}} \cdot T_{\text{synodic, Ma}} = \frac{2\pi}{365.3 \text{ days}} \cdot 780.1 \text{ days};$$

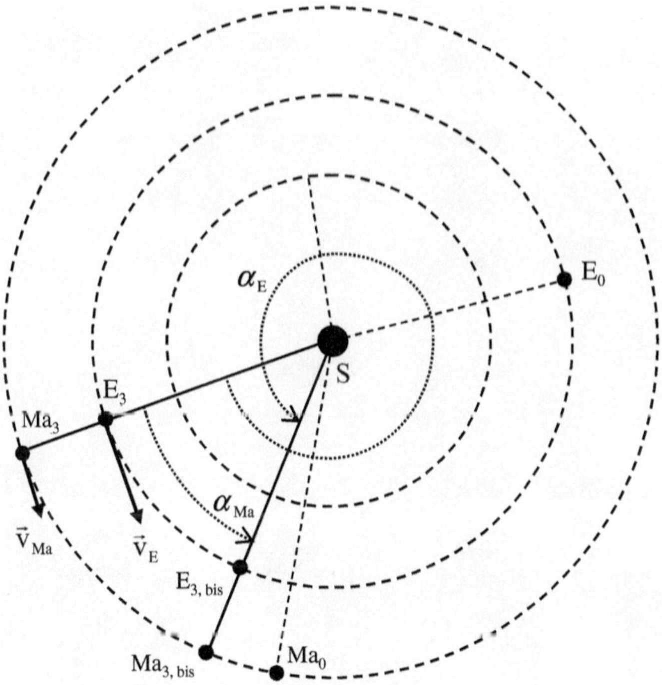

Fig. 16.11

$$\alpha_{\mathrm{E}} = 4.27 \cdot \pi = 768.6^\circ = 2 \cdot 360^\circ + 48.6^\circ;$$

$$\alpha_{\mathrm{E}} = \frac{2\pi}{T_{\text{sidereal, E}}} \cdot T_{\text{synodic, Ma}} = \frac{2\pi}{365.3 \text{ days}} \cdot 780.1 \text{ days};$$

$$\alpha_{\mathrm{Ma}} = \frac{2\pi}{T_{\text{sidereal, Ma}}} \cdot T_{\text{synodic, Ma}} = \frac{2\pi}{687.0 \text{ days}} \cdot 780.1 \text{ days};$$

$$\alpha_{\mathrm{Ma}} = 2.27 \cdot \pi = 408.6^\circ = 360^\circ + 48.6^\circ.$$

Under these conditions, the order of the three alignments is:

1) Earth, Mercury, Sun; $T_{\text{synodic, Mercury}} = 115.87$ days;
2) Earth, Venus, Sun; $T_{\text{synodic, Venus}} = 583.80$ days;
3) Earth, Mars, Sun; $T_{\text{synodic, Mars}} = 780.10$ days.

[illegible]

Problem 17

IOAA-jr Romania 2022 Zodiac

In their apparent motions, the Moon and the other large planets in our solar system do not stray far from the ecliptic plane. Their apparent trajectories described on the celestial sphere remain contained in a region that extends symmetrically on both sides of the ecliptic, having a total width of about 18°. In the drawing in Figure 17.1, the heliocentric orbit of the Earth is shown in the plane of the ecliptic, and in the drawing in Figure 17.2, the equivalent apparent geocentric orbit of the Sun in relation to the Earth is shown.

The time intervals of the evolution of the true sun in each of the 12 constellations of the zodiac are as follows: **1)** Aries, 21 III–20 IV; **2)** Taurus, 21 IV–20 V; **3)** Gemini, 21 V–20 VI; **4)** Cancer, 21 VI–22 VII; **5)** Leo, 23 VII–22 VIII; **6)** Virgo, 23 VIII–22 IX; **7)** Libra, 23 IX–22 X; **8)** Scorpio, 23 X–21 XI; **9)** Sagittarius, 22 XI–21 XII; **10)** Capricorn, 22 XII–21 I; **11)** Aquarius, 22 I–19 II; **12)** Pisces, 20 II–21 III.

The International Olympiad on Astronomy and Astrophysics for Juniors, 1st Edition, was to take place in Suceava, Romania, from March 28 to April 3, 2020. The event, proposed by Romania, was established at the International Olympiad on Astronomy and Astrophysics, 13th Edition, held in Hungary from August 2 to 10, 2019.

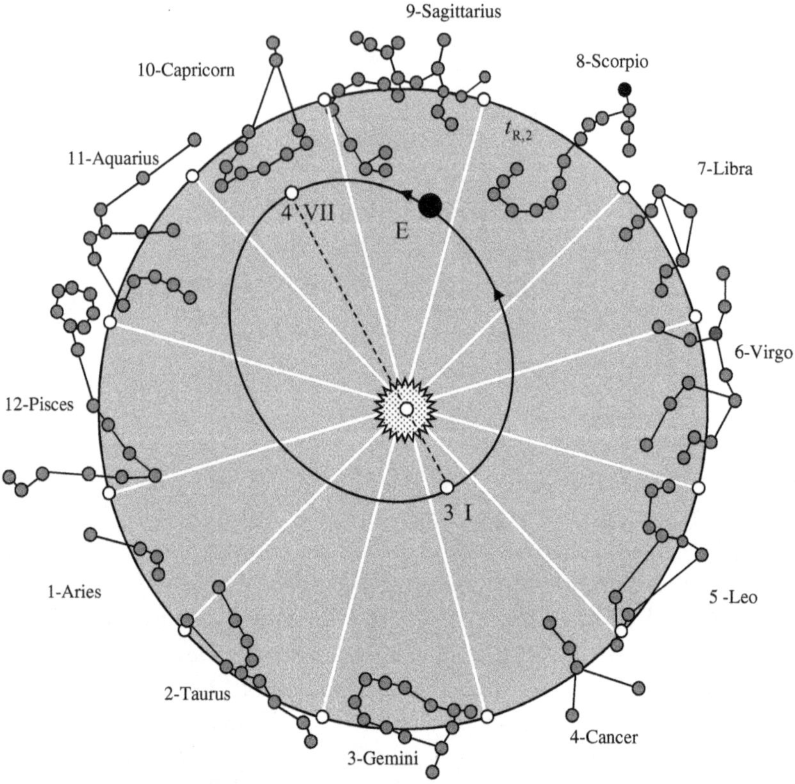

Fig. 17.1

But, given the known conditions, the OIAA for Juniors, 1st Edition, took place in Romania in November 2022!

a) *Identify*:

  1) the sign of the OIAA, 13th Edition, Hungary, August 2–10, 2019;
  2) the sign of the proposed date for the OIAA-jr 1st Edition, March 28–April 3, 2020, Romania;
  3) the sign of the OIAA-jr 1st Edition, November, 2022.

b) Each of the 3 specified Olympiad events can be considered to have taken place on the day when the Sun, viewed from Earth, was in the middle of the angular range corresponding to each of the 3 signs.

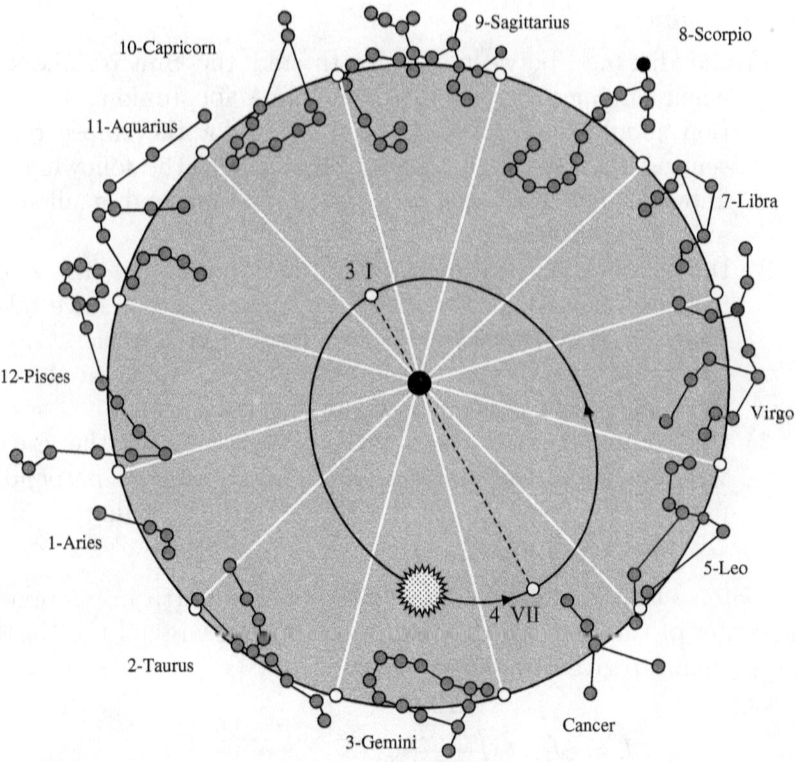

Fig. 17.2

1) *Determine* the time interval between any two of these three events:

$$\Delta t_{\text{H-R},1}; \quad \Delta t_{\text{R},1-\text{R},2}; \quad \Delta t_{\text{H-R},2}.$$

2) *Estimate*, by direct measurements on the drawing, using a protractor, the value of the angle between the position vector of the Earth in relation to the Sun, corresponding to each of the three specified moments, i.e. the directions of the vectors $\vec{r}_{\text{H}}$, $\vec{r}_{\text{R},1}$ and $\vec{r}_{\text{R},2}$, and the direction of the apsides line, Aph–Ph, i.e. the angles α_{H}, $\alpha_{\text{R},1}$ and $\alpha_{\text{R},2}$, respectively.

c) *Determine* the area of the surface described by the position vector of the center of the Earth, $\vec{r}$, in relation to the center of the Sun:

1) from position $\vec{r}_{\text{Hungary}}$ to position $\vec{r}_{\text{Romania},1}$;
2) from position $\vec{r}_{\text{Romania},1}$ to position $\vec{r}_{\text{Romania},2}$;
3) from position $\vec{r}_{\text{Hungary}}$ to position $\vec{r}_{\text{Romania},2}$.

d) *Determine*:

1) the distance between the Earth and the Sun on the day when, in Suceava, in 2020, the IOAA for Juniors, 1st Edition should have taken place, $r_{R,1}$, if, for the ellipse representing the Earth's orbit around the Sun, the following are known: large semi-axis, $a = 149\,597\,500$ km and small semi-axis, $b = 149\,580\,670$ km;

2) the acceleration of the center of the Earth on the day when, in Suceava, in 2020, the IOAA for Juniors should have taken place, $a_{E,1}$, compared with the gravitational acceleration in the gravitational field of the Sun, corresponding to the distance $r_{R,1}$ from the center of the Sun, $g_{S,1}$;

3) the components of the speed of the center of the Earth, $\vec{v}_R$, parallel to the major axis of the ellipse and perpendicular to the major axis of the ellipse during the International Olympiad in Romania, $v_{//}$ and $v_{\perp}$, respectively.

It is known that the angular momentum of the Earth in relation to the center of the Sun and the total mechanical energy of the Earth–Sun system are given by the expressions:

$$L = M_E b \cdot \sqrt{\frac{KM_S}{a}}; \quad E = -K\frac{M_E M_S}{2a}.$$

Given: the gravitational attraction constant, $K = 6.67 \cdot 10^{-11}$ $\mathrm{Nm^2kg^{-2}}$; the mass of the Sun, $M_S = 1.989 \cdot 10^{30}$ kg.

Solution

a) In the drawing in Figure 17.3 are represented the positions of the Earth on its elliptical orbit around the Sun, as well as the positions of the projections of the Sun in the zodiac belt, corresponding to the moments of the three events specified in the statement of the problem.

In the drawing in Figure 17.4, the positions of the Sun on its apparent elliptical orbit around the Earth and the positions

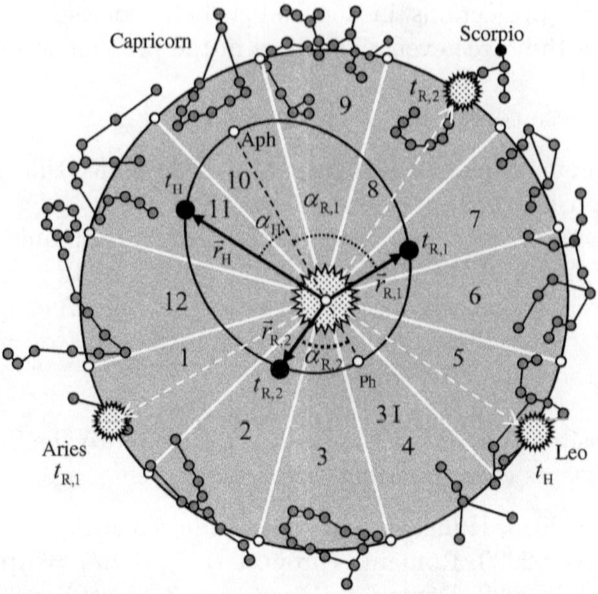

Fig. 17.3

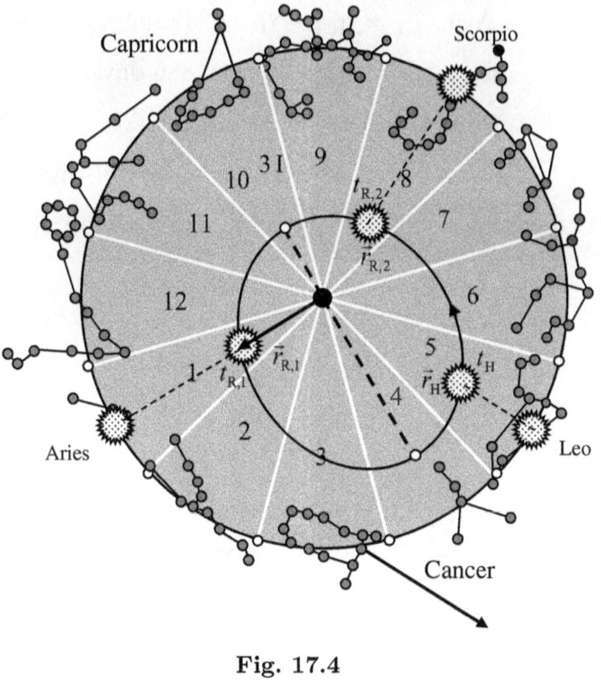

Fig. 17.4

of the Sun's projections in the zodiac belt, corresponding to the moments of the three events specified in the problem statement, are represented.

Under these conditions, we identify that:

1) the announcement in Hungary was made under the sign of Leo on August 6 2019;
2) the proposed event was due to take place in Romania under the sign of Aries, in the middle of it, on April 4 2020;
3) the International Olympiad on Astronomy and Astrophysics for Juniors, 1st Edition, took place in Romania in November 2022, under the sign of Scorpio.

b) The position vectors of the Earth, in relation to the Sun, as well as the moments corresponding to the three events, are:

- for IOAA 2019, Hungary: $\vec{r}_{\rm H}$; $t_{\rm H}$ = August 6 2019;
- for IOAA-jr 2020, Romania (proposed): $\vec{r}_{\rm R,1}$; $t_{\rm R,1}$ = April 4 2020;
- for IOAA-jr 2022, Romania: $\vec{r}_{\rm R,2}$; $t_{\rm R,2}$ = 3 November 2022.

Thus, the time intervals between any two of these three events are:

$$\Delta t_{\rm H-R,1} = t_{\rm R,1} - t_{\rm H} = 242 \text{ days};$$

$$\Delta t_{\rm R,1-R,2} = t_{\rm R,2} - t_{\rm R,1} = 886 \text{ days};$$

$$\begin{aligned}\Delta t_{\rm H-R,2} &= t_{\rm R,2} - t_{\rm H} = \Delta t_{\rm H-R,1} + \Delta t_{\rm R,1-R,2}\\ &= 223\,\text{days} + 886\,\text{days} = 1109\,\text{days}.\end{aligned}$$

From direct measurements performed on the drawing in Figure 17.3, it results that:

$$\alpha_{\rm H} \approx 30°;$$

$$\alpha_{\rm R,1} \approx 90°; \quad \vec{r}_{\rm R,1} \perp (\text{Aph–Ph});$$

$$\alpha_{\rm R,2} \approx 60°.$$

c)

1) It is known that during the evolution of the Earth in its elliptical orbit around the Sun, according to Kepler's third law, the areolar velocity of the Earth is constant, so that:

$$\Omega = \frac{\Delta A}{\Delta t} = \text{constant};$$

$$\frac{A_{\text{ellipse}}}{T} = \frac{\Delta A_{\text{H-R,1}}}{\Delta t_{\text{H-R,1}}};$$

$$\Delta A_{\text{H-R,1}} = \frac{\Delta t_{\text{H-R,1}}}{T} A_{\text{ellipse}} = \frac{\Delta t_{\text{H-R,1}}}{T} \cdot \pi \cdot a \cdot b;$$

$$\Delta t_{\text{H-R,1}} = 223\,\text{days}; \quad T = 365.256\,\text{days};$$

$$a = 149\,597\,500\text{ km}; \quad b = 149\,580\,670\text{ km};$$

$$\Delta A_{\text{H-R,1}} = \frac{223\text{ days}}{365.256\text{ days}} \cdot 3.14 \cdot 149\,597\,500\text{ km} \cdot 149\,580\,670\text{ km};$$

$$\Delta A_{\text{H-R,1}} = \frac{223}{365.256} \cdot 3.14 \cdot 149\,597\,500 \cdot 149\,580\,670\,\text{km}^2;$$

$$\Delta A_{\text{H-R,1}} = 4.2897 \cdot 10^{16}\text{ km}^2.$$

2)

$$\frac{A_{\text{ellipse}}}{T} = \frac{\Delta A_{\text{R,1-R,2}}}{\Delta t_{\text{R,1-R,2}}};$$

$$\Delta A_{\text{R,1-R,2}} = \frac{\Delta t_{\text{R,1-R,2}}}{T} A_{\text{ellipse}} = \frac{\Delta t_{\text{R,1-R,2}}}{T} \cdot \pi \cdot a \cdot b;$$

$$\Delta A_{\text{R,1-R,2}} = \frac{886\text{ days}}{365.256\text{ days}} \cdot 3.14 \cdot 149\,597\,500\text{ km} \cdot 149\,580\,670\text{ km};$$

$$\Delta A_{\text{R,1-R,2}} = \frac{886}{365.256} \cdot 3.14 \cdot 149\,597\,500 \cdot 149\,580\,670\text{ km}^2;$$

$$\Delta A_{\text{R,1-R,2}} = 1.704 \cdot 10^{17}\text{ km}^2.$$

3)

$$\frac{A_{\text{ellipse}}}{T} = \frac{\Delta A_{\text{H-R,2}}}{\Delta t_{\text{H-R,2}}}; \quad \frac{A_{\text{ellipse}}}{T} = \frac{\Delta A_{\text{H-R,1}}}{\Delta t_{\text{H-R,1}}};$$

$$\Delta A_{\text{H-R,2}} = \frac{\Delta t_{\text{H-R,2}}}{T} A_{\text{ellipse}} = \frac{\Delta t_{\text{H-R,2}}}{T} \cdot \pi \cdot a \cdot b;$$

$$\Delta t_{\text{H-R},2} = 1109 \text{ days}; \quad T = 365.256 \text{ days};$$

$$a = 149\,597\,500\,\text{km}; \quad b = 149\,580\,670\,\text{km};$$

$$\Delta A_{\text{H-R},2} = \frac{1109 \text{ days}}{365.256 \text{ days}} \cdot 3.14 \cdot 149\,597\,500 \text{ km} \cdot 149\,580\,670 \text{ km};$$

$$\Delta A_{\text{H-R},2} = \frac{1109}{365.256} \cdot 3.14 \cdot 149\,597\,500 \cdot 149\,580\,670\,\text{km}^2;$$

$$\Delta A_{\text{H-R},2} = 2.133 \cdot 10^{17}\,\text{km}^2.$$

d)

1) According to the notation in Figure 17.5, representing the position of the Earth in relation to the Sun on the day when, as we have shown, $\vec{r}_{\text{R},1} \perp (\text{Aph–Ph})$, the result is:

$$\frac{x^2}{a^2} + \frac{y^2}{b^2} = 1; \quad x = c; \quad y = r_{\text{R},1};$$

$$c^2 = a^2 - b^2;$$

$$\frac{c^2}{a^2} + \frac{r_{\text{R},1}^2}{b^2} = 1; \quad \frac{a^2 - b^2}{a^2} + \frac{r_{\text{R},1}^2}{b^2} = 1;$$

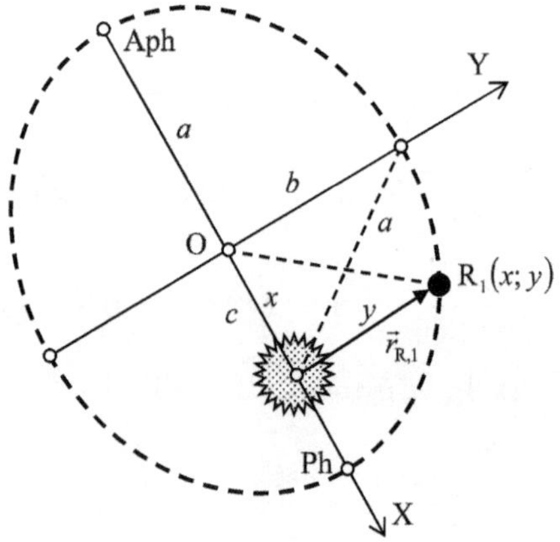

Fig. 17.5

$$1 - \frac{b^2}{a^2} + \frac{r_{\mathrm{R},1}^2}{b^2} = 1; \quad \frac{r_{\mathrm{R},1}^2}{b^2} = \frac{b^2}{a^2};$$

$$r_{\mathrm{R},1}^2 = \frac{b^4}{a^2};$$

$$r_{\mathrm{R},1} = \frac{b^2}{a},$$

representing the distance between the center of the Earth and the center of the Sun during the proposed date for the International Olympiad on Astronomy and Astrophysics for Juniors, 1st Edition, March 28–April 3 2020, in Romania;

$$a = 149\,597\,500\,\mathrm{km}; \quad b = 149\,580\,670\,\mathrm{km};$$

$$r_{\mathrm{R},1} = 149\,563\,841.9\,\mathrm{km}.$$

2)

$$K\frac{M_\mathrm{E} M_\mathrm{S}}{r_{\mathrm{R},1}^2} = M_\mathrm{E} a_{\mathrm{E},1}; \quad a_{\mathrm{E},1} = K\frac{M_\mathrm{S}}{r_{\mathrm{R},1}^2};$$

$$K = 6.67 \cdot 10^{-11}\,\mathrm{Nm^2kg^{-2}}; \quad M_\mathrm{S} = 1.989 \cdot 10^{30}\,\mathrm{kg};$$

$$a_{\mathrm{E},1} = \frac{6.67 \cdot 10^{-11} \cdot 1.989 \cdot 10^{30}}{(149\,563\,841.9)^2 \cdot 10^6}\,\frac{\mathrm{m}}{\mathrm{s}^2};$$

$$a_{\mathrm{E},1} \approx 0.006\,\frac{\mathrm{m}}{\mathrm{s}^2};$$

$$g_{\mathrm{S},1} = K \cdot \frac{M_\mathrm{S}}{r_{\mathrm{R},1}^2} = a_{\mathrm{E},1}.$$

3) Corresponding to the position R_1 of the Earth, represented in the drawing in Figure 17.6, when its angular momentum is

$$\vec{L}_{\mathrm{R},1} = \vec{r}_{\mathrm{R},1} \mathrm{x} M_\mathrm{E} \vec{\mathrm{v}}_{\mathrm{R},1},$$

$$L_{\mathrm{R},1} = r_{\mathrm{R},1} M_\mathrm{E} \mathrm{v}_{\mathrm{R},1} \cdot \sin(\vec{r}_{\mathrm{R},1}; \vec{\mathrm{v}}_{\mathrm{R},1}) = r_{\mathrm{R},1} M_\mathrm{E} \mathrm{v}_{\mathrm{R},1} \cdot \sin\alpha,$$

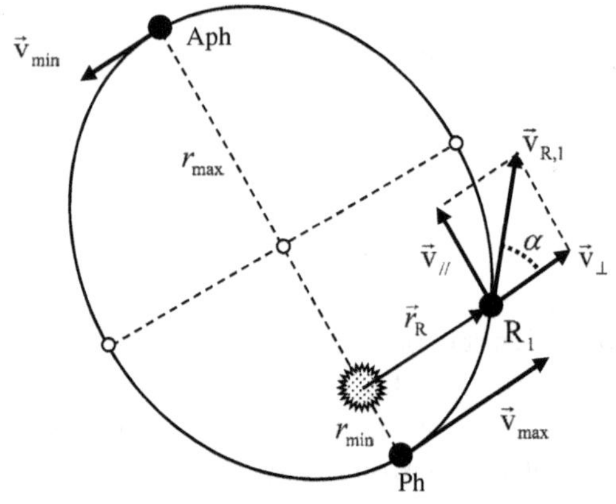

Fig. 17.6

according to the law of conservation of angular momentum, it follows that:

$$\mathrm{v}_{\mathrm{R},1} \cdot \sin\alpha = \mathrm{v}_{//};$$

$$L_{\mathrm{R},1} = r_{\mathrm{R},1} M_{\mathrm{E}} \mathrm{v}_{//}; \quad L = M_{\mathrm{E}} b \cdot \sqrt{\frac{K M_{\mathrm{S}}}{a}};$$

$$M_{\mathrm{E}} b \cdot \sqrt{\frac{K M_{\mathrm{S}}}{a}} = r_{\mathrm{R},1} M_{\mathrm{E}} \mathrm{v}_{//}; \quad r_{\mathrm{R}} = \frac{b^2}{a};$$

$$\mathrm{v}_{//} = \frac{a}{b} \cdot \sqrt{\frac{K M_{\mathrm{S}}}{a}};$$

$$K = 6.67 \cdot 10^{-11}\,\mathrm{Nm^2\,kg^{-2}}; \quad M_{\mathrm{S}} = 1.989 \cdot 10^{30}\,\mathrm{kg};$$

$$a = 149\,597\,500 \;\mathrm{km}; \quad b = 149\,580\,670 \;\mathrm{km};$$

$$\mathrm{v}_{//} = \frac{149\,597\,500\,\mathrm{km}}{149\,580\,670\,\mathrm{km}} \cdot \sqrt{\frac{6.67 \cdot 10^{-11}\,\mathrm{Nm^2\,kg^{-2}} \cdot 1.989 \cdot 10^{30}\,\mathrm{kg}}{149\,597\,500 \cdot 10^3\,\mathrm{m}}};$$

$$\mathrm{v}_{//} = \frac{149\,597\,500}{149\,580\,670} \cdot \sqrt{\frac{6.67 \cdot 10^{-11} \cdot 1.989 \cdot 10^{30}}{149\,597\,500 \cdot 10^3}}\,\frac{\mathrm{m}}{\mathrm{s}};$$

$$\mathrm{v}_{//} = \frac{149\,597\,500}{149\,580\,670} \cdot \sqrt{\frac{6.67 \cdot 1.989}{149\,597\,500}} \cdot 10^8\,\frac{\mathrm{m}}{\mathrm{s}};$$

$$\mathrm{v}_{//} = \frac{\sqrt{149\,597\,500 \cdot 6.67 \cdot 1.989}}{149\,580\,670} \cdot 10^8\,\frac{\mathrm{m}}{\mathrm{s}};$$

$$\mathrm{v}_{//} = 29\,782.9\,\frac{\mathrm{m}}{\mathrm{s}}; \quad \mathrm{v}_{//} = 29\,782.9 \cdot 10^{-3}\,\frac{\mathrm{km}}{\mathrm{s}};$$

$$\mathrm{v}_{//} = 29.7829\,\frac{\mathrm{km}}{\mathrm{s}}.$$

According to the law of conservation of mechanical energy, it results that:

$$E = -K\frac{M_{\mathrm{E}}M_{\mathrm{S}}}{2a};$$

$$E_{\mathrm{R},1} = \frac{M_{\mathrm{E}}\mathrm{v}_{\mathrm{R}}^2}{2} - K\frac{M_{\mathrm{E}}M_{\mathrm{S}}}{r_{\mathrm{R},1}};$$

$$\frac{M_{\mathrm{E}}\mathrm{v}_{\mathrm{R},1}^2}{2} - K\frac{M_{\mathrm{E}}M_{\mathrm{S}}}{r_{\mathrm{R},1}} = -K\frac{M_{\mathrm{E}}M_{\mathrm{S}}}{2a};$$

$$\frac{\mathrm{v}_{\mathrm{R},1}^2}{2} - K\frac{M_{\mathrm{S}}}{r_{\mathrm{R},1}} = -K\frac{M_{\mathrm{S}}}{2a};$$

$$\mathrm{v}_{\mathrm{R},1} = \sqrt{KM_{\mathrm{S}}\left(\frac{2}{r_{\mathrm{R},1}} - \frac{1}{a}\right)} = \sqrt{KM_{\mathrm{S}} \cdot \frac{2a - r_{\mathrm{R},1}}{r_{\mathrm{R},1}a}};$$

$$K = 6.67 \cdot 10^{-11}\,\mathrm{Nm}^2\,\mathrm{kg}^{-2}; \quad M_{\mathrm{S}} = 1.989 \cdot 10^{30}\,\mathrm{kg};$$

$$r_{\mathrm{R},1} = 149\,563\,841.9\ \mathrm{km}; \quad a = 149\,597\,500\,\mathrm{km};$$

$$\mathrm{v}_{\mathrm{R},1} = \sqrt{\begin{array}{c} 6.67 \cdot 10^{-11}\,\mathrm{Nm}^2\,\mathrm{kg}^{-2} \cdot 1.989 \cdot 10^{30}\mathrm{kg} \\ \cdot \dfrac{2 \cdot 149\,597\,500\,\mathrm{km} - 149\,563\,841.9\,\mathrm{km}}{149\,563\,841.9\ \mathrm{km} \cdot 149\,597\,500\,\mathrm{km}} \end{array}};$$

$$\mathrm{v}_{\mathrm{R},1} = \sqrt{\begin{array}{c} 6.67 \cdot 10^{-11}\,\mathrm{Nm}^2\,\mathrm{kg}^{-2} \cdot 1.989 \cdot 10^{30}\mathrm{kg} \\ \cdot \dfrac{2 \cdot 149\,597\,500\ - 149\,563\,841.9}{149\,563\,841.9 \cdot 149\,597\,500 \cdot 10^3\,\mathrm{m}} \end{array}};$$

$$\mathrm{v}_{\mathrm{R},1} = \sqrt{\begin{array}{c} 6.67 \cdot 10^{-11} \cdot 1.989 \cdot 10^{30} \\ \cdot \dfrac{2 \cdot 149\,597\,500\ - 149\,563\,841.9}{149\,563\,841.9 \cdot 149\,597\,500 \cdot 10^3} \end{array}} \cdot 10^8\,\frac{\mathrm{m}}{\mathrm{s}};$$

$$\mathrm{v}_{\mathrm{R},1} = 29\,786.25\,\frac{\mathrm{m}}{\mathrm{s}} = 29\,786.25 \cdot 10^{-3}\,\frac{\mathrm{km}}{\mathrm{s}};$$

$$\mathrm{v}_{\mathrm{R},1} = 29.7862\,\frac{\mathrm{km}}{\mathrm{s}};$$

$$\mathrm{v}_{//} = 29.7829\,\frac{\mathrm{km}}{\mathrm{s}};$$

$$\mathrm{v}_{\perp} = \sqrt{\mathrm{v}_{\mathrm{R},1}^2 - \mathrm{v}_{//}^2} = 0.4433\,\frac{\mathrm{km}}{\mathrm{s}}.$$

Problem 18

A Cosmonaut's Descent onto a Satellite of Saturn

An artist painted "*A Cosmonaut's Descent onto a Satellite of Saturn*" and represented, against the starry background of the sky, the disks of the Sun and Saturn as having the same dimensions.

Considering the representation of the artist to be correct, *specify* the distance from the planet Saturn to the satellite that the artist had in mind when he created the mentioned work.

It is known that: the distance between Saturn and the Sun is 9.54 times greater than the distance between Earth and the Sun; the angular diameter of the Sun's disk, seen from Saturn, is $32'$; Saturn's equatorial diameter is 120 000 km; the closest satellite to Saturn is Pan, located 133.570 km from Saturn; Saturn's farthest satellite is Phoebe, 13 000 000 km from Saturn.

The distances between the Sun, Saturn, and the satellite on which the cosmonaut landed will be approximately equal.

For the calculation of the equatorial diameter of a star, d, the following formula will be used: $d = \alpha r$, where α is the angle (expressed in radians; $1' = 1/3438$ radians) under which the disk of a star can be seen from the observation point (the angular diameter of the star), and r is the distance from the observation point to the star.

Solution

Since the distance between Saturn and the Sun is 9.54 times greater than the distance from Earth to the Sun, it turns out that the angular diameter of the Sun's disk, observed from Saturn, is 9.54 times smaller than the angular diameter of the Sun, observed from Earth. Therefore, using the drawing in Figure 18.1 results in:

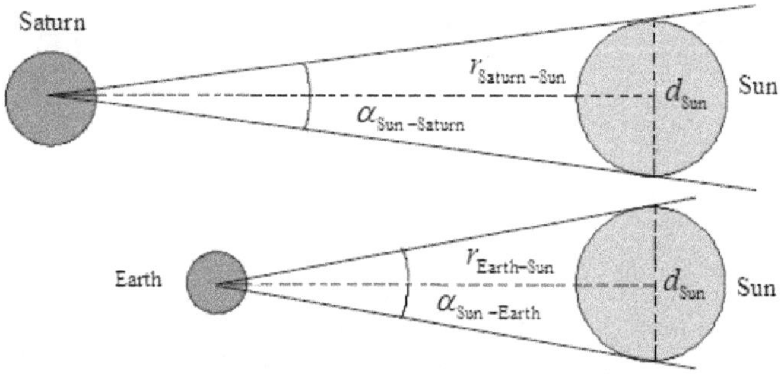

Fig. 18.1

$$d_{\text{Sun}} = r_{\text{Saturn-Sun}} \cdot \alpha_{\text{Sun-Saturn}};$$

$$d_{\text{Sun}} = r_{\text{Earth-Sun}} \cdot \alpha_{\text{Sun-Earth}};$$

$$r_{\text{Saturn-Sun}} = 9.54 \cdot r_{\text{Earth-Sun}};$$

$$\alpha_{\text{Sun-Saturn}} = \frac{1}{9.54} \cdot \alpha_{\text{Sun-Earth}} = \frac{32'}{9.54} \approx 3.35' \approx 0.00097 \text{ radians};$$

$$\alpha_{\text{Sun-Saturn}} \approx 0.001 \text{ rad},$$

representing the angular diameter of the Sun seen from Saturn.

The distance from the satellite of Saturn to the Sun is approximately equal to the distance from Saturn to the Sun, meaning that the angular diameter of the Sun, seen from the satellite of Saturn, is equal to the angular diameter of the Sun seen from Saturn, as indicated in the drawing in Figure 18.2, meaning:

$$\alpha_{\text{Sun-satellite}} \approx \alpha_{\text{Sun-Saturn}} = 0.001 \text{ radians}.$$

If, in the considered painting, the disk of Saturn projected on the celestial sphere has the same dimensions as the disk of the Sun projected on the celestial sphere, it means that the angular diameter

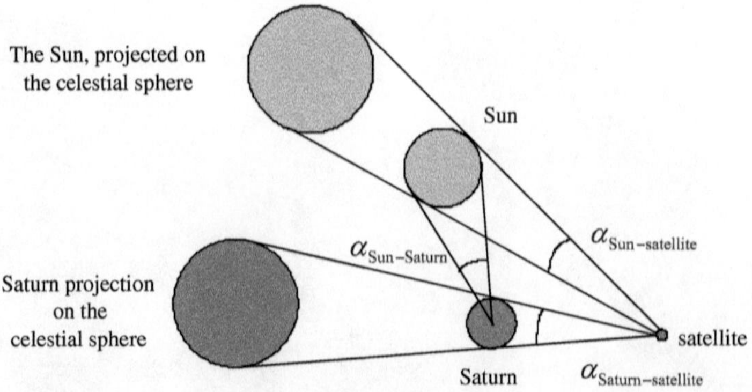

Fig. 18.2

of Saturn seen from its satellite is equal to the angular diameter of the Sun seen from the same satellite, meaning:

$$\alpha_{\text{Saturn-satellite}} = \alpha_{\text{Sun-satellite}} = 0.001\,\text{radians}.$$

Let us now calculate how far from Saturn the satellite onto which the artist's cosmonaut descends must be so that the angular diameter of Saturn, seen from that satellite, is the one calculated above (0.001 radians):

$$r_{\text{satellite-Saturn}} = \frac{d}{\alpha_{\text{Saturn-satellite}}} = \frac{120\,000\,\text{km}}{0.001} = 120\,000\,000\,\text{km}.$$

This is much greater than the distance between Saturn and Saturn's farthest known satellite.

Conclusion: in the artist's vision, the cosmonaut landed on a still undiscovered satellite of Saturn.

Problem 19

The Masses of the Components of Several Visual Binary Stellar Systems

Table 19.1 shows numerical information for 10 visual binary star systems.

Identify, without calculating the masses of the two components of each binary star system, the binary star system for which the difference in mass of its components, $M_2 - M_1$, has the smallest value. Then, *verify* the result by calculation, determining the masses of the components of all the visual binary star systems.

Notation: T – the period of rotation of each component around the CM; p – the parallax of the main star of each visual binary stellar system; α – the angular distance between the two components of each visual binary stellar system; α_1 – the angular distance of each secondary component from the CM of the system; α_2 – the angular distance of each main component from the CM of the system; the period of the Earth's rotation around the Sun, $T_0 = 1\,\text{year}$; the mass of the Earth, $M_E \ll M_S$.

The two components of each visual binary stellar system evolve in the plane of the sky, in circular orbits, around the common CM, at rest relative to the observer, as shown in the drawing in Figure 19.1.

Table 19.1

	Binary system	Period T (years)	Parallax p (arcsec)	α (arcsec)	α_1 (arcsec)	α_2 (arcsec)	M_1 (M_{Sun})	M_2 (M_{Sun})
1	η Cassiopeiae	480	0.184	11.09	8.54			
2	Sirius	49.94	0.379	7.62	5.33			
3	Procyon	40.65	0.287	4.55	3.32			
4	ξ Cnc A, B	59.7	0.047	0.88	0.46			
5	ξ Boötis	149.95	0.148	4.884	2.589			
6	ξ Herc.	34.42	0.104	1.35	0.78			
7	99 Herc.	56.40	0.058	1.03	0.62			
8	β 648	61.8	0.061	1.24	0.65			
9	K Peg.	11.405	0.036	0.336	0.187			
10	85 Peg.	26.27	0.086	0.83	0.42			

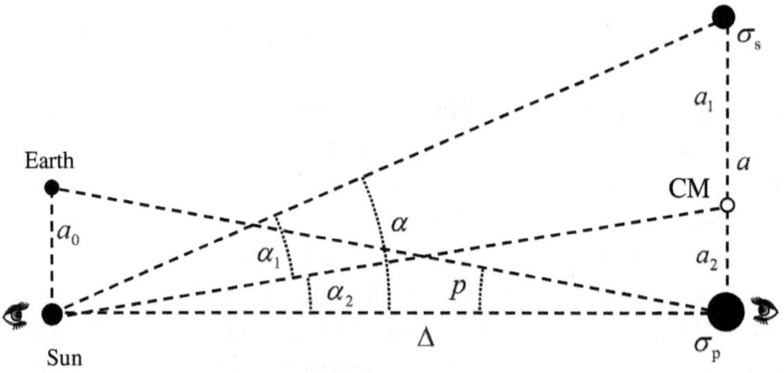

Fig. 19.1

Solution

According to the notation in Figure 19.2, it results that:

$$a \approx \alpha\Delta; \quad a_0 \approx p\Delta;$$

$$F = K\frac{M_1 M_2}{a^2} = M_1\frac{4\pi^2}{T^2}a_1 = M_2\frac{4\pi^2}{T^2}a_2;$$

$$a_1 + a_2 = a; \quad M_1 a_1 = M_2 a_2;$$

$$a_1 = \frac{aM_2}{M_1 + M_2}; \quad a_2 = \frac{aM_1}{M_1 + M_2};$$

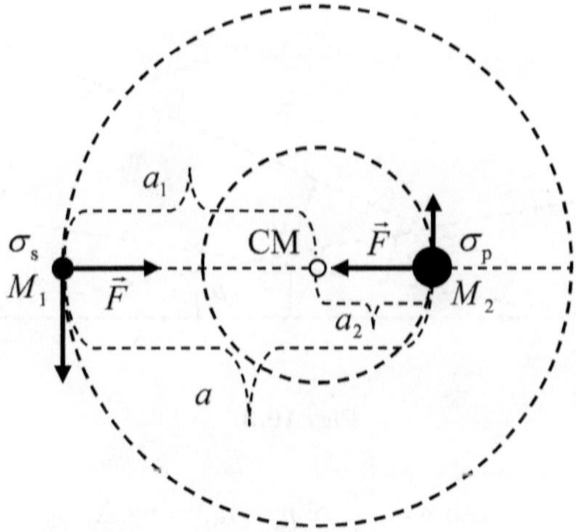

Fig. 19.2

$$K\frac{M_1 M_2}{a^2} = M_1 \frac{4\pi^2}{T^2} \cdot \frac{aM_2}{M_1 + M_2};$$

$$T^2 = \frac{4\pi^2}{K}\frac{a^3}{M_1 + M_2},$$

representing Kepler's third law for the visual binary stellar system.

Writing a similar expression for the Sun–Earth system, it follows that:

$$T_0^2 = \frac{4\pi^2}{K}\frac{a_0^3}{M_S + M_E};$$

$$\frac{T^2}{T_0^2} = \frac{a^3}{a_0^3} \cdot \frac{M_S + M_E}{M_1 + M_2};$$

$$M_1 + M_2 = \frac{a^3}{a_0^3}\frac{T_0^2}{T^2}(M_S + M_E); \quad M_E \ll M_S;$$

$$M_1 + M_2 = \frac{a^3}{a_0^3}\frac{T_0^2}{T^2} M_S.$$

According to the notation in Figure 19.3, it results that:

$$a \approx \alpha\Delta; \quad a_0 \approx p\Delta;$$

$$M_1 + M_2 = \frac{\alpha^3}{p^3}\frac{T_0^2}{T^2} M_S;$$

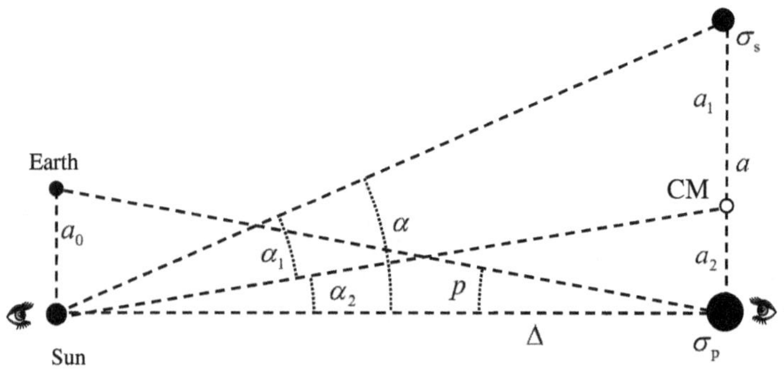

Fig. 19.3

$$\tan p = \frac{a_0}{\Delta} \approx p; \quad a_0 = p \cdot \Delta;$$

$$\tan\alpha = \frac{a}{\Delta} \approx \alpha; \quad a = \alpha \cdot \Delta; \quad \tan\alpha_2 = \frac{a_2}{\Delta} \approx \alpha_2; \quad a_2 = \alpha_2 \cdot \Delta;$$

$$a_1 + a_2 = a; \quad a_1 = a - a_2 = \alpha \cdot \Delta - \alpha_2 \cdot \Delta = (\alpha - \alpha_2) \cdot \Delta;$$

$$\alpha_1 = \alpha - \alpha_2; \quad a_1 = \alpha_1 \cdot \Delta;$$

$$a_1 + a_2 = a; \quad \alpha_1 + \alpha_2 = \alpha;$$

$$M_1 a_1 = M_2 a_2; \quad M_1 \alpha_1 \cdot \Delta = M_2 \alpha_2 \cdot \Delta; \quad M_1 \alpha_1 = M_2 \alpha_2;$$

$$\alpha_1 = \frac{\alpha M_2}{M_1 + M_2}; \quad \alpha_2 = \frac{\alpha M_1}{M_1 + M_2};$$

$$M_2 = \frac{\alpha_1}{\alpha}(M_1 + M_2);$$

$$M_1 = \frac{\alpha_2}{\alpha}(M_1 + M_2); \quad \alpha_2 = \alpha - \alpha_1;$$

$$M_1 + M_2 = \frac{\alpha^3}{p^3}\frac{T_0^2}{T^2} M_S;$$

$$M_1 = \frac{\alpha_2}{\alpha}\left(\frac{\alpha}{p}\right)^3 \left(\frac{T_0}{T}\right)^2 M_S = \frac{\alpha - \alpha_1}{\alpha}\left(\frac{\alpha}{p}\right)^3 \left(\frac{T_0}{T}\right)^2 M_S;$$

$$M_1 = \frac{\alpha - \alpha_1}{\alpha}\left(\frac{\alpha}{p}\right)^3 \left(\frac{T_0}{T}\right)^2 M_S;$$

$$M_2 = \frac{\alpha_1}{\alpha}\left(\frac{\alpha}{p}\right)^3 \left(\frac{T_0}{T}\right)^2 M_S.$$

Using the information in Table 19.1 in the problem statement, where the angular distances α and α_1 are given, we calculate

$$\alpha_2 = \alpha - \alpha_1.$$

The results are recorded in Table 19.2.

We have shown that:

$$M_1 = \frac{\alpha_2}{\alpha}(M_1 + M_2) = M_1(\sigma_{\text{secondary}});$$

$$M_2 = \frac{\alpha_1}{\alpha}(M_1 + M_2) = M_2(\sigma_{\text{principal}}).$$

Thus, we identify the binary stellar system with the minimum difference between the masses of the two components as the stellar system for which the difference in the angular distances from the CM of the system is the minimum. The results are recorded in Table 19.3.

Conclusion:

$$(\alpha_1 - \alpha_2)_{\text{min}} = (\alpha_1 - \alpha_2)_{85\text{ Peg.}} = 0.01;$$

$$(M_2 - M_1)_{\text{min}} = (M_2 - M_1)_{85\text{ Peg.}}.$$

That is, of the visual binary stellar systems analyzed, 85 Peg. has the smallest difference between its two components.

Table 19.2

	Binary system	Period T (years)	Parallax p (arcsec)	α (arcsec)	α_1 (arcsec)	α_2 (arcsec)	M_1 (M_{Sun})	M_2 (M_{Sun})
1	η Glass	480	0.184	11.09	8.54	2.55		
2	Sirius	49.94	0.379	7.62	5.33	2.29		
3	Procyon	40.65	0.287	4.55	3.32	1.23		
4	ξ Cnc A, B	59.7	0.047	0.88	0.46	0.42		
5	ξ Boötis	149.95	0.148	4.884	2.589	2.295		
6	ξ Herc.	34.42	0.104	1.35	0.78	0.57		
7	99 Herc.	56.40	0.058	1.03	0.62	0.41		
8	β 648	61.8	0.061	1.24	0.65	0.59		
9	K Peg.	11.405	0.036	0.336	0.187	0.149		
10	85 Peg.	26.27	0.086	0.83	0.42	0.41		

Table 19.3

	Binary system	Period T (years)	Parallax p (arcsec)	α (arcsec)	α_1 (arcsec)	α_2 (arcsec)	$\alpha_1 - \alpha_2$ (arcsec)
1	η Glass	480	0.184	11.09	8.54	2.55	5.99
2	Sirius	49.94	0.379	7.62	5.33	2.29	3.04
3	Procyon	40.65	0.287	4.55	3.32	1.23	2.09
4	ξ Cnc A, B	59.7	0.047	0.88	0.46	0.42	0.04
5	ξ Boötis	149.95	0.148	4.884	2.589	2.295	0.294
6	ξ Herc.	34.42	0.104	1.35	0.78	0.57	0.21
7	99 Herc.	56.40	0.058	1.03	0.62	0.41	0.21
8	β 648	61.8	0.061	1.24	0.65	0.59	0.06
9	K Peg.	11.405	0.036	0.336	0.187	0.149	0.038
10	85 Peg.	26.27	0.086	0.83	0.42	0.41	0.01

Table 19.4

	Binary system	Period T (years)	Parallax p (arcsec)	α_1 (arcsec)	α_2 (arcsec)	M_1 (M_{Sun})	M_2 (M_{Sun})	$M_2 - M_1$ (M_{Sun})
1	η Glass	480	0.184	8.54	2.55	0.47	0.73	0.26
2	Sirius	49.94	0.379	5.33	2.29	0.98	2.28	1.3
3	Procyon	40.65	0.287	3.32	1.23	0.65	1.76	1.11
4	ξ Cnc A, B	59.7	0.047	0.46	0.42	0.86	0.98	0.12
5	ξ Boötis	149.95	0.148	2.589	2.295	0.75	0.85	0.1
6	ξ Herc.	34.42	0.104	0.78	0.57	0.78	1.07	0.29
7	99 Herc.	56.40	0.058	0.62	0.41	0.69	1.07	0.38
8	β 648	61.8	0.061	0.65	0.59	1.03	1.17	0.14
9	K Peg.	11.405	0.036	0.187	0.149	2.75	3.49	0.74
10	85 Peg.	26.27	0.086	0.42	0.41	0.64	0.66	0.02

We have shown that:

$$M_1 = \frac{\alpha_2}{\alpha}\left(\frac{\alpha}{p}\right)^3\left(\frac{T_0}{T}\right)^2 M_S; \quad M_2 = \frac{\alpha_1}{\alpha}\left(\frac{\alpha}{p}\right)^3\left(\frac{T_0}{T}\right)^2 M_S.$$

The results of the calculations are recorded in Table 19.4.

Conclusion:

$$(M_2 - M_1)_{\min} = (M_2 - M_1)_{85\,\text{Peg.}} = 0.02.$$

Problem 20

Ice Particles Detached from a Comet's Core

A comet evolves around the Sun, moving on a very elongated ellipse. Under the action of solar radiation, ice particles of different diameters detach from the comet's ice core.

Determine the values of the diameters of the ice particles torn from the comet's core, for which, when the comet is in its orbit:

a) the extracted ice particles will continue their evolution around the Sun in the same orbit as the comet's core;
b) the extracted ice particles will move in the direction of the Sun, leaving the orbit of the comet and approaching the Sun;
c) the extracted ice particles will leave the comet's orbit and move away from the Sun.

It is known that the pressure due to solar radiation is given by the expression $p_{\text{rad}} = \varphi/c$, where: φ – the density of the energy flux of the Sun at a distance Δ from it; c – the speed of light in a vacuum.

Given: the mass of the Sun, $M = 1.9 \cdot 10^{30}\,\text{kg}$; the radius of the Sun, $R = 695 \cdot 10^6\,\text{m}$; the temperature of the Sun, $T = 6000\,\text{K}$; the density of ice, $\rho = 300\,\text{kg/m}^3$; the Stefan–Boltzman coefficient $\sigma = 5.67 \cdot 10^{-8}\,\text{J/m}^2\text{K}^4\,\text{s}$; the constant of gravitational attraction, $K = 6.67 \cdot 10^{-11}\text{Nm}^2/\text{kg}^2$; the speed of light in a vacuum, $c = 3 \cdot 10^8\,\text{m/s}$.

We consider negligible: the gravitational interaction between the comet's core and the ice particles extracted from it; the speed of the comet in its orbit.

Solution

$$p_{\text{rad}} = \frac{\varphi}{c};$$

$$\varphi = \frac{L}{4\pi\Delta^2} = \frac{4\pi R^2 \cdot Q}{4\pi\Delta^2} = \frac{4\pi R^2 \cdot \sigma T^4}{4\pi\Delta^2} = \frac{\sigma T^4 R^2}{\Delta^2};$$

$$p_{\text{rad}} = \frac{\sigma T^4 R^2}{c\Delta^2};$$

$$F_{\text{p,rad}} = s \cdot p_{\text{rad}} = \pi r^2 \cdot p_{\text{rad}} = \pi \left(\frac{d}{2}\right)^2 \cdot \frac{\sigma T^4 R^2}{c\Delta^2};$$

$$F_{\text{p,rad}} = \frac{\pi\sigma T^4 R^2 d^2}{4c\Delta^2};$$

$$F_{\text{grav}} = K\frac{mM}{\Delta^2} = K\frac{\rho \cdot \frac{4\pi}{3} \cdot \left(\frac{d}{2}\right)^3 M}{\Delta^2} = K\frac{4\pi\rho d^3 M}{24\Delta^2};$$

$$F_{\text{grav}} = K\frac{\pi\rho d^3 M}{6\Delta^2}.$$

a)

$$F_{\text{grav}} = F_{\text{p,rad}};$$

$$\frac{\pi\sigma T^4 R^2 d^2}{4c\Delta^2} = K\frac{\pi\rho d^3 M}{6\Delta^2};$$

$$\frac{\sigma T^4 R^2}{2c} = K\frac{\rho d M}{3};$$

$$d > \frac{3}{2} \cdot \frac{\sigma T^4 R^2}{c\rho K M};$$

$$d = \frac{3}{2} \cdot \frac{5.67 \cdot 10^{-8}\frac{\text{J}}{\text{m}^2\text{K}^4\text{s}} \cdot 6^4 \cdot 10^{12}\text{K}^4 \cdot 695^2 \cdot 10^{12}\text{m}^2}{3 \cdot 10^8\frac{\text{m}}{\text{s}} \cdot 3 \cdot 10^4\frac{kg}{\text{m}^3} \cdot 6.67 \cdot 10^{-11}\frac{\text{Nm}^2}{\text{kg}^2} \cdot 1.9 \cdot 10^{30}\text{kg}};$$

$$d = \frac{3}{2} \cdot \frac{567 \cdot 10^{-10}\frac{\text{J}}{\text{m}^2\text{K}^4\text{s}} \cdot 6^4 \cdot 10^{12}\text{K}^4 \cdot 695^2 \cdot 10^{12}\text{m}^2}{3 \cdot 10^8\frac{\text{m}}{\text{s}} \cdot 3 \cdot 10^4\frac{\text{kg}}{\text{m}^3} \cdot 667 \cdot 10^{-13}\frac{\text{Nm}^2}{\text{kg}^2} \cdot 19 \cdot 10^{29}\text{kg}};$$

$$d = \frac{3}{2} \cdot \frac{567 \cdot 6^4 \cdot 695^2 \cdot 10^{14}}{3 \cdot 3 \cdot 667 \cdot 19 \cdot 10^{28}} \cdot \text{m} \approx 88 \cdot 10^6 \cdot 10^{-14}\text{m};$$

$$d = 88 \cdot 10^{-8}\text{m} = 88 \cdot 10^{-6}\text{cm} = 88 \cdot 10^{-5}\text{mm}.$$

b)

$$F_{\text{grav}} > F_{\text{p,rad}};$$

$$d > \frac{3}{2} \cdot \frac{\sigma T^4 R^2}{c \rho K M}; \quad d > 88 \cdot 10^{-5}\text{mm}.$$

c)

$$F_{\text{grav}} < F_{\text{p,rad}};$$

$$d < \frac{3}{2} \cdot \frac{\sigma T^4 R^2}{c \rho K M}; \quad d < 88 \cdot 10^{-5}\text{mm}.$$

Problem 21

Dumbbell Suspended by a Spaceship

A dumbbell whose rigid linear bar has a length l and is very light is connected through a movable joint, S, by another light linear bar with a rigid length $L > l/2$, fixed at point N, to a spacecraft which orbits the Earth in a circular orbit with a large radius, r_0, in the variants represented in Figures 21.1 and 21.2. The figures show the dumbbell under and above a spacecraft, respectively.

The dumbbell rotates around its suspension point, S, remaining permanently in the plane of the spacecraft's orbit, without influencing the movement of the spacecraft. It is known that $l << r_0$ and $L << r_0$, and the two spherical dumbbell balls are identical, each being a material point.

a) *Determine*, in relation to the spacecraft, corresponding to each variant, the shape of the trajectory of the point at which the resultant gravitational pull force is exerted by the Earth on the dumbbell balls suspended from the spacecraft during a complete rotation of the dumbbell in the orbit of the spacecraft. It will be considered that the masses of the balls and the two bars are negligible in relation to the mass of the spacecraft, so the rotation of the dumbbell has no influence on the movement of the spacecraft. Any gravitational contraction of the dumbbell bar is also neglected.

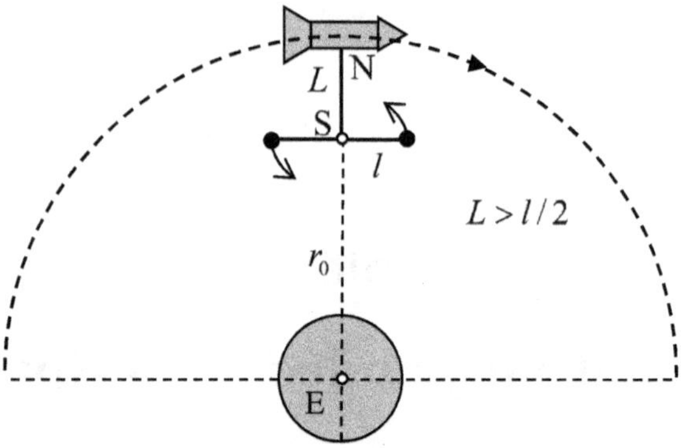

Fig. 21.1

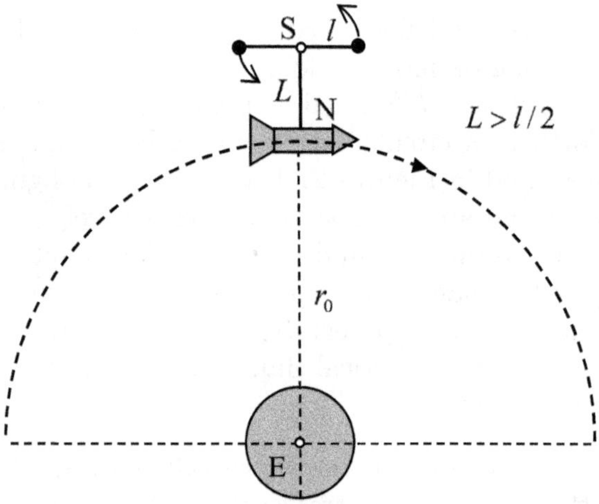

Fig. 21.2

b) *Compare* the geometric elements of the trajectories corresponding to the two variants.

It is known that

$$(1 + a)^n \approx 1 + na, \text{ if } a << 1.$$

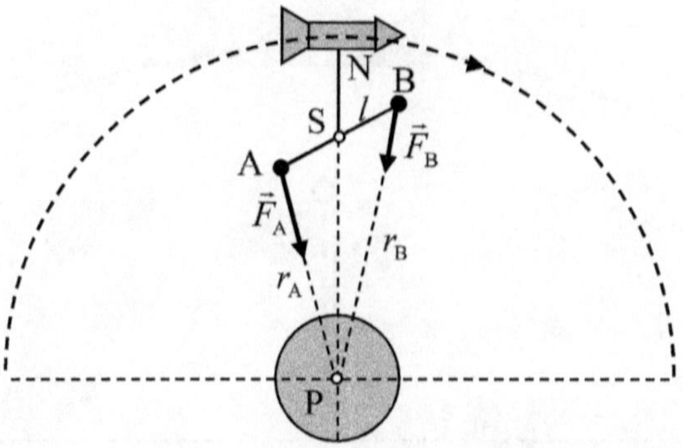

Fig. 21.3

Solution

a)
1) At some point during the dumbbell's rotation, corresponding to the variant shown in the drawing in Figure 21.1, the gravitational forces acting on the two dumbbell balls are represented in the drawing in Figure 21.3, where the distances from the two are also noted.

But because the distances between the elements suspended from the spacecraft, whose geometry is shown in the drawing in Figure 21.3, are small in relation to their distances to the center of the Earth, the suspended elements shown in the drawing in Figure 21.4 can be considered to be a good approximation.

Under these conditions, when

$$\vec{F}_A \| \vec{F}_B, \quad r_A < r_B, \quad F_A = K\frac{mM}{r_A^2}, \quad F_B = K\frac{mM}{r_B^2}, \quad F_A > F_B,$$

it turns out that the point at which the resultant of the two gravitational forces exerted by the Earth are applied to the two balls (on the dumbbell) will not coincide with the center of mass of the dumbbell (point S), but will be at point C, closer of the point of application of

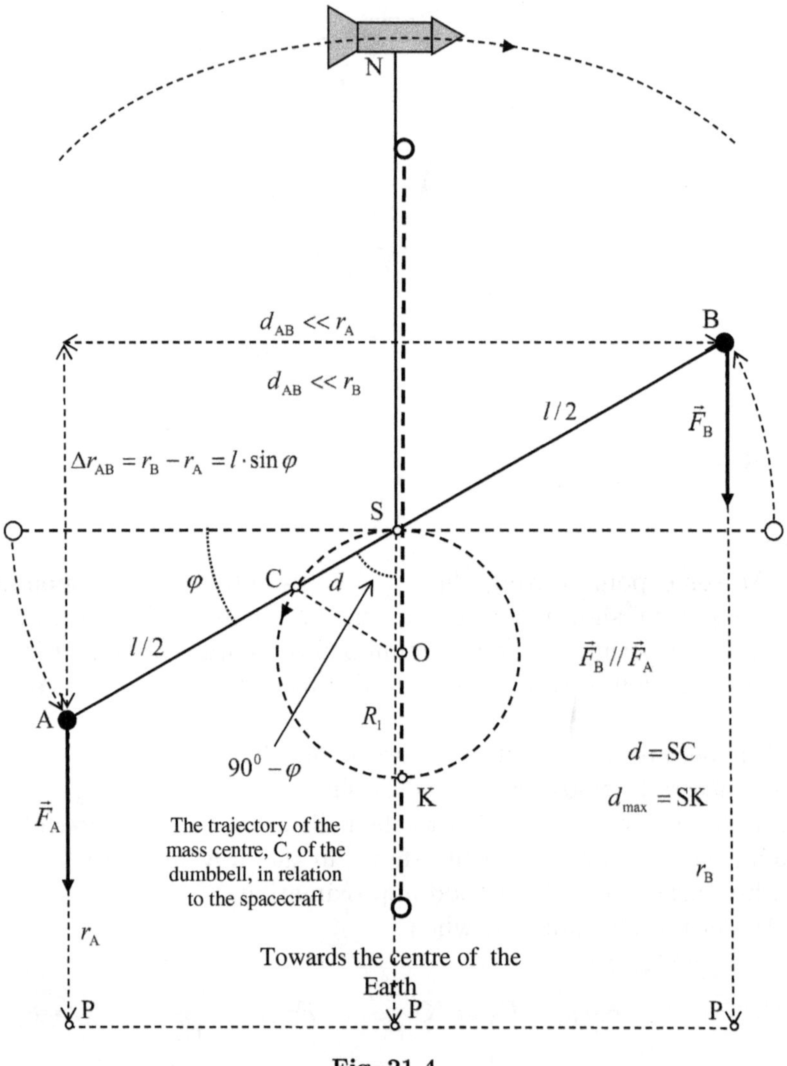

Fig. 21.4

the higher force, so that:

$$F_A \cdot AC = F_B \cdot BC; \quad CS = d;$$

$$\frac{F_A}{F_B} = \frac{BC}{AC} = \frac{\frac{l}{2} + d}{\frac{l}{2} - d};$$

$$F_{\mathrm{A}} = K \cdot \frac{mM}{r_{\mathrm{A}}^2}; \quad F_{\mathrm{B}} = K \cdot \frac{mM}{r_{\mathrm{B}}^2};$$

$$\frac{F_{\mathrm{A}}}{F_{\mathrm{B}}} = \frac{r_{\mathrm{B}}^2}{r_{\mathrm{A}}^2};$$

$$r_{\mathrm{B}} - r_{\mathrm{A}} = \Delta r_{\mathrm{AB}} = l \cdot \sin\varphi << r_{\mathrm{A}};$$

$$r_{\mathrm{B}} = r_{\mathrm{A}} + l \cdot \sin\varphi;$$

$$\frac{\frac{l}{2} + d}{\frac{l}{2} - d} = \frac{r_{\mathrm{B}}^2}{r_{\mathrm{A}}^2}; \quad \frac{l + 2d}{l - 2d} = \frac{(r_{\mathrm{A}} + l \cdot \sin\varphi)^2}{r_{\mathrm{A}}^2};$$

$$\frac{l + 2d}{l - 2d} = \frac{r_{\mathrm{A}}^2 \cdot \left(1 + \frac{l}{r_{\mathrm{A}}} \cdot \sin\varphi\right)^2}{r_{\mathrm{A}}^2} = \left(1 + \frac{l}{r_{\mathrm{A}}} \cdot \sin\varphi\right)^2;$$

$$\frac{l}{r_{\mathrm{A}}} \cdot \sin\varphi << 1;$$

$$\left(1 + \frac{l}{r_{\mathrm{A}}} \cdot \sin\varphi\right)^2 \simeq \left(1 + 2 \cdot \frac{l}{r_{\mathrm{A}}} \cdot \sin\varphi\right);$$

$$\frac{l + 2d}{l - 2d} = \left(1 + 2 \cdot \frac{l}{r_{\mathrm{A}}} \cdot \sin\varphi\right);$$

$$l + 2d = (l - 2d) \cdot \left(1 + 2 \cdot \frac{l}{r_{\mathrm{A}}} \cdot \sin\varphi\right);$$

$$l + 2d = l + 2 \cdot \frac{l^2}{r_{\mathrm{A}}} \cdot \sin\varphi - 2d - 4\frac{d \cdot l}{r_{\mathrm{A}}} \cdot \sin\varphi;$$

$$2d = 2 \cdot \frac{l^2}{r_{\mathrm{A}}} \cdot \sin\varphi - 2d - 4\frac{d \cdot l}{r_{\mathrm{A}}} \cdot \sin\varphi;$$

$$d = \frac{l^2}{r_{\mathrm{A}}} \cdot \sin\varphi - d - 2\frac{d \cdot l}{r_{\mathrm{A}}} \cdot \sin\varphi;$$

$$2d = \frac{l^2}{r_{\mathrm{A}}} \cdot \sin\varphi - 2\frac{d \cdot l}{r_{\mathrm{A}}} \cdot \sin\varphi;$$

$$2d + 2\frac{d \cdot l}{r_{\mathrm{A}}} \cdot \sin\varphi = \frac{l^2}{r_{\mathrm{A}}} \cdot \sin\varphi;$$

$$2d \cdot \left(1 + \frac{l}{r_{\mathrm{A}}} \cdot \sin\varphi\right) = \frac{l^2}{r_{\mathrm{A}}} \cdot \sin\varphi;$$

$$d = \frac{l^2 \cdot \sin\varphi}{2 \cdot (r_A + l \cdot \sin\varphi)};$$

$$l \cdot \sin\varphi << r_A;$$

$$d = \frac{l^2 \cdot \sin\varphi}{2 \cdot r_A}.$$

In particular, when the dumbbell is aligned with the direction of the center of the Earth, the directions of the two bars coincide, and the point at which the resultant of the two gravitational forces exerted by the Earth is applied to the two balls is in position K, the result is:

$$\varphi = \frac{\pi}{2};$$

$$\mathrm{SK} = d_{\max} = \frac{l^2}{2 \cdot r_A},$$

and point O divides the segment SK into two segments of equal length, OS = OK.

From the triangle OCS, it follows that:

$$\mathrm{OC} = \sqrt{(\mathrm{CS})^2 + (\mathrm{OS})^2 - 2 \cdot (\mathrm{CS}) \cdot (\mathrm{OS}) \cdot \cos(90^\circ - \varphi)};$$

$$\mathrm{OC} = \sqrt{d^2 + \frac{d_{\max}^2}{4} - 2 \cdot d \cdot \frac{d_{\max}}{2} \cdot \sin\varphi};$$

$$\mathrm{OC} = \sqrt{d^2 + \frac{d_{\max}^2}{4} - d \cdot d_{\max} \cdot \sin\varphi};$$

$$d = \frac{l^2 \cdot \sin\varphi}{2 \cdot r_A}; \quad SK = d_{\max} = \frac{l^2}{2 \cdot r_A};$$

$$\mathrm{OC} = \sqrt{\frac{l^4 \cdot \sin^2\varphi}{4 \cdot r_A^2} + \frac{l^4}{16 \cdot r_A^2} - \frac{l^2 \cdot \sin\varphi}{2 \cdot r_A} \cdot \frac{l^2}{2 \cdot r_A} \cdot \sin\varphi};$$

$$\mathrm{OC} = \sqrt{\frac{l^4 \cdot \sin^2\varphi}{4 \cdot r_A^2} + \frac{l^4}{16 \cdot r_A^2} - \frac{l^4 \cdot \sin^2\varphi}{4 \cdot r_A^2}};$$

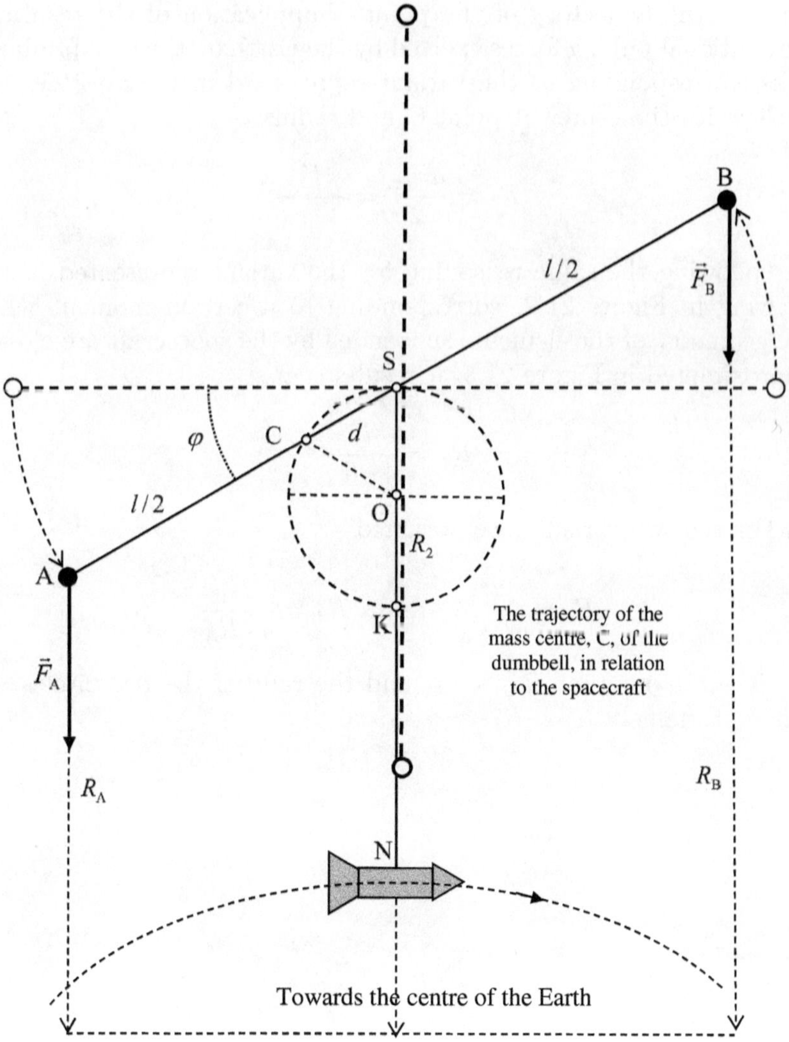

Fig. 21.5

$$\text{OC} = \sqrt{\frac{l^4}{16 \cdot r_{\text{A}}^2}};$$

$$\text{OC} = \frac{l^2}{4 \cdot r_{\text{A}}} = \frac{d_{\max}}{2}.$$

That is, the trajectory of the point of application of the resultant gravitational pulling forces exerted by the Earth on the two dumbbell balls, corresponding to the variant represented in Figure 21.1, is a circle, with the center at point O and radius

$$R_1 = \frac{d_{\max}}{2} = \frac{l^2}{4 \cdot r_{\mathrm{A}}}.$$

2) Following the same reasoning for the variant represented in the drawing in Figure 21.2, corresponding to a certain moment when the geometry of the elements suspended by the spacecraft are closely approximated in Figure 21.5, it results that

$$R_2 = \frac{l^2}{4 \cdot R_{\mathrm{A}}}.$$

b) For the two variants, we obtained

$$R_1 = \frac{l^2}{4 \cdot r_{\mathrm{A}}} \quad \text{and} \quad R_2 = \frac{l^2}{4 \cdot R_{\mathrm{A}}}.$$

So, it turns out that $R_{\mathrm{A}} > r_{\mathrm{A}}$, and the radii of the two circles are different, namely $R_2 < R_1$.

Problem 22

The Temperatures of the Planets in Our Solar System

a) Knowing the distances between the Sun and the planets of our solar system, presented in Table 22.1, *determine* the temperatures at the surfaces of the 8 planets, knowing: the temperature at the surface of the Sun, $T_S = 5\,780\,\mathrm{K}$; the radius of the Sun, $R_S = 696\,000\,\mathrm{km}$.

Both the Sun and the planets are considered perfect black bodies. It is also assumed that the planets do not reflect solar radiation.

b) Identifying that the planet at the shortest distance from Earth is Venus, *determine* the result of the radiative interaction between Earth and Venus on each of the two planets. The following are known: the radius of the Earth, $R_E = 6378.10\,\mathrm{km}$; the radius of Venus, $R_V = 6051.59\,\mathrm{km}$.

Solution

a) According to the Stefan–Boltzmann law, the *radiance* of the Sun (considered an absolute black body), i.e. the total energy for all

Table 22.1

Planet		Distance from the Sun $\Delta(km)$	Temperature $T(K)$
Mercury		57 909 175	
Venus		108 208 930	
Earth	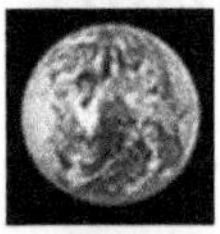	149 597 890	
Mars		227 936 640	
Jupiter	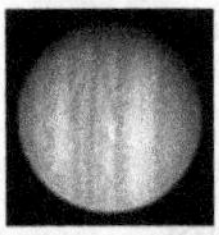	778 412 010	
Saturn		1 426 725 400	
Uranus		2 870 972 200	
Neptune	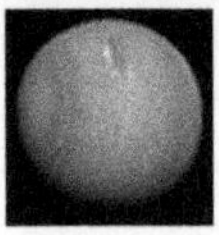	4 498 252 900	

wavelengths emitted in one unit of time by a sector of the Sun's surface, with an area of one unit, is directly proportional to the fourth power of the absolute temperature of its surface, meaning:

$$Q_{\text{Sun}} = \frac{E_{\text{emitted, Sun}}}{t \cdot S_{\text{Sun}}} = \sigma \cdot T_{\text{S}}^4,$$

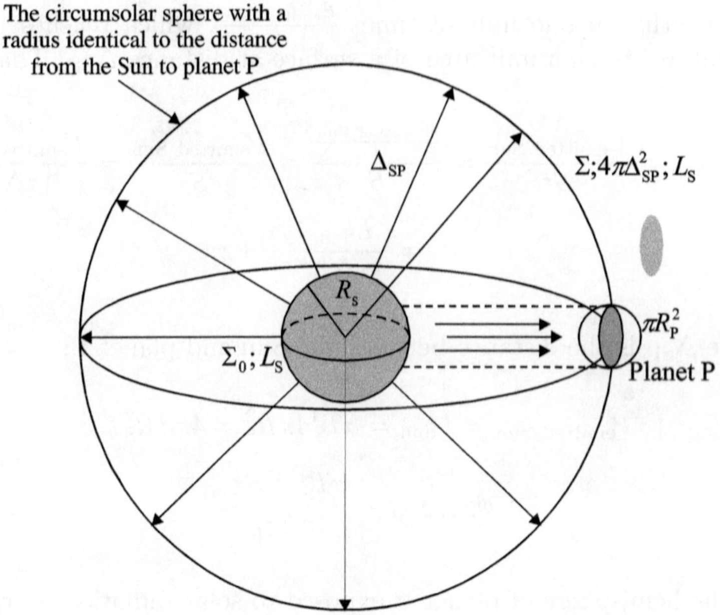

Fig. 22.1

where σ is a constant;

$$\frac{E_{\text{emitted, Sun}}}{t} = P_{\text{emitted, Sun}} = L_{\text{Sun}};$$

$$\sigma \cdot T_{\text{S}}^4 = \frac{P_{\text{emitted, Sun}}}{4\pi R_{\text{S}}^2} = \frac{L_{\text{Sun}}}{4\pi R_{\text{S}}^2},$$

where L_{S} is the *brightness* of the Sun;

$$P_{\text{emitted, Sun}} = L_{\text{Sun}} = \sigma T_{\text{S}}^4 4\pi R_{\text{S}}^2 = 4\pi\sigma R_{\text{S}}^2 T_{\text{S}}^4,$$

representing the total energy emitted by the Sun through its entire surface in one unit of time.

If Σ is the area of the circumsolar sphere whose radius, Δ_{SP}, is the instantaneous distance between the center of the Sun and the center of a planet, P, as shown in the drawing in Figure 22.1, then the energy of the solar radiation passing through the surface of Σ is equal to L_{S}.

The *density* of the Sun's *energy flow* at a distance Δ_{SP} from it refers to the fraction of the total energy emitted by the Sun, on all wavelengths, in one unit of time, $\frac{E_{\text{emitted,Sun}}}{t}$, which reaches under normal incidence a unit area of a surface at distance Δ_{SP}. That is:

$$\phi_{\text{Sun},\Delta_{\text{SP}}} = \frac{E_{\text{emitted, Sun}}}{St} = \frac{\frac{E_{\text{emitted, Sun}}}{t}}{S} = \frac{P_{\text{emitted, Sun}}}{S} = \frac{P_{\text{emitted, Sun}}}{4\pi\Delta_{\text{SP}}^2}$$
$$= \frac{L_{\text{Sun}}}{4\pi\Delta_{\text{SP}}^2},$$

where Δ_{SP} is the distance between the Sun and planet P;

$$P_{\text{emitted, Sun}} = L_{\text{Sun}} = \sigma T_{\text{S}}^4 4\pi R_{\text{S}}^2 = 4\pi\sigma R_{\text{S}}^2 T_{\text{S}}^4;$$

$$\phi_{\text{Sun},\Delta_{\text{SP}}} = \frac{\sigma T_{\text{S}}^4 4\pi R_{\text{S}}^2}{4\pi\Delta_{\text{SP}}^2}.$$

The hemisphere of planet P exposed to solar radiation is equivalent to a circular flat disk at a distance Δ_{SP} from the Sun, with the radius R_{P} and area πR_{P}^2, placed perpendicular to the Sun–Planet direction, so that the *energy flow* of incident solar radiation, F_{incident}, at planet P (a flat circular disk with radius R_{P}) per unit time on the flat circular disk with the radius R_{P}, is:

$$F_{\text{incident}} = \phi_{\text{Sun},\Delta_{\text{SP}}} \cdot \pi R_{\text{P}}^2 = P_{\text{incident}} = P_{\text{emitted, Sun}};$$

$$\phi_{\text{Sun},\Delta_{\text{SP}}} = \frac{\sigma T_{\text{S}}^4 4\pi R_{\text{S}}^2}{4\pi\Delta_{\text{SP}}^2};$$

$$F_{\text{incident}} = \frac{\sigma T_{\text{S}}^4 4\pi R_{\text{S}}^2}{4\pi\Delta_{\text{SP}}^2} \cdot \pi R_{\text{P}}^2 = P_{\text{incident}}.$$

Considering that planet P does not reflect any of the received radiation, it means that

$$P_{\text{reflected, Planet}} = 0.$$

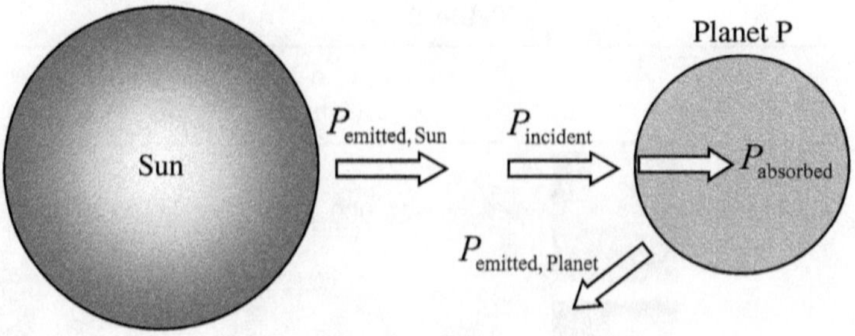

Fig. 22.2

Therefore, according to the drawing in Figure 22.2, the energy balance equation of the analyzed process is:

$$P_{\text{emtted, Sun}} = L_{\text{Sun}} = \sigma T_{\text{S}}^4 4\pi R_{\text{S}}^2 = 4\pi\sigma R_{\text{S}}^2 T_{\text{S}}^4;$$

$$P_{\text{emitted, Sun}} = P_{\text{incident,Planet}} = P_{\text{absorbed,Planet}};$$

$$P_{\text{emitted, Sun}} = 4\pi\sigma R_{\text{S}}^2 T_{\text{S}}^4;$$

$$P_{\text{absorbed,Planet}} = \frac{\sigma T_{\text{S}}^4 4\pi R_{\text{S}}^2}{4\pi\Delta_{\text{SP}}^2} \cdot \pi R_{\text{P}}^2.$$

In turn, the energy emitted by planet P, considered a black body, through its entire surface in one unit of time, its temperature being T_{P}, is:

$$P_{\text{emitted,Planet}} = \sigma T_{\text{P}}^4 \cdot 4\pi R_{\text{P}}^2.$$

When the process becomes stationary, the result is:

$$P_{\text{absorbed,Planet}} = P_{\text{emitted,Planet}};$$

$$P_{\text{emitted,Sun}} = 4\pi\sigma R_{\text{S}}^2 T_{\text{S}}^4; \quad P_{\text{incident}} = \frac{\sigma T_{\text{S}}^4 4\pi R_{\text{S}}^2}{4\pi\Delta_{\text{SP}}^2} \cdot \pi R_{\text{P}}^2;$$

$$P_{\text{absorbed,Planet}} = \frac{\sigma T_{\text{S}}^4 4\pi R_{\text{S}}^2}{4\pi\Delta_{\text{SP}}^2} \cdot \pi R_{\text{P}}^2;$$

$$P_{\text{absorbed,Planet}} = \frac{\sigma T_{\text{S}}^4 4\pi R_{\text{S}}^2}{4\pi\Delta_{\text{SP}}^2} \cdot \pi R_{\text{P}}^2; \quad P_{\text{emitted,Planet}} = \sigma T_{\text{P}}^4 \cdot 4\pi R_{\text{P}}^2;$$

Table 22.2

Planet		Distance from the Sun $\Delta(km)$	Temperature T (K)
Mercury		57 909 175	448.06
Venus		108 208 930	327.78
Earth	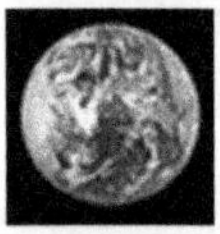	149 597 890	278.77
Mars		227 936 640	225.84
Jupiter	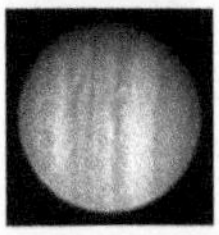	778 412 010	122.21
Saturn		1 426 725 400	90.27
Uranus		2 870 972 200	63.63
Neptune		4 498 252 900	50.83

$$\frac{\sigma T_{\mathrm{S}}^4 4\pi R_{\mathrm{S}}^2}{4\pi\Delta_{\mathrm{SP}}^2}\cdot\pi R_{\mathrm{P}}^2 = \sigma T_{\mathrm{P}}^4\cdot 4\pi R_{\mathrm{P}}^2; \qquad \frac{T_{\mathrm{S}}^4 R_{\mathrm{S}}^2}{\Delta_{\mathrm{SP}}^2} = T_{\mathrm{P}}^4\cdot 4;$$

$$\sqrt{\frac{T_{\mathrm{S}}^4 R_{\mathrm{S}}^2}{\Delta_{\mathrm{SP}}^2}} = \sqrt{4\cdot T_{\mathrm{P}}^4};$$

$$\frac{T_S^2 R_S}{\Delta_{SP}} = 2 \cdot T_P^2;$$

$$T_P = T_S \cdot \sqrt{\frac{R_S}{2\Delta_{SP}}},$$

representing the temperature of planet P.

The results of the determinations are presented in Table 22.2.

b) Particularizing the previously established relation for the radiative action of the Sun on a certain planet, P,

$$T_P = T_S \cdot \sqrt{\frac{R_S}{2\Delta_{SP}}},$$

we get:

1) For the radiative action of the Earth on Venus:

$$T_{\text{Venus}} = T_{\text{Earth}} \cdot \sqrt{\frac{R_{\text{Earth}}}{2\Delta_{\text{Earth}-\text{Venus}}}};$$

$$\Delta_{\text{Earth}-\text{Venus}} = \Delta_{\text{Sun}-\text{Earth}} - \Delta_{\text{Sun}-\text{Venus}} = 413\,88\,960\,\text{km};$$

$$T_{\text{Earth}} = 278.77\,\text{K};$$

$$T_{\text{Venus}} = 2.44\,\text{K};$$

2) For the radiative action of Venus on the Earth:

$$T_{\text{Earth}} = T_{\text{Venus}} \cdot \sqrt{\frac{R_{\text{Venus}}}{2\Delta_{\text{Earth}-\text{Venus}}}};$$

$$\Delta_{\text{Earth}-\text{Venus}} = \Delta_{\text{Sun}-\text{Earth}} - \Delta_{\text{Sun}-\text{Venus}} = 41\,388\,960\,\text{km};$$

$$T_{\text{Venus}} = 327.78\,\text{K};$$

$$T_{\text{Earth}} = 2.80\,\text{K}.$$

Problem 23

Binary Star System Eclipse

The two components, (σ_1; σ_2) of an eclipsing binary star system evolve in circular orbits around the common center of mass. The direction of the observation line is in the plane of the orbits of the centers of the two stars, as shown in the drawing in Figure 23.1. In relation to the observer on Earth, the system's center of mass is at rest. For the two stars, we know the radii, R_1 and R_2, and the temperatures at their surface, T_1 and T_2.

Determine the difference between the apparent visual magnitudes of this eclipsing binary star system corresponding to the following moments:

a) when the apparent disks of the two stars completely overlap, in variants a and b of Figure 23.2;
b) when the apparent disks of the two stars are tangent, in variants a and b of Figure 23.3.

It is known that the energy of radiation from the binary star system reaching the detector of the observer on Earth in one unit of time (total measured energy flow), is

$$\Phi_{\text{measured}} = k\left(\varphi_1 \Delta A_1 + \varphi_2 \Delta A_2\right),$$

where: φ_1 and φ_2 are the densities of the energy fluxes (radiation) at the surfaces of the two stars (the energy corresponding to all wavelengths released by each of the two stars in one unit of time, from a sector of one unit surface area; k is a dimensionless

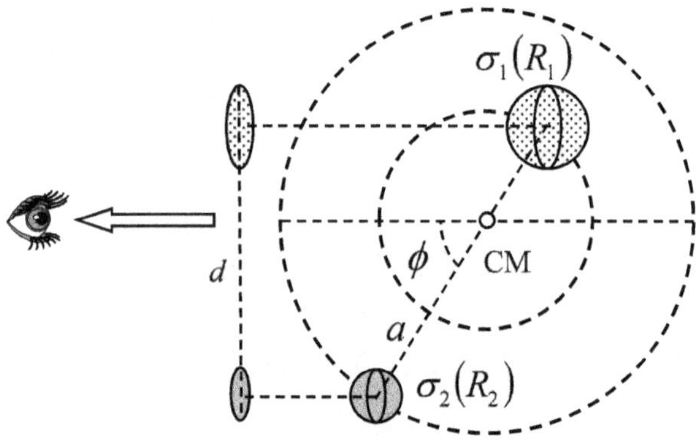

Fig. 23.1

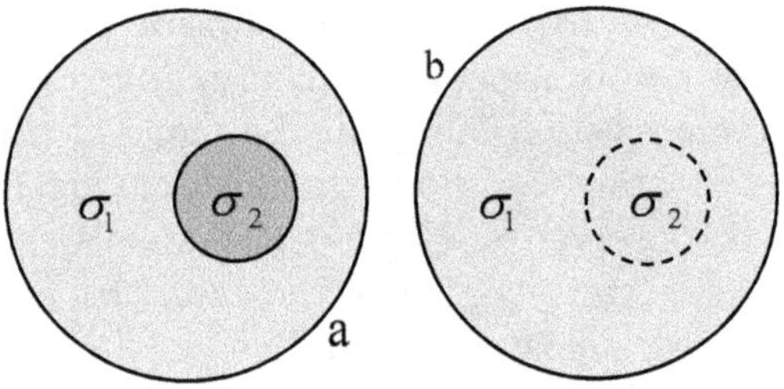

Fig. 23.2

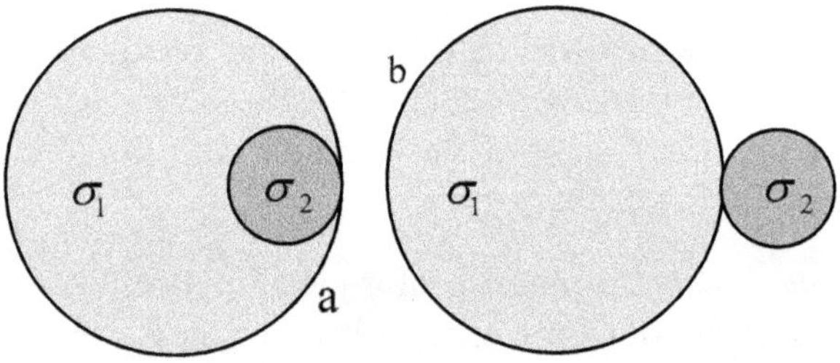

Fig. 23.3

proportionality constant, the value of which depends on the distance between the observer and the star, the area of the detector surface and the efficiency of the detector; ΔA_1 and ΔA_2 are the areas of the free (non-eclipsed) sectors of stellar disks visible to the Earth observer.

Solution

Using Pogson's formula, it follows that:

$$\log \frac{\Phi_{\text{measured, a}}}{\Phi_{\text{measured, b}}} = -0.4\,(m_{\text{a}} - m_{\text{b}});$$

$$m_{\text{a}} - m_{\text{b}} = 2.5 \log \frac{\Phi_{\text{measured, b}}}{\Phi_{\text{measured, a}}}.$$

When the components of the binary stellar system eclipse each other, the apparent circular disks of the two stars overlap, more or less, so that the areas of the free (non-eclipsed) sectors of the stellar disks visible to the Earth observer are, respectively, ΔA_1 and ΔA_2.

The energy of radiation from the binary star system reaching the detector of the observer on Earth, at a distance D, in one unit of time (total measured energy flow), is

$$\Phi_{\text{measured}} = k\,(\varphi_1 \Delta A_1 + \varphi_2 \Delta A_2).$$

This results in:

$$\Phi_{\text{measured}} = k\left(\frac{L_1}{4\pi R_1^2}\Delta A_1 + \frac{L_2}{4\pi R_2^2}\Delta A_2\right);$$

$$\Phi_{\text{measured}} = k'\left(\frac{L_1}{R_1^2}\Delta A_1 + \frac{L_2}{R_2^2}\Delta A_2\right); \quad k' = \frac{k}{4\pi};$$

$$L_1 = 4\pi R_1^2 \sigma T_1^4; \quad L_2 = 4\pi R_2^2 \sigma T_2^4;$$

$$\Phi_{\text{measured}} = k''\left(T_1^4 \Delta A_1 + T_2^4 \Delta A_2\right); \quad k'' = k\sigma.$$

a) According to the details in Figure 23.4, it results that:

$$\begin{aligned}\Phi_{\text{measured,a}} &= k''\left(T_1^4 \Delta A_{1,\text{a}} + T_2^4 \Delta A_{2,\text{a}}\right)\\ &= k''\left[T_1^4 \pi \left(R_1^2 - R_2^2\right) + T_2^4 \pi R_2^2\right];\end{aligned}$$

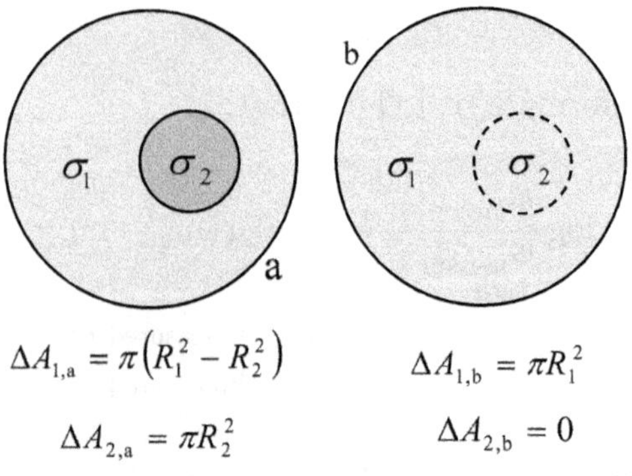

Fig. 23.4

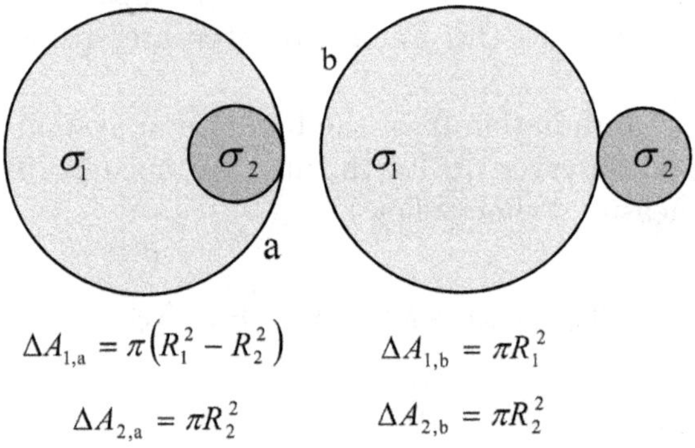

Fig. 23.5

$$\Phi_{\text{measured,b}} = k'' \left(T_1^4 \Delta A_{1,\text{b}} + T_2^4 \Delta A_{2,\text{b}}\right) = k'' T_1^4 \pi R_1^2;$$

$$m_\text{a} - m_\text{b} = 2.5 \log \frac{T_1^4 R_1^2}{T_1^4 \left(R_1^2 - R_2^2\right) + T_2^4 R_2^2}.$$

b) According to the details in Figure 23.5:

$$\begin{aligned}\Phi_{\text{measured,a}} &= k'' \left(T_1^4 \Delta A_{1,\text{a}} + T_2^4 \Delta A_{2,\text{a}}\right)\\ &= k'' \left[T_1^4 \pi \left(R_1^2 - R_2^2\right) + T_2^4 \pi R_2^2\right];\end{aligned}$$

$$\Phi_{\text{measured,b}} = k'' \left(T_1^4 \Delta A_{1,b} + T_2^4 \Delta A_{2,b}\right) = k'' \left(T_1^4 \pi R_1^2 + T_2^4 \pi R_2^2\right);$$

$$m_\text{a} - m_\text{b} = 2.5 \log \frac{T_1^4 \left(R_1^2 - R_2^2\right) + T_2^4 R_2^2}{T_1^4 R_1^2 + T_2^4 R_2^2}.$$

Problem 24

Oblique Connecting Pyramids

For the motion of a planet on an ellipse in the gravitational field of the Sun, it is *shown that*:

1) The *average distance* from the planet to the Sun during a complete revolution of the planet around the Sun is equal to the distance from the Sun (located in one of the foci of the ellipse) to one of the extreme points of the small axis of the ellipse (B, one of the minor points of the ellipse), as shown in the drawing in Figure 24.1:

$$\langle r \rangle = \mathrm{FB} = a.$$

2) The *average orbital speed* of the planet during a complete revolution around the Sun is equal to the instantaneous speed of the planet at one of the minor vertices of the ellipse:

$$\langle \mathrm{v} \rangle = \mathrm{v_B} = \sqrt{KM\left(\frac{2}{\mathrm{FB}} - \frac{1}{a}\right)} = \sqrt{KM\left(\frac{2}{a} - \frac{1}{a}\right)} = \sqrt{\frac{KM}{a}},$$

where M is the mass of the Sun.

3) The *average distance* from the planet to the Sun and the *average orbital speed* of the planet in its motion on the ellipse in whose focus the Sun is located are the same as, respectively, the *radius of the confocal circle* of the ellipse (constructed as shown in the drawing in Figure 24.1) and the speed of the planet if it evolved around the Sun on the specified confocal circle:

$$\langle \mathrm{v} \rangle = \mathrm{v_B} = \frac{2\pi a}{T},$$

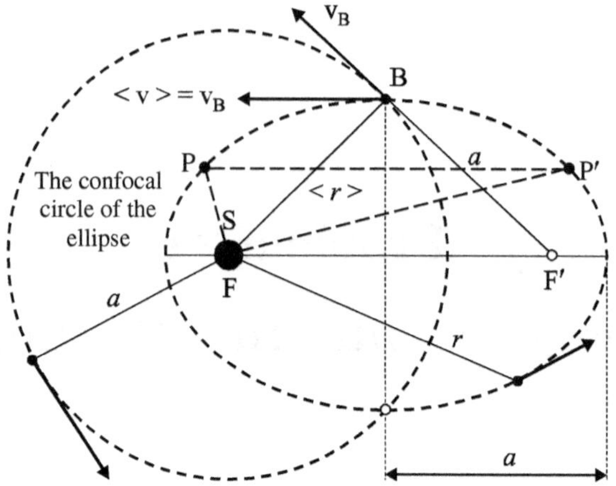

Fig. 24.1

where T is the period of the planet's motion on the confocal circle, the same as the period of the planet's motion on the ellipse with the major semi-axis a.

The drawing in Figure 24.2 shows two pyramids, called *oblique connecting pyramids*, each with its apex in the center of the ellipse (center of the Sun) and its base a square centered on the minor apex of each ellipse, so that the base plane of each pyramid is perpendicular to each direction. The minor apex of the ellipse and the height of each pyramid is equal to the major half-axis of each ellipse (average distance between the Sun and each planet).

The two pyramids create an *oblique connection* between the foci of the ellipses and their minor peaks. The length of the sides of the base of each pyramid is equal to the elementary tangential displacement of each planet, near the minor apex of the corresponding ellipse, in a time $\Delta t << T_1$ and $\Delta t << T_2$, during which the speed can be considered constant, $\vec{v}_B = \text{constant}$.

a) *Calculate* and *compare* the volumes of each planet's two obliquely connected pyramids and draw a conclusion.

b) Four planets (P_1, P_2, P_3, P_4) evolve around a star in concentric and coplanar circular orbits with radii R, $2R$, $3R$, and $4R$, respectively, as shown in the drawing in Figure 24.3, being at time $t = 0$ in the same direction as the star Σ.

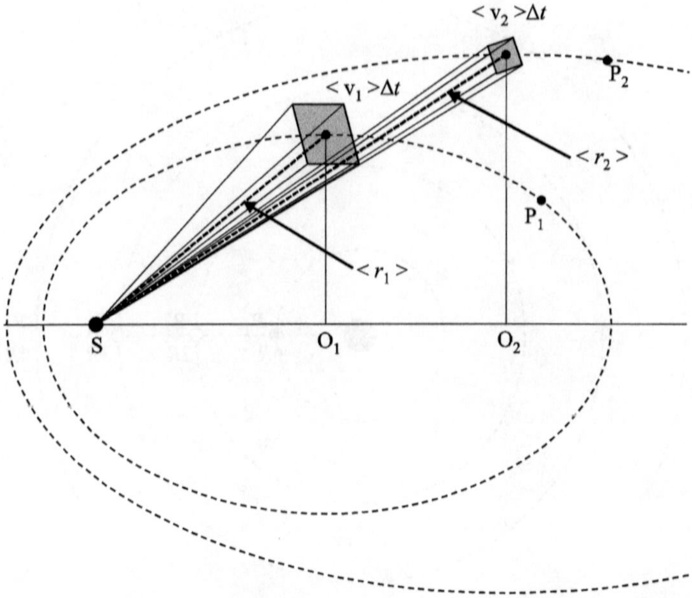

Fig. 24.2

Mark, on the orbits of planets P_2, P_3, and P_4, the arcs that indicate their positions at time $t = T_1$, representing the duration of a complete revolution of planet P_1.

The intuitive graphical presentation of Kepler's third law proposed by Manfred Bucher of the Department of Physics at California State University will allow a precise answer to the proposed test.

Solution

a) The volume of each connecting pyramid is:

$$\Delta V = \frac{1}{3}(\Delta l)^2 \langle r \rangle;$$

$$\Delta l = v_B \Delta t = \langle v \rangle \Delta t;$$

$$\Delta V = \frac{1}{3}(\langle v \rangle \Delta t)^2 \langle r \rangle.$$

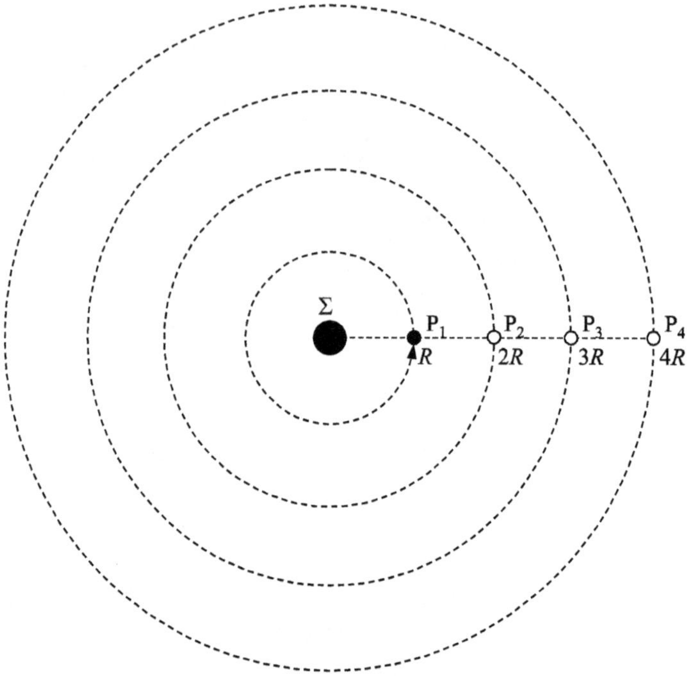

Fig. 24.3

This results in:

$$\Delta V_1 = \frac{1}{3}\langle v_1\rangle^2(\Delta t)^2\langle r_1\rangle = \frac{1}{3}\left(\frac{2\pi a_1}{T_1}\right)^2 \cdot (\Delta t)^2 \cdot a_1 = \frac{1}{3}\,\frac{4\pi^2 a_1^3}{T_1^2}(\Delta t)^2;$$

$$\langle v_1\rangle = v_{B,1} = \sqrt{\frac{KM}{a_1}};\quad \langle r_1\rangle = a_1;\quad \Delta V_1 = \frac{1}{3}\cdot\frac{KM}{a_1}\cdot a_1\cdot(\Delta t)^2;$$

$$\Delta V_1 = \frac{KM}{3}\cdot(\Delta t)^2;$$

$$\Delta V_2 = \frac{1}{3}\langle v_2\rangle^2(\Delta t)^2\langle r_2\rangle = \frac{1}{3}\left(\frac{2\pi a_2}{T_2}\right)^2\cdot(\Delta t)^2$$

$$\cdot a_2 = \frac{1}{3}\,\frac{4\pi^2 a_2^3}{T_2^2}\frac{1}{3}\,\frac{4\pi^2 a_2^3}{T_2^2}(\Delta t)^2;$$

$$\langle \mathrm{v}_2 \rangle >= \mathrm{v}_{\mathrm{B},2} = \sqrt{\frac{KM}{a_2}}; \quad \langle r_2 \rangle = a_2; \quad \Delta \mathrm{V}_2 = \frac{1}{3} \cdot \frac{KM}{a_2} \cdot a_1 \cdot (\Delta t)^2;$$

$$\Delta \mathrm{V}_2 = \frac{KM}{3} \cdot (\Delta t)^2;$$

$$\Delta \mathrm{v}_1 = \Delta \mathrm{v}_2.$$

Hence:

$$\frac{a_1^3}{T_1^2} = \frac{a_2^3}{T_2^2},$$

representing Kepler's third law in its original form.

Referring to these geometric constructions, Kepler's third law is reformulated as follows: *The volumes of the oblique connecting pyramids corresponding to the same time interval are the same for all planets evolving around the Sun.*

Note: In any of the proposed formulations, Kepler's third law correlates elements of one planet's motion with elements of another planet's motion, highlighting an "invariant" of them.

b) If sectors with equal areas can be constructed anywhere on the surface of the ellipse in the graphical presentation of Kepler's second law, in the graphical presentation of Kepler's third law, oblique pyramids with equal volumes can be constructed only with the bases centered on the minor vertices of the ellipses, the planes of these bases being perpendicular to the inclined Sun–minor peak axes.

The only exceptions to these restrictions are circular orbits, whose points are equivalent, so that the connecting pyramids can be oriented such that one of the edges of the base is tangent to the circle.

According to the new way of presenting Kepler's third law, the average orbital velocities of any two planets whose orbits are concentric circles have the relation

$$\frac{\langle \mathrm{v}_2 \rangle}{\langle \mathrm{v}_1 \rangle} = \sqrt{\frac{\langle r_1 \rangle}{\langle r_2 \rangle}}.$$

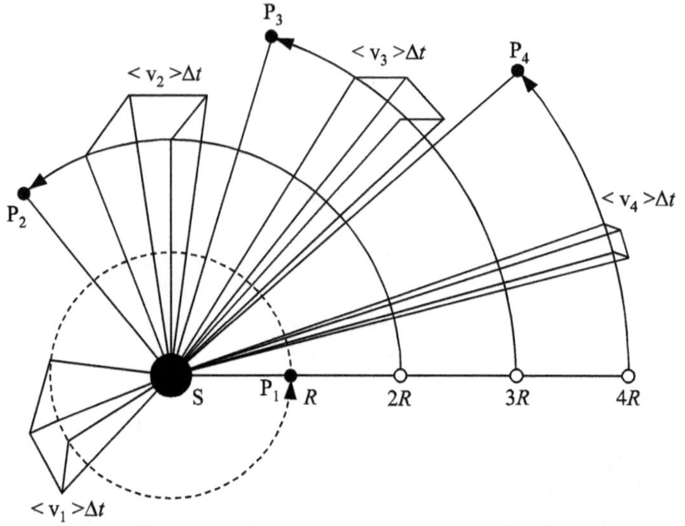

Fig. 24.4

Since the orbits of the four planets are circular, any points on the orbits can be used to construct the corresponding connecting pyramids, as shown in the drawing in Figure 24.4. The pyramids are oriented so that an edge of the base, representing the movement of that planet in a time $\Delta t << \mathrm{T}_1$, is tangent to the circular orbit.

As a result of the new way of presenting Kepler's third law, if the heights of the connecting pyramids of planets P_2, P_3, and P_4 are

$$\langle r_2 \rangle = 2\langle r_1 \rangle, \quad \langle r_3 \rangle = 3\langle r_1 \rangle, \quad \langle r_4 \rangle = 4\langle r_1 \rangle,$$

then the surface areas of their bases must be:

$$\langle \mathrm{v}_2 \rangle^2 (\Delta t)^2 = \frac{1}{2} < \mathrm{v}_1 \rangle^2 (\Delta t)^2;$$

$$\langle \mathrm{v}_3 \rangle^2 (\Delta t)^2 = \frac{1}{3} \langle \mathrm{v}_1 \rangle^2 (\Delta t)^2;$$

$$\langle \mathrm{v}_4 \rangle^2 (\Delta t)^2 = \frac{1}{4} \langle \mathrm{v}_1 \rangle^2 (\Delta t)^2.$$

Thus:

$$\langle \mathrm{v}_2 \rangle = \frac{\langle \mathrm{v}_1 \rangle}{\sqrt{2}}; \quad \langle \mathrm{v}_3 \rangle = \frac{\langle \mathrm{v}_1 \rangle}{\sqrt{3}}; \quad \langle \mathrm{v}_4 \rangle = \frac{\langle \mathrm{v}_1 \rangle}{\sqrt{4}}.$$

It follows that when the planet P_1 has returned to its original position (after a time T_1), the positions of the planets P_2, P_3, and P_4 on their circular orbits are given by the angles:

$$\alpha_2 = 2\pi \frac{\langle r_1 \rangle}{\langle r_2 \rangle} \frac{\langle \mathrm{v}_2 \rangle}{\langle \mathrm{v}_1 \rangle} \approx 127^{\mathrm{o}};$$

$$\alpha_3 = 2\pi \frac{\langle r_1 \rangle}{\langle r_3 \rangle} \frac{\langle \mathrm{v}_3 \rangle}{\langle \mathrm{v}_1 \rangle} \approx 69^{\mathrm{o}};$$

$$\alpha_4 = 2\pi \frac{\langle r_1 \rangle}{\langle r_4 \rangle} \frac{\langle \mathrm{v}_4 \rangle}{\langle \mathrm{v}_1 \rangle} = 45^{\mathrm{o}}.$$

Problem 25

Eclipses of a Planet's Satellite

In a certain galaxy, it is known that a planet, P, makes a complete rotation around a star, Z, considered fixed, during the time T_1. During a time $T_2 > T_1$, the planet P performs a complete rotation around its own axis. The satellite S of planet P (a moon of planet P) makes a complete rotation around planet P during a time $T_3 < T_1$.

A certain point, which we will consider as the initial moment, $t_0 = 0$, is shown in the drawing in Figure 25.1. For an observer on the surface of planet P, on its equator, in position O_0, the satellite in position S_0 is eclipsed.

a) *Determine* the time intervals, τ, after which, for the observer on the surface of planet P, the eclipse of satellite S will be repeated.

b) *Determine* the time intervals, t, after which, for the observer on the surface of planet P, the eclipse of satellite S will be repeated in exactly the same alignment as the initial eclipse. The angular velocity vectors of the three specified motions, $\vec{\omega}_1, \vec{\omega}_2$ and $\vec{\omega}_3$, are parallel and of the same direction, perpendicular to the plane of the orbits, and moving away from it.

c) *Prove*, without demonstration, the possibility/impossibility of the results obtained for this problem being true for the Sun–Earth–Moon alignments, which would represent the repetition of lunar eclipses for an observer on the Earth's equator.

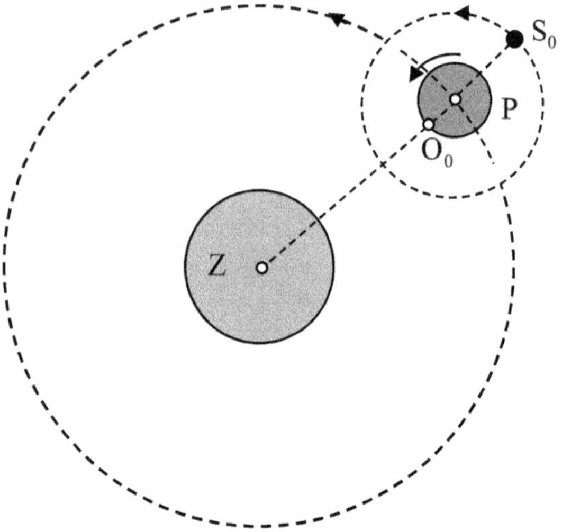

Fig. 25.1

Solution

a) In a coordinate system fixed in relation to the center of the star Z, as shown in the drawing in Figure 25.2, whose plane is that of the orbits of the center of planet P around the center of star Z and the satellite S (considered a material point) around the centre of planet P: $\omega_1 = \frac{2\pi}{T_1}$ is the angular velocity of the center of planet P in its orbit around the center of star Z; $\omega_2 = \frac{2\pi}{T_2} < \omega_1$ the angular velocity of the rotation of the planet P around its own axis; and $\omega_3 = \frac{2\pi}{T_3} > \omega_2$ is the angular velocity of satellite S in its orbit around the center of planet P. Thus,

$$\omega_2 < \omega_1 < \omega_3.$$

The three vectors, $\vec{\omega}_1, \vec{\omega}_2$ and $\vec{\omega}_3$, are parallel, of the same direction, and perpendicular to the common plane of the orbits emerging from the plane of the drawing.

The drawing in Figure 25.3 shows the circle C_{PZ}, representing the orbit of the absolute motion of the center of planet P relative to the center of star Z, as well as the identical circle, C_{ZP}, representing

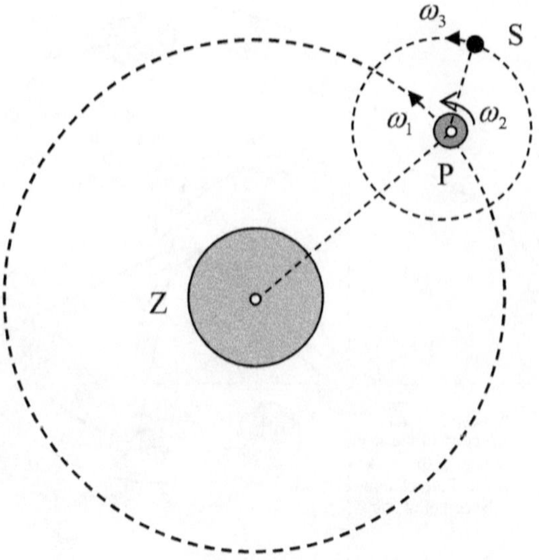

Fig. 25.2

the orbit of the relative motion of the center of star Z in relation to the center of planet P. Both have circular motions in the same direction and have equal angular velocities, $\vec{\omega}_1$.

In a coordinate system fixed in relation to the center of planet P, against which planet P does not rotate, as shown in the drawing in Figure 25.4, the center of star Z rotates around the center of planet P at an angular velocity ω_1, as shown in the drawing in Figure 25.3. The satellite S rotates around the center of the planet P with angular velocity ω_3. The vectors of the angular velocities $\vec{\omega}_1$ and $\vec{\omega}_3$ are parallel and of the same direction, perpendicular to the plane from which the orbits are emerging.

In a coordinate system fixed in relation to the center of planet P, as we proposed in the drawing in Figure 25.4, planet P now rotates around its own axis at an angular velocity $\vec{\omega}_2$, and the center of star Z rotates at angular velocity $\vec{\omega}_1$ relative to the center of the planet P, as shown in the drawing in Figure 25.5. Here, we show the possible alignment $Z_0O_0PS_0$ corresponding to the initial moment, $t_0 = 0$, representing the eclipse of the satellite S with respect to the observer on the surface of the planet P when in the position O_0.

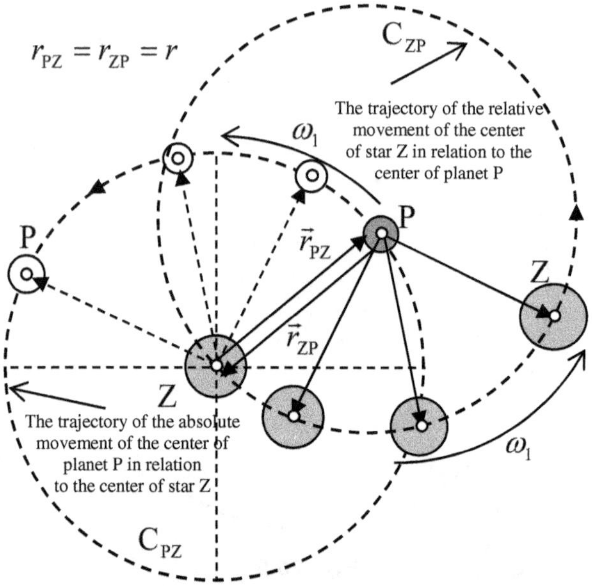

Fig. 25.3

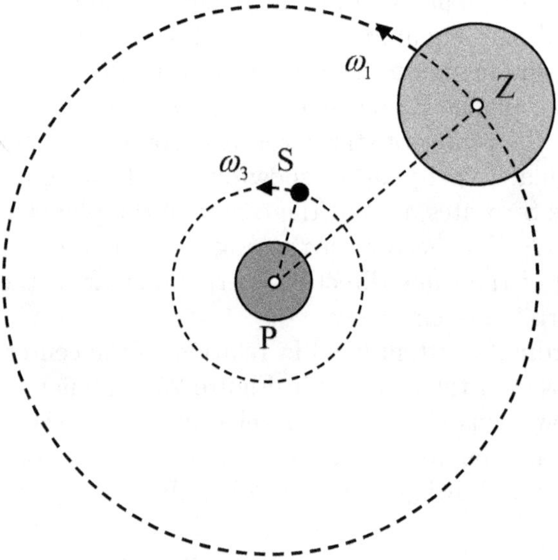

Fig. 25.4

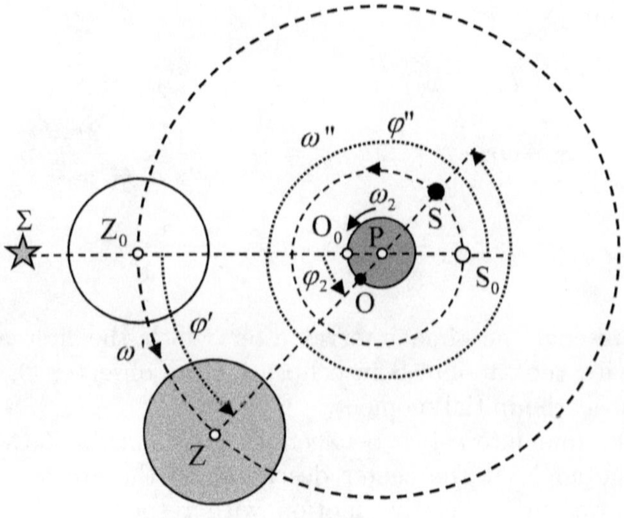

Fig. 25.5

Under these conditions, the eclipse of satellite S for the observer O will be repeated, as indicated by the notation in Figure 25.5, i.e., the alignment ZOPS, given that:

$$\vec{\omega}' = \vec{\omega}_1 - \vec{\omega}_2; \quad \omega' = \omega_1 - \omega_2; \quad \omega_1 > \omega_2,$$

representing the relative angular velocity of the center of the star Z relative to the observer on the surface of the planet;

$$\vec{\omega}'' = \vec{\omega}_3 - \vec{\omega}_2; \quad \omega'' = \omega_3 - \omega_2; \quad \omega_3 > \omega_2; \quad \omega_2 < \omega_1 < \omega_3,$$

representing the relative angular velocity of the satellite S relative to the observer on the planet's surface. Thus, for the angles at the center of the planet P described by the vector radii of the star Z and the satellite S, respectively, in their motion relative to the surface of the planet, are:

$$\phi' = \omega' \cdot \tau = (\omega_1 - \omega_2) \cdot \tau; \quad \omega' = \omega_1 - \omega_2;$$
$$\phi'' = \omega'' \cdot \tau = (\omega_3 - \omega_2) \cdot \tau; \quad \omega'' = \omega_3 - \omega_2.$$

The relationship between these two angles is

$$\phi'' = 2\pi + \phi',$$

resulting in:

$$(\omega_3 - \omega_2) \cdot \tau = 2\pi + (\omega_1 - \omega_2) \cdot \tau;$$

$$(\omega_3 - \omega_1) \cdot \tau = 2\pi; \quad \tau = \frac{2\pi}{\omega_3 - \omega_1} = \frac{2\pi}{\frac{2\pi}{T_3} - \frac{2\pi}{T_1}};$$

$$\tau = \frac{1}{\frac{1}{T_3} - \frac{1}{T_1}}; \quad \tau = \frac{T_1 T_3}{T_1 - T_3}.$$

This represents the time interval after which the first alignment occurs, with the satellite S in eclipse for the observer O, from the alignment at the initial moment.

In the time interval τ between the alignments $Z_0O_0PS_0$ and ZOPS, the angle at the center described by the arc on which the star Z moves in its relative motion with respect to the center of planet P is:

$$\phi' = \omega' \cdot \tau = (\omega_1 - \omega_2) \cdot \tau;$$

$$\phi' = \omega' \cdot \tau = \left(\frac{2\pi}{T_1} - \frac{2\pi}{T_2}\right) \cdot \frac{T_1 T_3}{T_1 - T_3};$$

$$\phi' = 2\pi \cdot \left(\frac{1}{T_1} - \frac{1}{T_2}\right) \cdot \frac{T_1 T_3}{T_1 - T_3}; \quad \phi' = 2\pi \cdot \frac{T_2 - T_1}{T_1 T_2} \cdot \frac{T_1 T_3}{T_1 - T_3};$$

$$\varphi' = 2\pi \cdot \frac{T_2 - T_1}{T_1 - T_3} \cdot \frac{T_3}{T_2}; \quad T_3 < T_1 < T_2.$$

On the other hand, in Figure 25.5, we notice that:

$$\phi_2 = \phi'; \quad \omega_2 \cdot \tau = (\omega_1 - \omega_2) \cdot \tau;$$

$$\omega_2 = (\omega_1 - \omega_2); \quad \omega_1 = 2\omega_2; \quad \frac{2\pi}{T_1} = 2\frac{2\pi}{T_2}; \quad \frac{1}{T_1} = \frac{2}{T_2};$$

$$2 \cdot T_1 = T_2; \quad T_1 = \frac{T_2}{2};$$

$$T_3 < T_1 < T_2;$$

$$\frac{T_3}{T_2} < 1;$$

$$\tau = \frac{T_1 T_3}{T_1 - T_3};$$

$$\phi' = 2\pi \cdot \frac{T_2 - T_1}{T_1 - T_3} \cdot \frac{T_3}{T_2};$$

$$\phi' = 2\pi \cdot \frac{T_1 \cdot \left(\frac{T_2}{T_1} - 1\right)}{T_1 \cdot \left(1 - \frac{T_3}{T_1}\right)_1} \cdot \frac{T_3}{T_2}; \quad \phi' = 2\pi \cdot \frac{\left(\frac{T_2}{T_1} - 1\right)}{\left(1 - \frac{T_3}{T_1}\right)_1} \cdot \frac{T_3}{T_2}; \quad 2 \cdot T_1 = T_2;$$

$$\phi' = 2\pi \cdot \frac{\left(\frac{2T_1}{T_1} - 1\right)}{\left(1 - \frac{T_3}{T_1}\right)_1} \cdot \frac{T_3}{T_2}; \quad \phi' = 2\pi \cdot \frac{(2-1)}{\left(1 - \frac{T_3}{T_1}\right)_1} \cdot \frac{T_3}{T_2};$$

$$\phi' = 2\pi \cdot \frac{1}{\left(1 - \frac{T_3}{T_1}\right)_1} \cdot \frac{T_3}{T_2};$$

$$\phi' = 2\pi \cdot \frac{1}{\left(1 - \frac{T_3}{T_1}\right)} \cdot \frac{T_3}{2T_1};$$

$$\phi' = 2\pi \cdot \frac{T_1}{T_1 - T_3} \cdot \frac{T_3}{2T_1}; \quad \phi' = \pi \cdot \frac{T_3}{T_1 - T_3}; \quad T_3 < T_1 < T_2.$$

b) In the first alignment after the initial alignment, occurring after a time τ, when the center of the star Z moves, relative to the center of planet P, on an arc of a circle whose angle at the center is ϕ', it follows that the alignments

$$2, 3, 4, \ldots\ldots\ldots\ldots, N$$

will be realized when the vector radius of the center of the star Z, in its apparent motion with respect to the center of the planet P, has rotated by an angle of

$$2\phi',\ 3\phi',\ 4\phi', \ldots\ldots\ldots\ldots,\ N\phi'.$$

If, for the N-th eclipse, the condition

$$N\phi' = 2\pi \cdot m$$

is met, then the angle $N\phi'$ is an integer (m) multiple of 2π, and the N-th eclipse can occur on exactly the same alignment with the fixed star Σ. This is the alignment $\Sigma Z_0 O_0 P S_0$, represented in the drawing in Figure 25.5, which occurred at the moment t_0. Thus, we calculate the ratio:

$$\frac{\phi'}{2\pi} = \frac{m}{N};$$

$$\phi' = \omega' \cdot \tau = (\omega_1 - \omega_2) \cdot \tau;$$

$$\phi'' = 2\pi + \phi'; \quad 2\pi = \phi'' - \phi';$$

$$\phi'' = \omega'' \cdot \tau = (\omega_3 - \omega_2) \cdot \tau; \quad \omega'' = \omega_3 - \omega_2;$$

$$\frac{\varphi'}{2\pi} = \frac{m}{N}; \quad \frac{\phi'}{\phi'' - \phi'} = \frac{m}{N};$$

$$\frac{\phi'}{2\pi} = \frac{(\omega_1 - \omega_2) \cdot \tau}{(\omega_3 - \omega_2) \cdot \tau - (\omega_1 - \omega_2) \cdot \tau} = \frac{m}{N};$$

$$\frac{\phi'}{2\pi} = \frac{(\omega_1 - \omega_2) \cdot \tau}{(\omega_3 - \omega_1) \cdot \tau} = \frac{m}{N}; \quad \frac{\phi'}{2\pi} = \frac{(\omega_1 - \omega_2)}{(\omega_3 - \omega_1)} = \frac{m}{N};$$

$$\frac{\phi'}{2\pi} = \frac{(\omega_1 - \omega_2)}{(\omega_3 - \omega_2) + (\omega_2 - \omega_1)} = \frac{m}{N};$$

$$\frac{\phi'}{2\pi} = \frac{(\omega_1 - \omega_2)}{(\omega_3 - \omega_2) - (\omega_1 - \omega_2)} = \frac{m}{N};$$

$$\frac{\phi'}{2\pi} = \frac{\omega'}{\omega'' - \omega'} = \frac{m}{N};$$

$$\frac{\phi'}{2\pi} = \frac{\left(\frac{2\pi}{T_1} - \frac{2\pi}{T_2}\right)}{\left(\frac{2\pi}{T_3} - \frac{2\pi}{T_1}\right)} = \frac{m}{N}; \quad \frac{\phi'}{2\pi} = \frac{\left(\frac{1}{T_1} - \frac{1}{T_2}\right)}{\left(\frac{1}{T_3} - \frac{2}{T_1}\right)} = \frac{m}{N};$$

$$\frac{\phi'}{2\pi} = \frac{\frac{T_2 - T_1}{T_1 T_2}}{\frac{T_1 - T_3}{T_1 T_3}} = \frac{m}{N}; \quad \frac{\phi'}{2\pi} = \frac{T_2 - T_1}{T_1 - T_3} \cdot \frac{T_1 T_3}{T_1 T_2} = \frac{m}{N}.$$

Finding that this ratio is in the form of an irreducible fraction, we get

$$\frac{\phi'}{2\pi} = \frac{\omega'}{\omega'' - \omega'} = \frac{T_2 - T_1}{T_1 - T_3} \cdot \frac{T_3}{T_2} = \frac{m}{N},$$

meaning that the N-th alignment corresponds exactly to the original alignment, $\Sigma Z_0 O_0 P S_0$, after a time interval:

$$t = m \cdot \tau;$$

$$\tau = \frac{T_1 T_3}{T_1 - T_3}; \quad t = m \cdot \frac{T_1 T_3}{T_1 - T_3}.$$

If the ratio

$$\frac{\phi'}{2\pi} = \frac{m}{N}$$

is an irrational number, then the angle $N\phi'$, for $N = 1, 2, 3, \ldots, \ldots, \ldots$, cannot be an integer multiple of 2π. In this case, the N-th eclipse will not occur at the original alignment.

In practice, the ratio

$$\frac{\phi'}{2\pi} = \frac{\omega'}{\omega'' - \omega'} = \frac{T_2 - T_1}{T_1 - T_3} \cdot \frac{T_3}{T_2} = \frac{m}{N}$$

can only be determined by approximation, so it does not make sense to say that it is a rational or irrational number.

As a result, only if the number $\frac{\phi'}{2\pi}$ closely approximates the ratio $\frac{m}{N}$ will the satellite's eclipse occur on one of the N alignments.

c) No! Arguments:

1) The orbits of the Earth around the Sun and the Moon around the Earth are not coplanar.

2) The axis of the Earth's own rotation is not perpendicular to the plane of its orbit.

3) The relation that determines the order of the period of the Earth's rotation around the Sun, T_{ES}, the period of the Earth's rotation around its own axis, T_{EE}, and the period of the Moon's rotation around the Earth, T_{ME}, is:

$$T_{\mathrm{EE}} < T_{\mathrm{ME}} < T_{\mathrm{ES}},$$

$$T_{\mathrm{EE}}(T_2) < T_{\mathrm{ME}}(T_3) < T_{\mathrm{ES}}(T_1),$$

which is different from the order of the periods of rotation of the elements of the system in the analyzed problem:

$$T_3 < T_1 < T_2.$$

Problem 26

Balls Suspended Inside a Terrestrial Satellite

An artificial satellite revolves around the Earth in a circular orbit with radius r, permanently maintaining the same orientation towards the Earth, as shown in the drawing in Figure 26.1. Inside the satellite, four identical spherical masses (each of mass m) are suspended by very light wires. The arrangement is such that balls (b) and (c) are symmetrically positioned relative to ball (a), while balls (a) and (d) are separated by a distance $\Delta r = 2d$.

Determine the tension in each suspension wire. *Draw a conclusion* about the existence or non-existence of a state of weightlessness in each of the four points corresponding to the positions of the spheres inside the satellite.

Given: $M-$ the mass of the Earth; $K-$ the gravitational attraction constant.

Solution

The orientation of the spacecraft being constant, and the spacecraft being a rigid solid body, all points belonging to the spacecraft move around the Earth at the same angular velocity, ω, whose expression is obtained as follows:

$$F_{\text{g(spacecraft)}} = K\frac{m_{\text{(spacecraft)}}M}{r^2} = m_{\text{(spacecraft)}}\omega^2 r; \quad \omega^2 = K\frac{M}{r^3}.$$

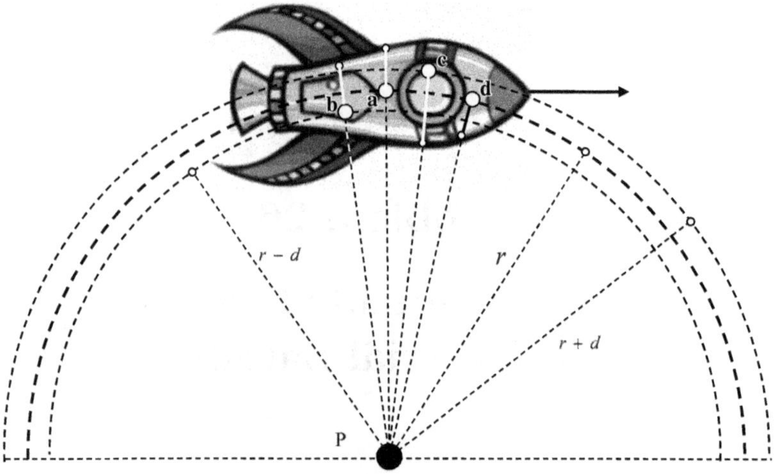

Fig. 26.1

The centripetal accelerations of the three suspended balls are:

$$a_{\text{cp(a)}} = \omega^2 r; \quad a_{\text{cp(b)}} = \omega^2 (r-d); \quad a_{\text{cp(c)}} = \omega^2 (r+d);$$
$$a_{\text{cp(d)}} = \omega^2 r = a_{\text{cp(a)}}.$$

Assuming that all the suspension wires are tensioned, then, for each ball, the resultant gravitational pull force from the tension in the respective suspension wire is the centripetal force responsible for the circular motion of each ball.

This results in:

For ball (a):

$$\vec{F}_{\text{cp(a)}} = \vec{F}_{\text{g(a)}} + \vec{T}_{\text{(a)}}; \quad F_{\text{cp(a)}} = F_{\text{g(a)}} - T_{\text{(a)}};$$
$$T_{(a)} = F_{\text{g(a)}} - F_{\text{cp(a)}};$$

$$F_{\text{g(a)}} = K\frac{mM}{r^2}; \quad F_{\text{cp(a)}} = m a_{\text{cp(a)}} = m\omega^2 r; \quad \omega^2 = K\frac{M}{r^3};$$

$$F_{\text{cp(a)}} = m a_{\text{cp(a)}} = K\frac{mM}{r^2};$$

$$T_{(a)} = K\frac{mM}{r^2} - K\frac{mM}{r^2};$$

$$T_{(a)} = 0,$$

which means that the suspension wire of ball (a) is not tensioned, so that ball (a) is in a weightless state.

For ball (b):

$$\vec{F}_{\text{cp(b)}} = \vec{F}_{\text{g(b)}} + \vec{T}_{(b)}; \quad F_{\text{cp(b)}} = F_{\text{g(b)}} - T_{(b)};$$

$$T_{(b)} = F_{\text{g(b)}} - F_{\text{cp(b)}};$$

$$F_{\text{g(b)}} = K\frac{mM}{(r-d)^2} = K\frac{mM}{r^2\left(1-\frac{d}{r}\right)^2} = K\frac{M}{r^2}\left(1-\frac{d}{r}\right)^{-2};$$

$$d << r; \quad \left(1-\frac{d}{r}\right)^{-2} \approx 1 + 2\cdot\frac{d}{r};$$

$$F_{\text{g(b)}} = K\frac{M}{r^2}\left(1+2\cdot\frac{d}{r}\right);$$

$$F_{\text{cp(b)}} = ma_{\text{cp(b)}}; \quad a_{\text{cp(b)}} = \omega^2(r-d); \quad F_{\text{cp(b)}} = m\omega^2(r-d);$$

$$\omega^2 = K\frac{M}{r^3}; \quad F_{\text{cp(b)}} = K\frac{mM}{r^3}(r-d);$$

$$T_{(\text{b})} = K\frac{M}{r^2}\left(1+2\cdot\frac{d}{r}\right) - K\frac{mM}{r^3}(r-d)$$

$$= K\frac{mM}{r^2} + 2K\frac{mMd}{r^3} - K\frac{mM}{r^2} + K\frac{mMd}{r^3};$$

$$T_{(\text{b})} = 3K\frac{mMd}{r^3} \neq 0; \quad T_{(b)} = 3K\frac{mMd}{r^3} > 0,$$

which means that the suspension wire of ball (b) is tensioned, so that ball (b) is not weightless.

For ball (c)

$$\vec{F}_{\text{cp(c)}} = \vec{F}_{\text{g(c)}} + \vec{T}_{(\text{c})}; \quad F_{\text{cp(c)}} = F_{\text{g(c)}} + T_{(\text{c})}; \quad T_{(\text{c})} = F_{\text{cp(c)}} - F_{\text{g(c)}};$$

$$F_{\text{g(c)}} = K\frac{mM}{(r+d)^2} = K\frac{mM}{r^2(1+\frac{d}{r})^2} = K\frac{mM}{r^2}\left(1+\frac{d}{r}\right)^{-2};$$

$$d << r; \quad \left(1+\frac{d}{r}\right)^{-2} \approx 1 - 2\cdot\frac{d}{r};$$

$$F_{\text{g(c)}} = K\frac{mM}{r^2}\left(1-2\cdot\frac{d}{r}\right);$$

$$F_{\text{cp(c)}} = ma_{\text{cp(c)}}; \quad a_{\text{cp(c)}} = \omega^2(r+d); \quad F_{\text{cp(c)}} = m\omega^2(r+d);$$

$$\omega^2 = K\frac{M}{r^3}; \quad F_{\text{cp(c)}} = K\frac{mM}{r^3}(r+d);$$

$$T_{\text{(c)}} = K\frac{mM}{r^3}(r+d) - K\frac{mM}{r^2}\left(1 - 2\cdot\frac{d}{r}\right)$$

$$= K\frac{mM}{r^2} + K\frac{mMd}{r^3} - K\frac{mM}{r^2} + 2K\frac{mMd}{r^3};$$

$$T_{\text{(c)}} = -3K\frac{mMd}{r^3} \neq 0; \quad T_{\text{(c)}} = 3K\frac{mMd}{r^3} > 0,$$

which means that the suspension wire of ball (c) is tensioned, so that ball (c) is not weightless.

For ball (d)

$$\vec{F}_{\text{cp(d)}} = \vec{F}_{\text{g(d)}} + \vec{T}_{\text{(d)}}; \quad F_{\text{cp(d)}} = F_{\text{g(d)}} + T_{\text{(d)}};$$

$$T_{\text{(d)}} = F_{\text{cp(d)}} - F_{\text{g(d)}};$$

$$F_{\text{g(d)}} = K\frac{mM}{r^2}; \quad F_{\text{cp(d)}} = ma_{\text{cp(d)}} = m\omega^2 r; \quad \omega^2 = K\frac{M}{r^3};$$

$$F_{\text{cp(d)}} = ma_{\text{cp(d)}} = K\frac{mM}{r^2};$$

$$T_{\text{(d)}} = K\frac{mM}{r^2} - K\frac{mM}{r^2};$$

$$T_{\text{(d)}} = 0,$$

which means that the suspension wire of ball (d) is not tensioned, so that ball (d) is in a weightless state.

Problem 27

Maximum Radial Velocities

For the elliptical orbit of the star σ around the star Σ, the drawing in Figure 27.1 represents the system's line of observation, located in the plane of the orbit.

a) *Justify qualitatively* (without mathematical proof) the existence of a moment when in the evolution of the star σ, on the elliptical orbit around the star Σ, there is a point (a position) at which the radial velocity of the star σ has the maximum value.

b) *Determine* the maximum radial velocity of the star σ, around the star Σ, relative to the observer located as indicated by the problem statement, $\vec{v}_{\text{rad,max}}$.

Given: the semi-axes of the ellipse, a and b; α – the angle between the aspides line and the observation line, measured clockwise; K – the universal attraction constant; M – the mass of star Σ.

Particular cases: **1)** $\alpha = 0$; **2)** $\alpha = 90°$; **3)** $\alpha = 180°$; **4)** $\alpha = 270°$.

Solution

a) For a star σ, moving at a speed $\vec{v}$, relative to an observer O, as indicated in the drawing in Figure 27.2, we define the radial velocity, $\vec{v}_{\text{radial}}$, as the component of the velocity oriented along the direction of observation of the star and the transverse velocity, $\vec{v}_{\text{transverse}}$, as the component of the velocity perpendicular to the direction of view.

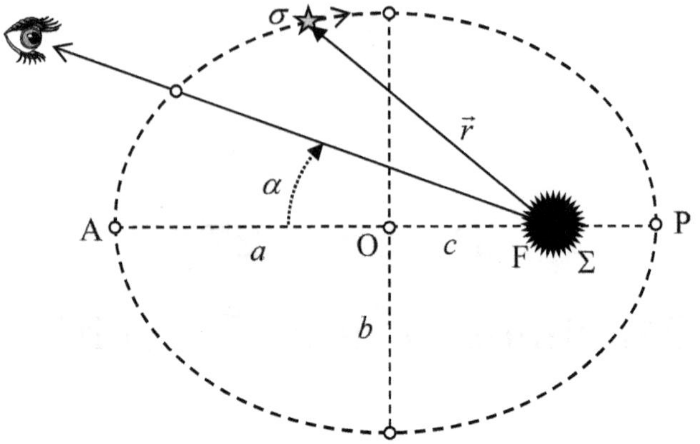

Fig. 27.1

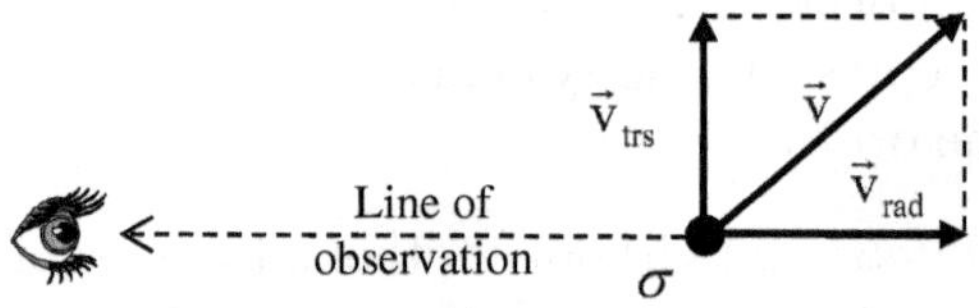

Fig. 27.2

At any point in its evolution, as shown in the drawing in Figure 27.3, the motion of the star σ is characterized by a velocity vector, $\vec{v}$, tangent to the ellipse, having the components $\vec{v}_{\text{radial}}$ and $\vec{v}_{\text{transverse}}$, so that

$$\vec{v} = \vec{v}_{\text{rad}} + \vec{v}_{\text{trs}}.$$

At any point in its evolution, as shown in the drawing in Figure 27.3, the motion of the star is also characterized by an acceleration vector:

$$\vec{a} = \frac{\vec{F}}{m} = \frac{F \cdot \text{vers}\vec{F}}{m} = K\frac{mM}{r^2} \cdot \frac{1}{m} \cdot \text{vers}\vec{F} = K\frac{M}{r^2} \cdot \text{vers}\vec{F}$$

$$= -K\frac{M}{r^2} \cdot \text{vers}\vec{r} = -K\frac{M}{r^2} \cdot \frac{\vec{r}}{r};$$

$$\vec{a} = -K\frac{M}{r^3} \cdot \vec{r},$$

whose components, in relation to the specified observer, are $\vec{a}_{\text{radial}}$ and $\vec{a}_{\text{transverse}}$. Thus,

$$\vec{a} = \vec{a}_{\text{rad}} + \vec{a}_{\text{trs}},$$

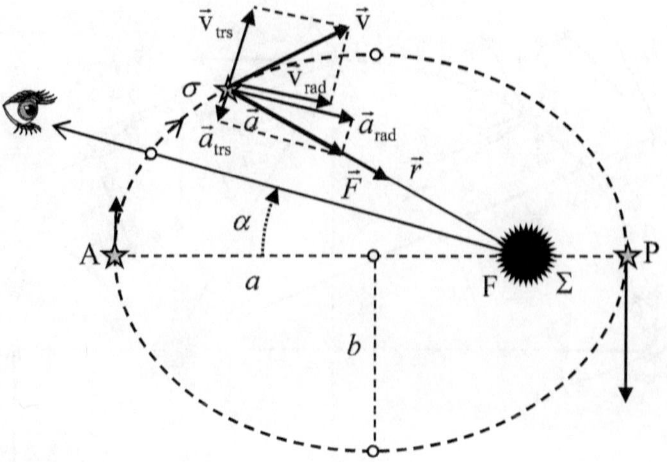

Fig. 27.3

of which the radial component, $\vec{a}_{rad}$, determines the variation of the radial component, $\vec{v}_{rad}$, of the speed of the star σ, and the transverse component, $\vec{a}_{trs}$, determines the variation of the transverse component, $\vec{v}_{trs}$, of the speed of the star σ.

Under these conditions, the evolution of the moduli of the two-speed components (ascending or descending) will be determined by the evolution of the orientations of the acceleration vector components in relation to the orientations of the velocity vector components, as shown in the drawing in Figure 27.4.

There is a moment in time, corresponding to the E position of the star σ, highlighted in the drawings in Figures 27.3 and 27.4, when:

$$\vec{a}_{rad} = 0; \quad \vec{a} \equiv \vec{a}_{trs} \perp \text{line of sight.}$$

Until then, the components $\vec{v}_{rad}$ and $\vec{a}_{rad}$ have identical orientations, which means that, until then, the radial motion of the star σ, relative to the observer, is an accelerated motion.

After exceeding the E position in Figure 27.5, the orientations of the components $\vec{v}_{rad}$ and $\vec{a}_{rad}$ are opposite, which proves that after exceeding the E position, the radial motion of the star relative to the observer becomes slower.

As a result, when the star σ reaches position E, the radial component of the velocity vector is at its maximum, $\vec{v}_{rad,max}$.

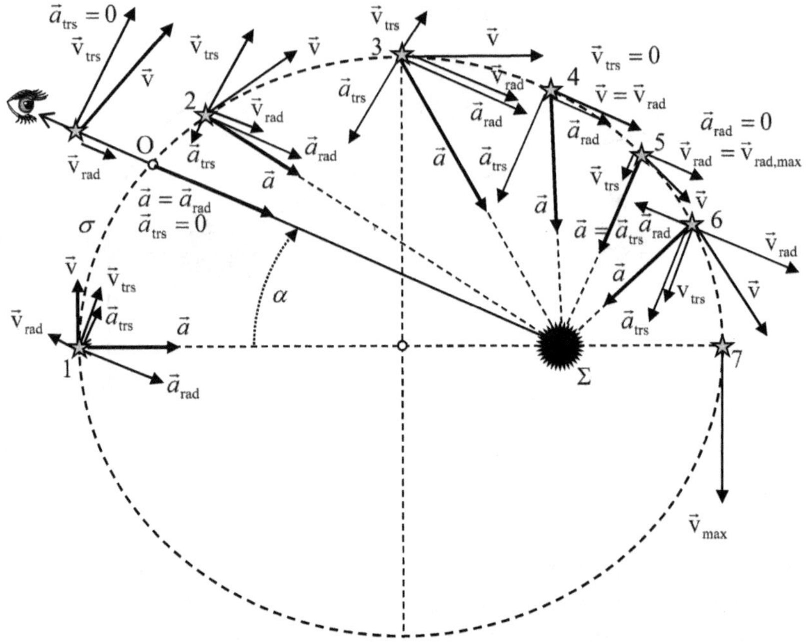

Fig. 27.4

b) Corresponding to position E, on the elliptical orbit around the star Σ, the angular momentum of the star σ is:

$$\vec{L}_E = m\vec{v}_E \times \vec{r}_E = m\left(\vec{v}_{rad,max} + \vec{v}_{trs}\right) \times \vec{r}_E = m\vec{v}_{rad,max} \times \vec{r}_E;$$

$$L_E = m v_{rad,max}\, r_E \cdot \sin 90^\circ = m v_{rad,max} \cdot r_E.$$

Knowing that in its motion on the elliptical orbit, the angular momentum of the star σ is conserved, having the value

$$L = mb\sqrt{\frac{KM}{a}} = \text{constant},$$

this results in:

$$m v_{rad,max} \cdot r_E = mb\sqrt{\frac{KM}{a}};$$

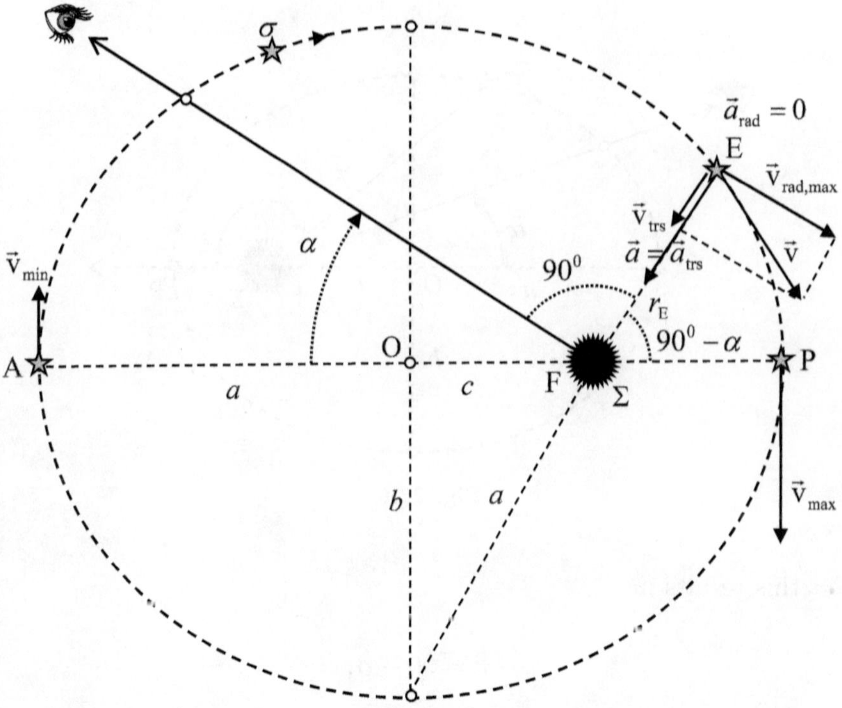

Fig. 27.5

$$\mathrm{v}_{\mathrm{rad,max}} \cdot r_{\mathrm{E}} = b \cdot \sqrt{\frac{KM}{a}};$$

$$\mathrm{v}_{\mathrm{rad,max}} = \frac{b}{r_{\mathrm{E}}} \cdot \sqrt{\frac{KM}{a}}.$$

Also, we know that in Cartesian coordinates and in plane polar coordinates, according to the notation in Figure 27.6, the equations of the elliptic orbit are given by the expressions:

$$\frac{x^2}{a^2} + \frac{y^2}{b^2} = 1; \quad r = \frac{p}{1 + e\cos\theta}; \quad p = a\left(1 - e^2\right); \quad p = b\sqrt{1 - e^2};$$

$$e = \sqrt{1 - \frac{b^2}{a^2}}; \quad e^2 = 1 - \frac{b^2}{a^2}; \quad 1 - e^2 = \frac{b^2}{a^2},$$

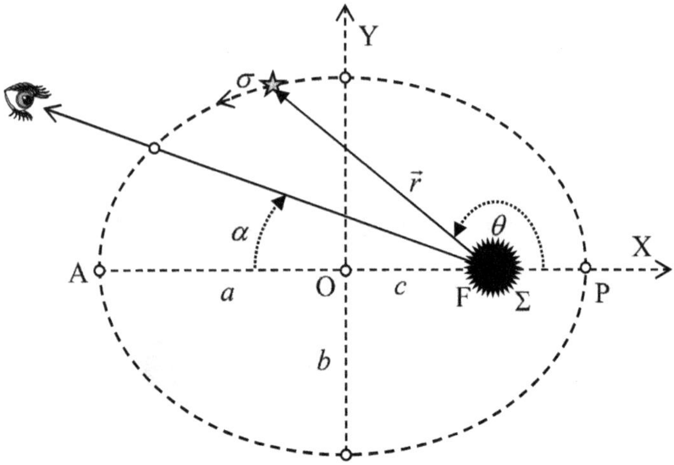

Fig. 27.6

this results in:

$$\theta = 90 - \alpha;$$

$$r_E = \frac{p}{1 + e\cos(90^\circ - \alpha)} = \frac{p}{1 + e\sin\alpha};$$

$$v_{rad,max} = \frac{b}{r_E} \cdot \sqrt{\frac{KM}{a}};$$

$$v_{rad,max} = \frac{b}{\frac{p}{1+e\sin\alpha}} \cdot \sqrt{\frac{KM}{a}} = \frac{b}{p} \cdot (1 + e\sin\alpha)\sqrt{\frac{KM}{a}};$$

$$v_{rad,max} = \frac{b}{a(1 - e^2)} \cdot (1 + e\sin\alpha)\sqrt{\frac{KM}{a}};$$

$$v_{rad,max} = \frac{b}{a \cdot \frac{b^2}{a^2}} \cdot (1 + e\sin\alpha) \cdot \sqrt{\frac{KM}{a}};$$

$$v_{rad,max} = \frac{a}{b} \cdot (1 + e\sin\alpha) \cdot \sqrt{\frac{KM}{a}};$$

$$v_{rad,max} = \frac{1 + e\sin\alpha}{b} \cdot \sqrt{aKM}.$$

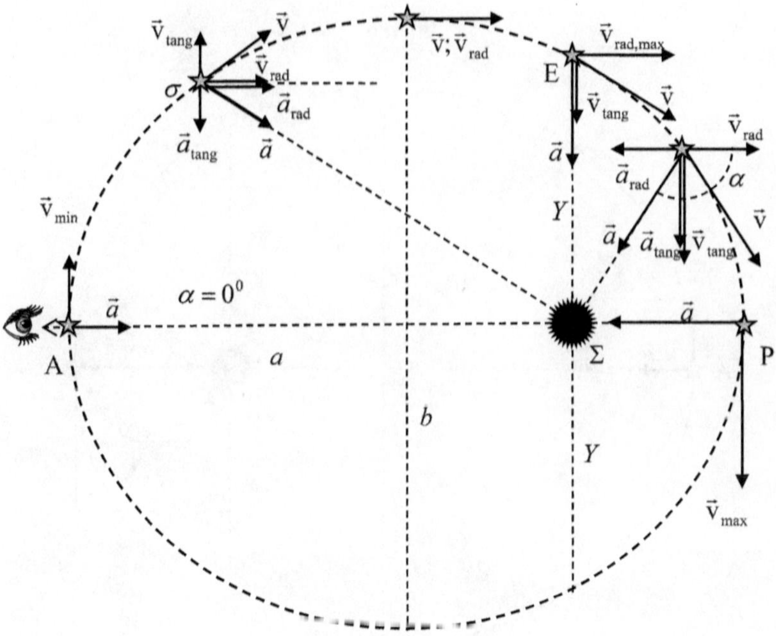

Fig. 27.7

Particular cases:

1) $\alpha = 0°$, Figure 27.7.

The drawing in Figure 27.7 shows the evolution of the radial and tangential components of the velocity and acceleration vectors relative to the observer when the line of sight is in the plane of the orbit and $\alpha = 0$:

$$\mathrm{v}_{\mathrm{rad,max}} = \frac{1}{b} \cdot \sqrt{aKM}.$$

Corresponding to the E position of the star σ, when its angular momentum is

$$L = Ym\mathrm{v}_{\mathrm{rad,max}},$$

it results that:

$$Ym\mathrm{v}_{\mathrm{rad,max}} = mb\sqrt{\frac{KM}{a}}; \quad \mathrm{v}_{\mathrm{rad,max}} = \frac{b}{Y} \cdot \sqrt{\frac{KM}{a}};$$

$$\frac{x^2}{a^2} + \frac{y^2}{b^2} = 1;$$

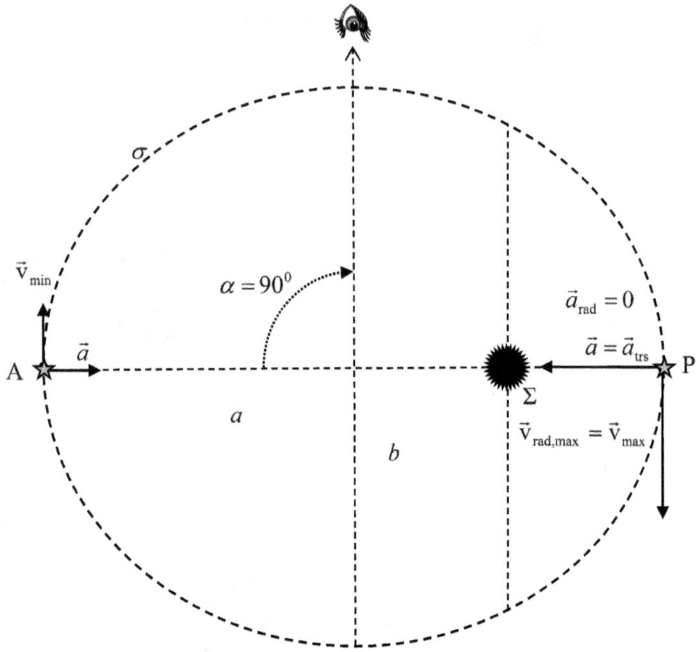

Fig. 27.8

$$x = c; \quad y = Y; \quad \frac{c^2}{a^2} + \frac{Y^2}{b^2} = 1; \quad c^2 = a^2 - b^2; \quad Y = \frac{b^2}{a};$$

$$\mathrm{v}_{\mathrm{rad,max}} = \frac{a}{b} \cdot \sqrt{\frac{KM}{a}}.$$

2) $\alpha = 90°$, Figure 27.8.

$$\mathrm{v}_{\mathrm{rad,max}} = \frac{1 + e\sin\alpha}{b} \cdot \sqrt{aKM};$$

$$\alpha = 90°;$$

$$\mathrm{v}_{\mathrm{rad,max}} = \frac{1+e}{b} \cdot \sqrt{aKM};$$

$$1 - e^2 = \frac{b^2}{a^2} = (1-e) \cdot (1+e); \quad b = a \cdot \sqrt{(1-e) \cdot (1+e)};$$

$$\mathrm{v}_{\mathrm{rad,max}} = \frac{1+e}{a \cdot \sqrt{(1-e) \cdot (1+e)}} \cdot \sqrt{aKM};$$

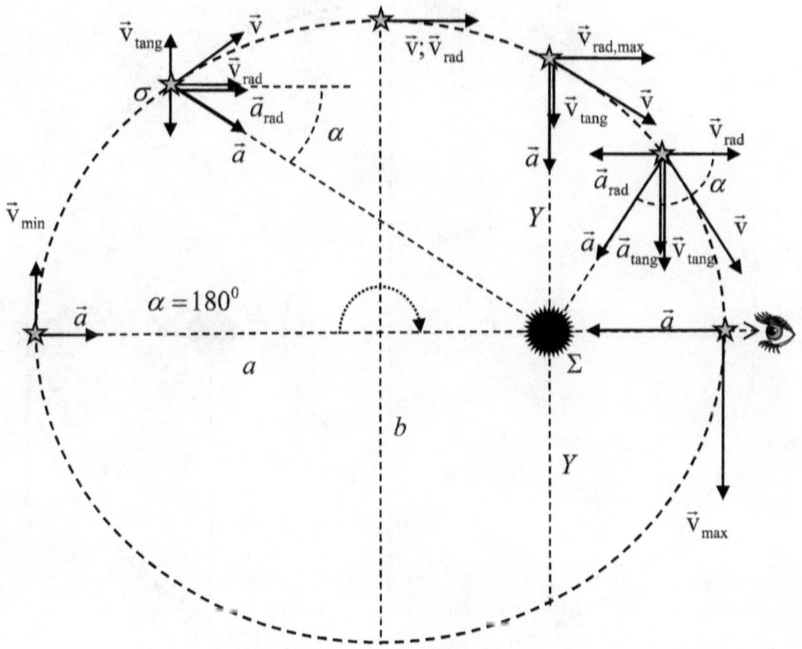

Fig. 27.9

$$v_{rad,max} = \frac{\sqrt{1+e}}{a \cdot \sqrt{1-e}} \cdot \sqrt{aKM};$$

$$v_{rad,max} = \sqrt{\frac{KM(1+e)}{a(1-e)}} = v_E = v_{max}.$$

3) $\alpha = 180°$, Figure 27.9.

$$\frac{1 + e\sin\alpha}{b} \cdot \sqrt{aKM}.$$

4) $\alpha = 270°$, Figure 27.10.

$$v_{rad,max} = \frac{1 + e\sin\alpha}{b} \cdot \sqrt{aKM};$$

$$\sin 270° = -1;$$

$$v_{rad,max} = \frac{1-e}{b} \cdot \sqrt{aKM};$$

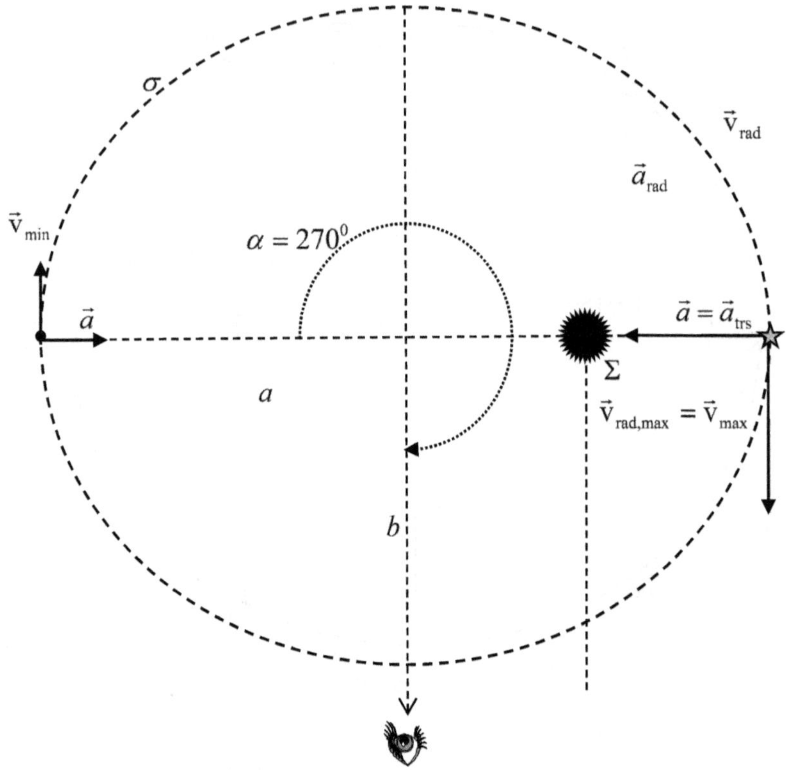

Fig. 27.10

$$1 - e^2 = \frac{b^2}{a^2} = (1-e)\cdot(1+e); \quad b = a\cdot\sqrt{(1-e)\cdot(1+e)};$$

$$v_{rad,max} = \frac{1-e}{a\cdot\sqrt{(1-e)\cdot(1+e)}}\cdot\sqrt{aKM};$$

$$v_{rad,max} = \frac{\sqrt{1-e}}{a\cdot\sqrt{1+e}}\cdot\sqrt{aKM};$$

$$v_{rad,max} = \sqrt{\frac{KM(1-e)}{a(1+e)}} = v_A = v_{min}.$$

Conclusion: The drawing in Figure 27.11 shows the evolution of the radial and tangential components of the velocity and acceleration

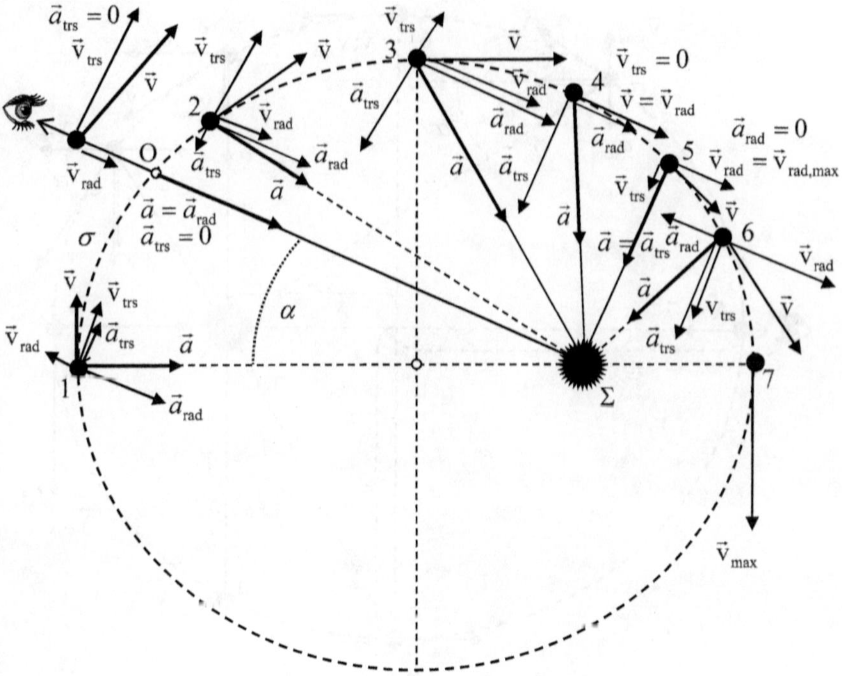

Fig. 27.11

vectors relative to the observer when the line of sight is in the plane of the orbit.

Particular direction of observation

The drawing in Figure 27.12 shows the evolution of the radial and tangential components of the velocity and acceleration vectors relative to the observer when the line of sight is in the plane of the orbit, along the AP direction.

On the 2–6–3 sector of the ellipse, when the angle α between the acceleration vector, $\vec{a}$, and the radial velocity vector, $\vec{v}_{\text{rad}}$, is

$$0 \leq \alpha \leq 90^\circ,$$

the effect of acceleration (the effect of gravitational pull) on the radial component of the velocity is an accelerating effect. As a result, in this

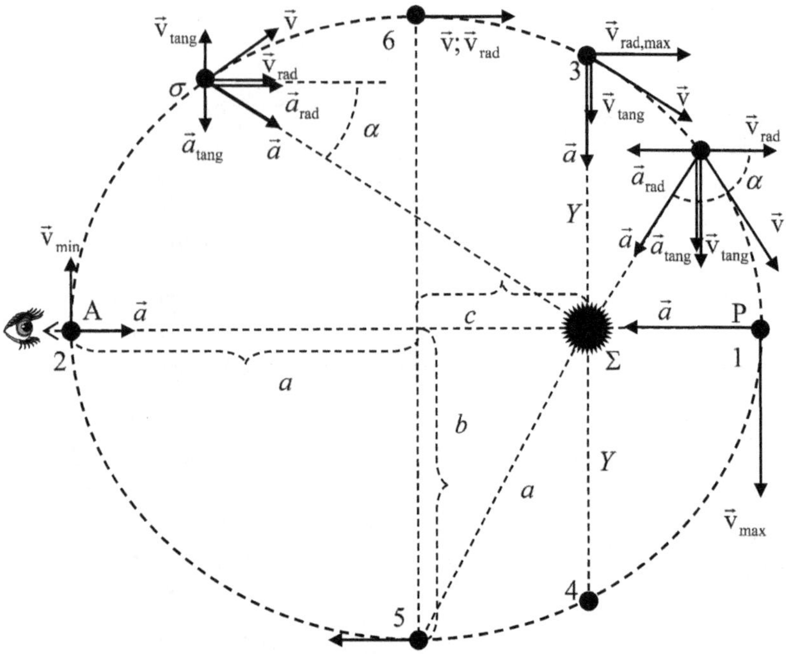

Fig. 27.12

sector, the radial component of the speed is

$$0 \leq v_{\text{rad}} \leq v_{\text{rad,max}}.$$

On sector 3–1 of the ellipse, when the angle α between the acceleration vector, $\vec{a}$, and the radial velocity vector, $\vec{v}_{\text{rad}}$, is

$$90 \leq \alpha \leq 180°,$$

the effect of acceleration (the effect of gravitational pull) on the radial component of the velocity is a slowing effect. As a result, in this sector, the radial component of the speed is

$$0 \leq v_{\text{rad}} \leq v_{\text{rad,max}}.$$

Conclusions:

- For $\alpha = 90°$, when the star σ is in position 3, the radial component of its velocity has the maximum value.

- In the sector 2–3 of the ellipse, the radial effect of the gravitational pull is an accelerating effect, and in the sector 3–1 of the ellipse, the radial effect of the gravitational pull is a slowing effect.
- In the sector 2–3 of the ellipse, the radial effect of the gravitational pull is a slowing effect, and in the sector 3–1 of the ellipse, the radial effect of the gravitational pull is an accelerating effect.
- In the sector 2–3 of the ellipse, the vectors $\vec{v}_{rad}$ and $\vec{a}_{rad}$ have identical orientations, and in sector 3–1 the vectors $\vec{v}_{rad}$ and $\vec{a}_{rad}$ have opposite orientations.
- At point 3, where the direction of the vector $\vec{a}_{rad}$ changes, the vector $\vec{v}_{rad}$ has an extreme (maximum) value.

At some point on the ellipse, the angular momentum of the star σ is:

$$\vec{L} = \vec{r} \times m\vec{v}; \quad L = rmv \cdot \sin(\vec{r}; \vec{v}).$$

Thus, according to the law of conservation of angular momentum, it follows that:

$$\vec{L} = \text{constant}; \quad L = rmv \cdot \sin(\vec{r};\ \vec{v}) = r_{min} m v_{max} = r_{max} m v_{min};$$

$$r = \frac{p}{1 + e\cos\theta}; \quad v = \sqrt{KM\left(\frac{2}{r} - \frac{1}{a}\right)};$$

$$r_{min} = \frac{p}{1+e}; \quad v_{max} = \sqrt{\frac{KM(1+e)}{1-e}}; \quad r_{max} = \frac{p}{1-e};$$

$$v_{min} = \sqrt{\frac{KM(1-e)}{1+e}};$$

$$p = a\left(1 - e^2\right); \quad p = b\sqrt{1-e^2};$$

$$L = m\sqrt{\frac{KMa^2(1-e)^2(1+e)}{a(1-e)}} = mb\sqrt{\frac{KM}{a}}.$$

Corresponding to position 3 of the star σ, when its angular momentum is

$$L = Y m v_{rad,max},$$

it results that:

$$Ym\mathrm{v}_{\mathrm{rad,max}} = mb\sqrt{\frac{KM}{a}}; \quad \mathrm{v}_{\mathrm{rad,max}} = \frac{b}{Y}\cdot\sqrt{\frac{KM}{a}};$$

$$\frac{x^2}{a^2} + \frac{y^2}{b^2} = 1;$$

$$x = c; \quad y = Y; \quad \frac{c^2}{a^2} + \frac{Y^2}{b^2} = 1; \quad c^2 = a^2 - b^2; \quad Y = \frac{b^2}{a};$$

$$\mathrm{v}_{\mathrm{rad,max}} = \frac{a}{b}\cdot\sqrt{\frac{KM}{a}};$$

$$\Delta\mathrm{v}_{\mathrm{rad,23}} = \mathrm{v}_{\mathrm{rad,max}} - 0 = \mathrm{v}_{\mathrm{rad,max}}; \quad \Delta t_{23} = \tau_1;$$

$$\Delta\mathrm{v}_{\mathrm{rad,31}} = 0 - \mathrm{v}_{\mathrm{rad,max}} = -\mathrm{v}_{\mathrm{rad,max}}; \quad \Delta t_{31} = \tau_2;$$

$$\frac{\Delta S_1}{\tau_1} = \frac{\Delta S_2}{\tau_2} = \frac{S_{\mathrm{ellipse}}}{T} = \frac{\pi ab}{T}; \quad \Delta S_1 = \pi ab\cdot\frac{\tau_1}{T}; \quad \Delta S_2 = \pi ab\cdot\frac{\tau_2}{T};$$

$$\Delta S_1 + \Delta S_2 = \frac{1}{2}\cdot S_{\mathrm{ellipse}}; \quad \pi ab\cdot\frac{\tau_1}{T} + \pi ab\cdot\frac{\tau_2}{T} = \frac{1}{2}\cdot\pi ab;$$

$$\tau_1 + \tau_2 = \frac{1}{2}T; \quad \frac{\tau_1}{\tau_2} = n;$$

$$\tau_1 = \frac{nT}{2\,(n+1)}; \quad \tau_2 = \frac{T}{2\,(n+1)}; \quad T = 2\pi\sqrt{\frac{a^3}{KM}},$$

where: τ_2 is the time interval in which the radial component of star σ's velocity increases from zero (position 1) to the maximum value (position 4); τ_1 is the time interval during which the radial component of the star's velocity decreases from the maximum value (position 4) to zero (position 2); $\tau_1/\tau_2 = n$.

Problem 28

Cosmic Spaceship

A. Spaceship to the Sun

In a future program, NASA plans to launch a spacecraft aimed directly at the Sun, without a human crew, to gather information on its way to the Sun about both the inner planets and, in particular, the Sun.

a) *Determine* the approximate duration of the Earth–Sun flight, if the spacecraft is launched in such a way that its movement in relation to the Sun is a free fall. Determine the speed, in relation to Earth, at which a spacecraft must leave Earth for it to fall to the surface of the Sun, following a rectilinear path.

The following are known: the distance between Earth and the Sun, $r_{\mathrm{ES}} = 1.5 \cdot 10^{11}$ m; the period of the Earth's rotation around the Sun, $T_{\mathrm{E}} = 3.15 \cdot 10^{7}$ s; the radius of the Sun, $R_{\mathrm{S}} \approx \frac{r_{\mathrm{ES}}}{200}$.

B. Cosmic Ship to the Moon

A spacecraft is approaching the Moon on a parabolic trajectory, almost tangent to the Moon's surface. A brake motor is engaged when approaching the Moon to pass into a low circular orbit (very close to the Moon's surface).

b) *Determine* the variation of the speed of the ship for the success of this maneuver and the proportion of the initial mass of the ship represented by the burned fuel if the engine ejects the products of combustion at relative speed v. The acceleration of free fall on the Moon is $g_{0\text{M}}$, and the radius of the Moon is R_{L}.

C. Cosmic Spacecraft Around the Moon

A spacecraft rotates around the Moon in a circular orbit with radius R.

c) *Determine*: **1)** the speed with which a body must be ejected from the ship so that it falls on the other side of the Moon, at the point diametrically opposite to the launch point, if the ship is launched on the tangent to the trajectory; **2)** after how long the body falls to the surface of the Moon. The acceleration of the free fall on the surface of the Moon is $g_{0\text{L}}$. The Moon's radius is R_{L}.

Solution

a) Imagine that a spacecraft is launched from the Earth whose speed in relation to the Sun is slightly lower than the Earth's orbital speed in relation to the Sun. The ship will evolve around the Sun in an elliptical trajectory, with the Sun in the focus opposite the launch point. The lower the initial speed of the ship, the longer the ellipse on which the spacecraft will evolve around the Sun. For a certain value of this speed, the spacecraft, in its elliptical motion around the Sun, will touch the surface of the Sun, as shown in Figure 28.1. In this case, the height of the ship's orbit is equal to the Sun's radius, which we know is

$$r_{\text{min}} = R_{\text{S}} \approx \frac{r_{\text{ES}}}{200}.$$

As a result, the semi-axes of the ship's elliptical orbit, tangent to the Sun's surface, are a and $b << a$, respectively, so that the small half-axis of this elliptical orbit can be neglected ($b \approx 0$). Under these conditions, the ship's trajectory in relation to the Sun can be estimated to be rectilinear, joining the launching point of the ship with the Sun so that the ship will reach the surface of the Sun at point B after a free fall towards it.

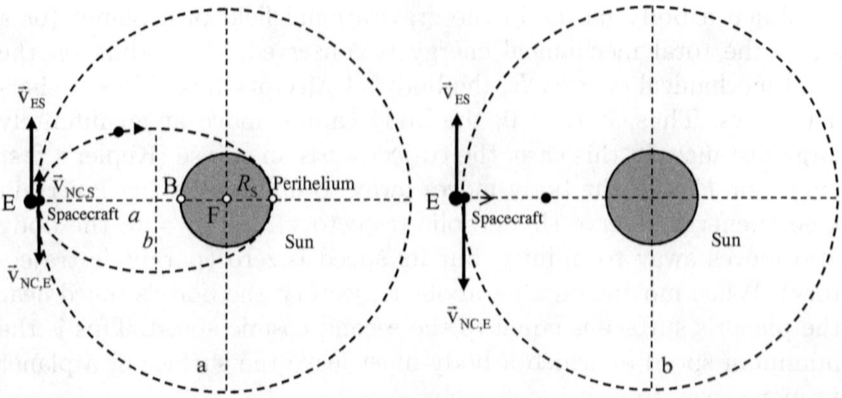

Fig. 28.1

Conclusion: For the ship to fall freely to the surface of the Sun, at point B, after a rectilinear course in relation to it, at the initial moment of its launch from the surface of the Earth, the speed of the ship in relation to the Sun must be zero.

To meet this condition, at the time of launching the ship from the Earth's surface, a certain speed relative to the Earth must be imparted to the ship.

The Earth's launch velocity relative to Earth must be inversely related to the orientation of the Earth's orbital velocity relative to the Sun, as shown in Figure 28.1b, and their moduli must be equal:

$$\mathrm{v} = \frac{2\pi r_{\mathrm{ES}}}{T_{\mathrm{E}}} = \frac{2 \cdot 3.14 \cdot 1.5 \cdot 10^{11}\,\mathrm{m}}{3.15 \cdot 10^{7}\,\mathrm{s}} \approx 3 \cdot 10^{4}\,\mathrm{m/s} = 30\,\mathrm{km/s}.$$

According to Kepler's third law, we can write:

$$\frac{T_{\mathrm{E}}^2}{r_{\mathrm{ES}}^3} = \frac{T_{\mathrm{NC}}^2}{\left(\frac{r_{\mathrm{ES}}}{2}\right)^3},$$

where T_{NC} is the period of the spacecraft's motion on the elongated elliptical trajectory around the Sun;

$$T_{\mathrm{NC}} = \frac{T_{\mathrm{E}}}{2\sqrt{2}}.$$

Thus, the duration of the free fall of the ship to the surface of the Sun is

$$t = \frac{T_{\mathrm{NC}}}{2} = \frac{T_{\mathrm{E}}}{4\sqrt{2}} \approx 5.6 \cdot 10^{6}\,\mathrm{s}.$$

b) When a body moves in the gravitational field of a planet (or a star), the total mechanical energy is conserved. Depending on the total mechanical energy, E, the body's trajectory has different characteristics. Thus, if $E < 0$, the body cannot move at an infinitely large distance. In this case, the trajectory is an ellipse (Kepler's first law). For $E > 0$, the body moves away to infinity, having a certain kinetic energy reserve (hyperbolic trajectory). For $E = 0$, the body also moves away to infinity, but its speed is zero (parabolic trajectory). When moving on a parabolic trajectory, the body's speed near the planet's surface is equal to the second cosmic speed. This is the minimum speed at which a body must leave the surface of a planet to move away from it indefinitely.

For the point closest to the Moon (the drawing in Figure 28.2), corresponding to the evolution on the parabolic trajectory, it can be written that:

$$E = \frac{m\mathrm{v}_{\mathrm{II}}^2}{2} - K\frac{mM_{\mathrm{L}}}{R_{\mathrm{L}}} = 0;$$

$$\mathrm{v}_{\mathrm{II}} = \sqrt{2g_{0\mathrm{L}}R_{\mathrm{L}}}.$$

During the braking process, the ship's speed must be reduced to the value of the first cosmic speed so that the ship can be transferred to

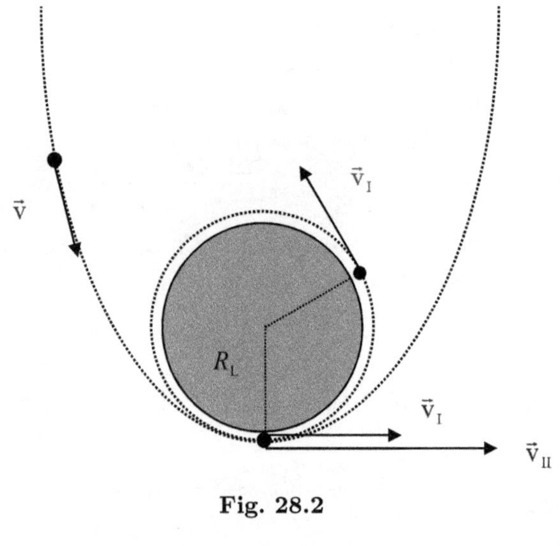

Fig. 28.2

a low circular orbit around the Moon:

$$\mathrm{v_I} = \sqrt{g_{0\mathrm{L}} R_\mathrm{L}}.$$

The transfer of the ship from the parabolic orbit to the low circular orbit is performed if, at the point closest to the surface of the Moon, the modulus of the ship's velocity vector decreases by the amount

$$\Delta \mathrm{v} = \mathrm{v_{II}} - \mathrm{v_I} = \sqrt{g_{0\mathrm{L}} R_\mathrm{L}}(\sqrt{2} - 1).$$

To determine the mass m of fuel burned for the transfer of the spacecraft from one orbit to another, we use the law of conservation of momentum. For simplicity, suppose that the products of combustion, having a total mass m, are ejected from the missile nozzle in the form of a single body, with relative velocity v. Writing the law of conservation of momentum in a reference system fixed in relation to the ship with $\mathrm{v_{II}}$, it results that:

$$(M_0 - m)\Delta \mathrm{v} = m\mathrm{v}; \quad m = M_0 \frac{\Delta \mathrm{v}}{\mathrm{v} + \Delta \mathrm{v}},$$

where M_0 is the initial mass of the ship.

c) The body thrown from the ship must move in an elliptical orbit, tangent to the surface of the Moon, as indicated in the drawing in Figure 28.3.

When the body thrown from the ship is already evolving in the elliptical orbit with the center of the Moon in the focus opposite to the launch point, its velocities at the extreme points of the ellipse are deduced using the conservation laws of mechanical energy and angular momentum (area law):

$$\frac{m\mathrm{v}_{\mathrm{min}}^2}{2} - K\frac{mM_\mathrm{L}}{r_{\mathrm{max}}} = \frac{m\mathrm{v}_{\mathrm{max}}^2}{2} - K\frac{mM_\mathrm{L}}{r_{\mathrm{min}}};$$

$$g_{0\mathrm{L}} = K\frac{M_\mathrm{L}}{R_\mathrm{L}^2};$$

$$\mathrm{v}_{\mathrm{min}}^2 - 2\frac{g_{0\mathrm{L}} R_\mathrm{L}^2}{r_{\mathrm{max}}} = \mathrm{v}_{\mathrm{max}}^2 - 2\frac{g_{0\mathrm{L}} R_\mathrm{L}^2}{r_{\mathrm{min}}};$$

$$\mathrm{v}_{\mathrm{min}} r_{\mathrm{max}} = \mathrm{v}_{\mathrm{max}} r_{\mathrm{min}};$$

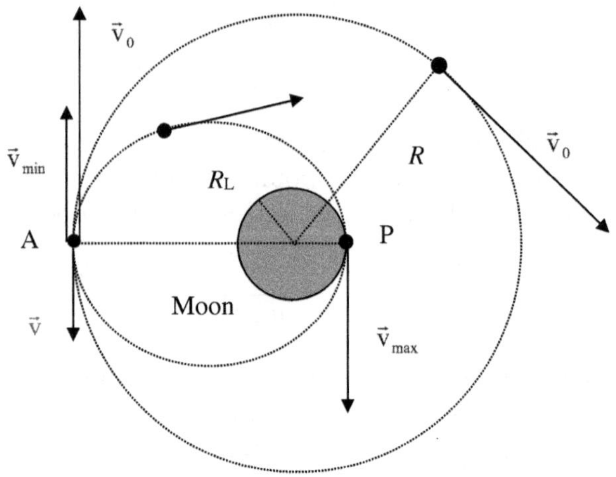

Fig. 28.3

$$r_{max} = R; \quad r_{min} = R_L; \quad v_{min} R = v_{max} R_L;$$

$$v_{max} = \frac{R}{R_L} v_{min}; \quad v_{min}^2 = 2\frac{g_{0L} R_L^3}{R(R+R_L)};$$

$$v_{min} = R_L \sqrt{\frac{2g_{0L} R_L}{R(R+R_L)}}; \quad v_{max} = R\sqrt{\frac{2g_{0L} R_L}{R(R+R_L)}}.$$

Under these conditions, the speed of the body at the time of launching from the ship, relative to the ship, for it to then evolve around the Moon in elliptical orbit, must be:

$$v = v_0 - v_{min} = \sqrt{\frac{KM_L}{R}} - v_{min} = R_L \sqrt{\frac{g_{0L}}{R}} - v_{min};$$

$$v_0 = R_L \sqrt{\frac{g_{0L}}{R}} \left(1 - \sqrt{\frac{2R_L}{R+R_L}}\right).$$

If T_0 is the period of rotation of the spaceship around the Moon in the circular orbit,

$$T_0 = 2\pi \frac{R}{R_L} \sqrt{\frac{R}{g_{0L}}},$$

then, according to Kepler's third law, we deduce the period of the body's motion in the elliptical orbit around the Moon as follows:

$$T^2 = k\left(\frac{R+R_{\rm L}}{2}\right)^3; \quad T_0^2 = kR^3;$$

$$\frac{T^2}{T_0^2} = \frac{1}{8}\left(\frac{R+R_{\rm L}}{R}\right)^3; \quad T = \frac{\pi(R+R_{\rm L})}{R_{\rm L}}\sqrt{\frac{R+R_{\rm L}}{2g_{0\rm L}}}.$$

Thus, the time after which the body thrown from the ship falls to the surface of the Moon is

$$t_{\rm fall} = \frac{T}{2} = \frac{\pi(R+R_{\rm L})}{2R_{\rm L}}\sqrt{\frac{R+R_{\rm L}}{2g_{0\rm L}}}.$$

Problem 29

The Apparent Magnitude of the Moon, Seen on the Way to the Moon

The Moon, whose linear diameter is $d = 3436\,\mathrm{km}$, and whose distance from Earth is $r_0 = 348\,000\,\mathrm{km}$, has the apparent magnitude $m_0 = -12.7$. The apparent magnitude of the Sun seen from Earth is $m_\mathrm{S} = -26.84$. The angular diameter of the Moon seen from Earth is $\delta_0 = 30\,\mathrm{arcmin}$. *Determine* the distance r from the Moon at which, for a cosmonaut on his way to the Moon in a spaceship, the brightness of the Moon is the same as the brightness of the Sun seen from Earth, and, corresponding for the distance r, the approximate angular diameter of the Moon, δ.

It will be considered that the astronaut's eye can perceive the full image of the Moon from any distance.

Solution

a) The approach of the cosmonaut to the Moon must be such that it determines the following variation of the apparent magnitude of the Moon:

$$\Delta m = m - m_0 = m_\mathrm{S} - m_0 = -26.84 - (-12.7) = -14.14.$$

Thus, the apparent magnitude of the Moon, viewed from this new distance, is:

$$m = m_0 + \Delta m = m_S = -26.84;$$

$$\Delta m = m_S - m_0 = -26.84 + 12.7 = -14.14.$$

Knowing that the luminous intensity, I, of a star, or, in this case, the Moon is

$$I = \frac{\phi}{\Omega} = \frac{L}{\Omega} = \frac{W}{t \cdot \Omega},$$

the energy flow emitted by the star, or the Moon, per unit solid angle, is:

$$\Omega = \frac{A}{\Delta^2}; \quad A = \Omega \cdot \Delta^2; \quad \phi = I \cdot \Omega; \quad E = \frac{\phi}{A} = \frac{I \cdot \Omega}{\Omega \cdot \Delta^2} = \frac{I}{\Delta^2};$$

$$E = \frac{L}{4\pi\Delta^2}.$$

According to Pogson's formula, if E is the brightnesses of the Moon, corresponding to the initial (r_0) and the final (r) distances shown in the drawing in Figure 29.1, it results that:

$$\log \frac{E}{E_0} = -0.4 \cdot (m - m_0) = \log 10^{-0.4(m-m_0)}; \quad E_0 = \frac{I}{r_0^2}; \quad E = \frac{I}{r^2},$$

where I is the luminous intensity of the Moon;

$$\log \frac{r_0^2}{r^2} = \log 10^{-0.4(m-m_0)};$$

$$2 \log \frac{r_0}{r} = -0.4 \cdot \Delta m; \quad \log \frac{r_0}{r} = 0.2 \cdot 14.14 = 2.828 = \log 10^{2.828};$$

$$\frac{r_0}{r} = 10^{2.828}; \quad r = \frac{r_0}{10^{2.828}};$$

$$x = 10^{2.828}; \quad \log x = 2.828; \quad x \approx 675;$$

$$r = \frac{r_0}{10^{2.828}}; \quad 10^{2.828} \approx 675;$$

$$r = \frac{r_0}{675} = \frac{384\,000\,\text{km}}{675} \approx 568.88\,\text{km}.$$

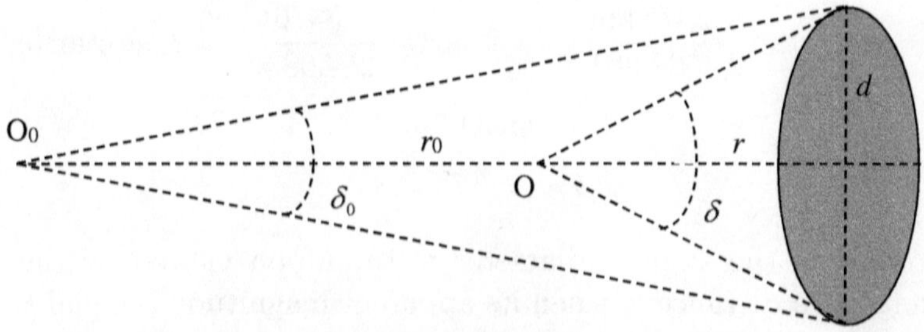

Fig. 29.1

Thus, we get the distance from which the Moon should be viewed so that its brightness is the same as the brightness of the Sun seen from Earth.

b) For the calculation of the angular diameter of the Moon viewed from distance r, using the drawing in Figure 29.1 where the linear diameter of the Moon is $d = 3476\,\text{km}$ and the angular diameter of the Moon viewed from Earth is $\delta_0 \cong 30\,\text{arcmin}$, the result is:

$$\tan\frac{\delta_0}{2} = \frac{d/2}{r_0}; \quad \frac{d}{2} = r_0 \cdot \tan\frac{\delta_0}{2};$$

$$\tan\frac{\delta_0}{2} = \frac{d}{2r_0} = \frac{3476\,\text{km}}{2 \cdot 384\,000\,\text{km}} = 0.004526041;$$

$$\tan\frac{\delta_0}{2} \approx \frac{\delta_0}{2}; \quad \frac{\delta_0}{2} = 0.004526041\,\text{rad}; \quad \delta_0 = 0.009052083\,\text{rad};$$

$$1\,\text{rad} = 206\;265\,\text{arcsec}; \quad \delta_0 = 1867.12\,\text{arcsec} = 31.11\,\text{arcmin}.$$

From the same drawing, when the cosmonaut approaches the Moon at a distance r, where the angular diameter of the Moon, δ, no longer has a small value, it results in:

$$\tan\frac{\delta}{2} = \frac{d/2}{r} = \frac{d}{2r};$$

$$\frac{\delta}{2} = \arctan\frac{d}{2r}; \quad \delta = 2 \cdot \arctan\frac{d}{2r}; \quad d = 3476\,\text{km}; \quad r \approx 568.88\,\text{km};$$

$$\delta = 2 \cdot \arctan \frac{3476\,\text{km}}{2 \cdot 568.88\,\text{km}} = 2 \cdot \text{artan} \frac{3476}{2 \cdot 568.88} = 2 \cdot \arctan 3.05;$$

$$\arctan 3.05 \approx 17°;$$

$$\delta = 2 \cdot 17° = 34°,$$

representing the angular diameter of the Moon viewed by the cosmonaut at a distance r, when its apparent magnitude is equal to the apparent magnitude of the Sun seen from Earth.

Problem 30

A Galaxy of Stars with the Same Brightness

In a limited region of a given galaxy where the distribution of stars in space is uniform, all stars have the same brightness, so their apparent magnitudes differ only because of their distances from the observer.

Notation:

N_m – the total number of stars whose apparent magnitude is $m_a \leq m$, or the total number of stars with apparent magnitudes $(m, m-1, m-2, \ldots, 0)$;

N_{m-1} – the total number of stars whose apparent magnitude is $m_a \leq m-1$, or the total number of stars with apparent magnitudes $(m-1, m-2, \ldots, 0)$; etc.

Determine the total number of stars existing in the considered region of the galaxy, corresponding to each of the magnitudes specified in Table 30.1, knowing that the number of stars with the apparent magnitude "zero" in that region is $N_0 = 3$.

Solution

If an apparent magnitude value of "zero" was given by convention to the star Vega ($m_{\text{Vega}} = m_0 = 0$), the illumination given by this star being E_0, then the illumination given by a star with the apparent

Table 30.1

m	0	1	2	3	....	13	14	15
N_m	3							

magnitude m is:

$$E_m = E_0 \cdot 10^{-0.4m} = E_0 \cdot (10^{-0.4})^m$$
$$= E_0 \cdot \left(\frac{1}{10^{0.4}}\right)^m = E_0 \cdot \left(\frac{1}{10^{2/5}}\right)^m;$$
$$E_m = E_0 \cdot \left(\frac{1}{\sqrt[5]{100}}\right)^m = E_0 \cdot \left(\frac{1}{2.512}\right)^m = E_0 \cdot 2.512^{-m},$$

representing Pogson's formula.

If W is the total energy of the radiation emitted by a star, and $S = 4\pi D^2$ is the area of the surface of the sphere with radius D that surrounds the star, then the density (intensity) of the energy flow of the star (apparent brightness of the star) at distance D from it (per unit area of a normal surface, per unit time), is

$$\phi = I = E = \frac{W}{St} = \frac{\frac{W}{t}}{S} = \frac{L}{S} = \frac{L}{4\pi D^2},$$

where L is the brightness (absolute brightness) of the star (the energy of the total radiation emitted by the star in one unit of time, through its entire surface, on all wavelengths, in all directions).

Let N_m be the total number of stars whose apparent magnitude is less than or equal to m (the total number of stars with magnitudes $m, m-1, m-2, \ldots, 0$).

Under these conditions, if E_m and E_0 are the apparent brightnesses (illuminations) of stars with magnitudes m and 0, respectively it results that:

$$E_m = \frac{L}{4\pi D_m^2},$$

where D_m is the distance at which the star with magnitude m is located;

$$E_0 = \frac{L}{4\pi D_0^2},$$

where D_0 is the distance at which the star with magnitude 0 is located;

$$\frac{E_m}{E_0} = \frac{D_0^2}{D_m^2};$$

$$D_m^2 = D_0^2 \cdot \frac{E_0}{E_m};$$

$$D_m = D_0 \cdot E_0^{1/2} \cdot E_m^{-1/2};$$

$$E_m = E_0 \cdot 2.512^{-m};$$

$$D_m = D_0 \cdot E_0^{1/2} \cdot E_0^{-1/2} \cdot (2.512^{-m})^{-1/2};$$

$$D_m = D_0 \cdot 2.512^{0.5m}.$$

Let us admit that the distribution of stars in space, regardless of their magnitudes, is uniform, i.e., the number of stars per unit volume (numerical density of stars), n, is constant.

This results in:

$$n = \frac{N_m}{\frac{4\pi}{3} \cdot D_m^3} = \text{constant};$$

$$N_m = n \cdot \frac{4\pi}{3} \cdot D_m^3 = n \cdot \frac{4\pi}{3} \cdot D_0^3 \cdot (2.512^{0.5m})^3;$$

$$N_m = n \cdot \frac{4\pi}{3} \cdot D_0^3 \cdot (2.512^{3/2})^m;$$

$$2.512^{3/2} = \sqrt{2.512^3} = 3.981;$$

$$N_m = n \cdot \frac{4\pi}{3} \cdot D_0^3 \cdot (3.981)^m,$$

representing the total number of stars whose apparent magnitude is less than or equal to m (the total number of stars with magnitudes $m, m-1, m-2, \ldots, 0$);

$$N_{m-1} = n \cdot \frac{4\pi}{3} \cdot D_0^3 \cdot (3.981)^{m-1},$$

representing the total number of stars whose apparent magnitude is less than or equal to $m-1$ (total number of stars with magnitudes $m-1, m-2, \ldots, 0$);

$$N_{m-2} = n \cdot \frac{4\pi}{3} \cdot D_0^3 \cdot (3.981)^{m-2},$$

representing the total number of stars whose apparent magnitude is less than or equal to $m-2$ (total number of stars with magnitudes $m-2, \ldots, 0$).

This number, called the star ratio, is

$$R = \frac{N_m}{N_{m-1}} = \frac{N_{m-1}}{N_{m-2}} = \frac{N_m - N_{m-1}}{N_{m-1} - N_{m-2}} = 3.981 = \text{constant}.$$

Conclusion: Assuming a uniform distribution of stars with the same brightness (absolute brightness), the ratio of stars, as defined above, is a constant number, $R = 3.981$.

The results are listed in Table 30.2.

Table 30.2

m	R	$R = N_m/N_{m-1}$	$N_m = RN_{m-1}$	N_m
0				3
1	3.981	$R = N_1/N_0$	$N_1 = RN_0$	11 943 **12**
2	3.981	$R = N_2/N_1$	$N_2 = RN_1$	47 545 **48**
3	3.981	$R = N_3/N_2$	$N_3 = RN_2$	189 276 **189**
4	3.981	$R = N_4/N_3$	$N_4 = RN_3$	753 511 **754**
5	3.981	$R = N_5/N_4$	$N_5 = RN_4$	2 999 729 **3 000**
6	3.981	$R = N_6/N_5$	$N_6 = RN_5$	11 941 924 **11 942**
7	3.981	$R = N_7/N_6$	$N_7 = RN_6$	47 540 801 **47 541**
8	3.981	$R = N_8/N_7$	$N_8 = RN_7$	189 259 930 **189 260**
9	3.981	$R = N_9/N_8$	$N_9 = RN_8$	753 443 781 **754 444**
10	3.981	$R = N_{10}/N_9$	$N_{10} = RN_9$	2 999 459 695 **2 999 460**
11	3.981	$R = N_{11}/N_{10}$	$N_{11} = RN_{10}$	11 940 849 050 **11 940 849**
12	3.981	$R = N_{12}/N_{11}$	$N_{12} = RN_{11}$	47 536 520 060 **47 536 520**
13	3.981	$R = N_{13}/N_{12}$	$N_{13} = RN_{12}$	189 242 886 3... **189 242 886**
14	3.981	$R = N_{14}/N_{13}$	$N_{14} = RN_{13}$	753 375 930 5... **753 375 931**
15	3.981	$R = N_{15}/N_{14}$	$N_{15} = RN_{14}$	2 999 189 579 ... **2 999 189 580**

Problem 31

Destroying a Threatening Asteroid with a Rocket

An asteroid is approaching threateningly towards the center of the Earth, as shown in Figure 31.1. With the help of a rocket, R, launched from a spaceship, N, it must be destroyed. The figure shows the positions of the spacecraft N and the asteroid A at the time of the launch of the rocket R to meet the asteroid. At these points, the speed of the spacecraft, N, relative to Earth, is constant, $\vec{v}_N$, oriented along the spacecraft–asteroid line, and the speed of asteroid A relative to Earth is considered constant, $\vec{v}_A$, its direction forming an angle α with the direction of the spacecraft, N.

During the rocket's flight, from its launch to the impact with the asteroid, the Earth, relative to the Sun, is considered to be at rest, and the movements of the rocket, R, the spacecraft, N, and the asteroid, A, are rectilinear and uniform.

Knowing that the direction of launch of rocket R was chosen in such a way that the duration of the rocket's movement from launch to impact with asteroid A is minimal, *determine* the angle β at which the rocket R was launched, relative to the direction of movement of the spacecraft, N.

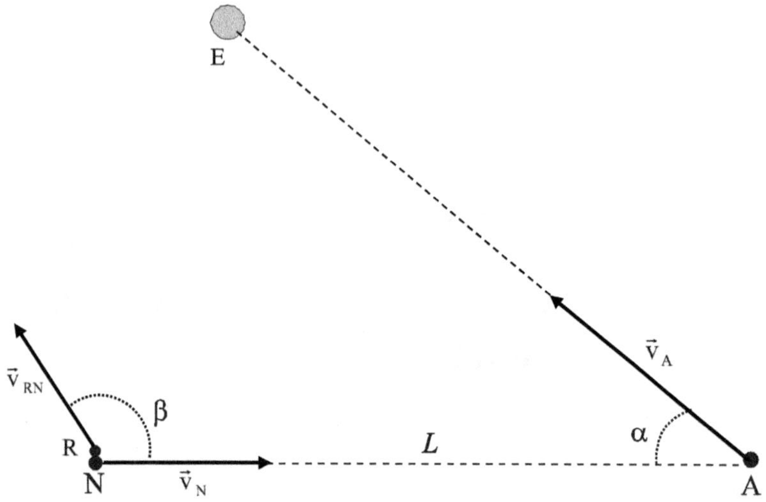

Fig. 31.1

Solution

The rocket R reaches the target asteroid A in the minimum time if the direction of absolute displacement of the rocket R, i.e. the direction of movement NC, is perpendicular to the direction of absolute movement of asteroid A, i.e., the direction of movement AC, as indicated in the drawing in Figure 31.2:

$$D = \mathrm{v_A} \cdot t_{\min} = L \cdot \cos\alpha;$$

$$d = \mathrm{v_R} \cdot t_{\min} = L \cdot \sin\alpha,$$

where $\mathrm{v_R}$ is the absolute speed of the rocket, from the moment of its launch;

$$t_{\min} = \frac{L \cdot \cos\alpha}{\mathrm{v_A}}; \quad t_{\min} = \frac{L \cdot \sin\alpha}{\mathrm{v_R}};$$

$$\frac{L \cdot \cos\alpha}{\mathrm{v_A}} = \frac{L \cdot \sin\alpha}{\mathrm{v_R}}; \quad \frac{\cos\alpha}{\mathrm{v_A}} = \frac{\sin\alpha}{\mathrm{v_R}};$$

$$\mathrm{v_R} = \mathrm{v_A} \cdot \tan\alpha,$$

where $\mathrm{v_R}$ is the absolute speed of the rocket after launch;

$$\vec{\mathrm{v}}_{\mathrm{RN}} = \vec{\mathrm{v}}_{\mathrm{R}} - \vec{\mathrm{v}}_{\mathrm{A}},$$

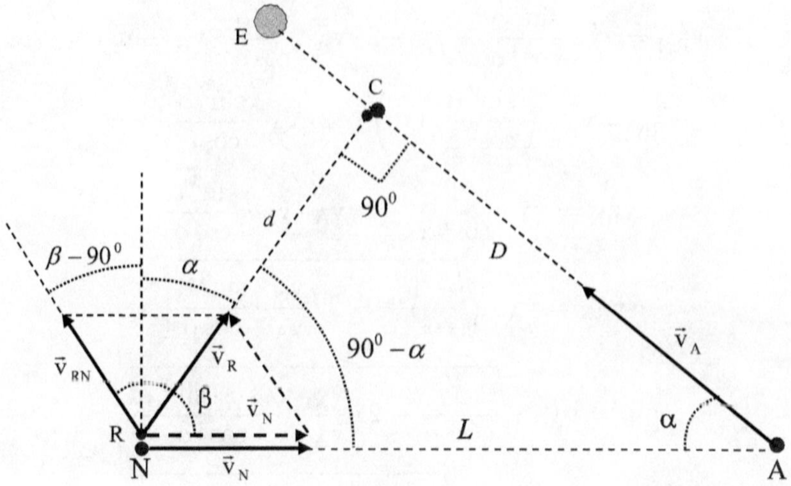

Fig. 31.2

where $\vec{v}_{RN}$ is the relative speed of the rocket R relative to the spacecraft N;

$$v_{RN}^2 = v_R^2 + v_A^2 - 2 \cdot v_R \cdot v_N \cdot \cos(90^\circ - \alpha);$$

$$v_{RN}^2 = v_R^2 + v_A^2 - 2 \cdot v_R \cdot v_N \cdot \sin\alpha;$$

$$v_R \cdot \cos\alpha = v_{RN} \cdot \cos(\beta - 90^\circ),$$

where β is the angle at which the rocket R is launched from the spacecraft N, relative to the flight direction of the spacecraft N;

$$v_R \cdot \cos\alpha = v_{RN} \cdot \sin\beta;$$

$$\sin\beta = \frac{v_R}{v_{RN}} \cdot \cos\alpha;$$

$$v_R = v_A \cdot \tan\alpha;$$

$$\sin\beta = \frac{v_A \cdot \tan\alpha}{v_{RN}} \cdot \cos\alpha = \frac{v_A}{v_{RN}} \cdot \frac{\sin\alpha}{\cos\alpha} \cdot \cos\alpha;$$

$$\sin\beta = \frac{v_A}{v_{RN}} \cdot \sin\alpha;$$

$$v_{RN}^2 = v_R^2 + v_A^2 - 2 \cdot v_R \cdot v_N \cdot \sin\alpha;$$

$$v_R = v_A \cdot \tan\alpha;$$

$$v_{RN}^2 = v_A^2 \cdot \frac{\sin^2\alpha}{\cos^2\alpha} + v_A^2 - 2 \cdot v_A \cdot \frac{\sin\alpha}{\cos\alpha} \cdot v_N \cdot \sin\alpha;$$

$$v_{RN}^2 = v_A^2 \cdot \left(\frac{\sin^2\alpha}{\cos^2\alpha} + 1\right) - 2 \cdot v_A \cdot \frac{\sin^2\alpha}{\cos\alpha} \cdot v_N;$$

$$v_{RN}^2 = v_A^2 \cdot \frac{1}{\cos^2\alpha} - 2 \cdot v_A \cdot v_N \cdot \frac{\sin^2\alpha}{\cos\alpha};$$

$$v_{RN} = v_A \cdot \sqrt{\frac{1}{\cos^2\alpha} - 2 \cdot \frac{v_N}{v_A} \cdot \frac{\sin^2\alpha}{\cos\alpha}};$$

$$v_{RN} = v_A \cdot \sqrt{\frac{1}{\cos^2\alpha} - 2 \cdot \frac{v_N}{v_A} \cdot \frac{\cos\alpha \cdot \sin^2\alpha}{\cos^2\alpha}};$$

$$v_{RN} = \frac{v_A}{\cos\alpha} \cdot \sqrt{1 - 2 \cdot \frac{v_N}{v_A} \cdot \sin^2\alpha \cdot \cos\alpha},$$

representing the speed of the rocket R relative to the spacecraft N;

$$\sin\beta = \frac{v_R}{v_{RN}} \cdot \cos\alpha;$$

$$v_R = v_A \cdot \tan\alpha;$$

$$v_{RN} = \frac{v_A}{\cos\alpha} \cdot \sqrt{1 - 2 \cdot \frac{v_N}{v_A} \cdot \sin^2\alpha \cdot \cos\alpha};$$

$$\sin\beta = \frac{v_A \cdot \frac{\sin\alpha}{\cos\alpha}}{\frac{v_A}{\cos\alpha} \cdot \sqrt{1 - 2 \cdot \frac{v_N}{v_A} \cdot \sin^2\alpha \cdot \cos\alpha}} \cdot \cos\alpha;$$

$$\sin\beta = \frac{\sin\alpha \cdot \cos\alpha}{\sqrt{1 - 2 \cdot \frac{v_N}{v_A} \cdot \sin^2\alpha \cdot \cos\alpha}},$$

where β is the angle at which the rocket R is launched from the spacecraft, N, relative to the direction of flight of the spacecraft N, to destroy the asteroid.

Problem 32

The Apparent Shape of the Celestial Vault and the Apparent Linear Diameter of the Moon

It is known that particles in the atmosphere partially diffuse the light we receive from the Sun and other stars. Since these particles are very fine and cannot be seen with the naked eye, we have the feeling that the light from the stars reaches our eyes as if coming from sources at the same distances. As a result, the celestial vault appears to us as a hemisphere with its center at the point of observation whose edges are supported on the contour of our horizon. However, if we examine carefully the celestial vault from a flat, clear place, we find that its apparent shape is not a hemisphere with the center at the observation point. Still, it appears to us as flattened at the zenith.

The causes of this apparent form must be sought in the physiological peculiarities of visual perception, i.e., how our eye appreciates distances and aerial perspectives.

We will now put forward arguments supporting the claim that the celestial vault, under direct visual observation, has an apparent flattened shape at the zenith, and explore some of the consequences of this fact.

To highlight the aforementioned phenomenon, let's observe the celestial vault on a clear night from a flat place, constituting the observation point, O, on the Earth's surface. We attempt to locate in the sky, by visual observation, a solitary star, S′, located at a point whose position, appreciated visually, is in the middle of the

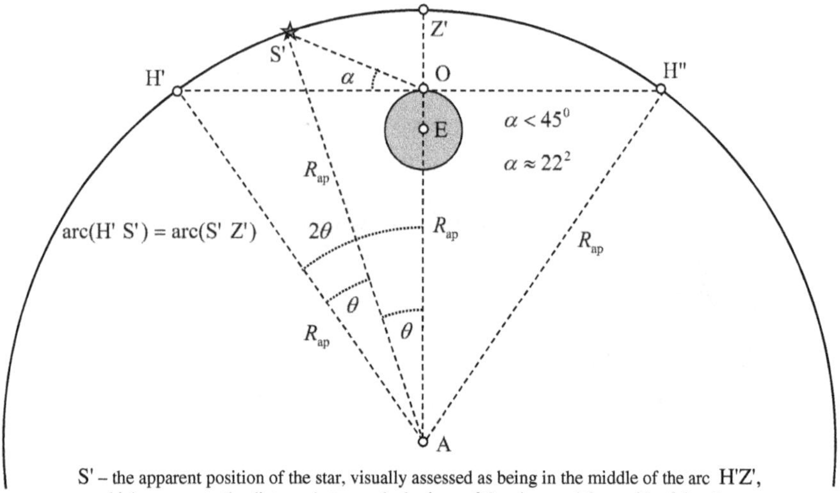

S' – the apparent position of the star, visually assessed as being in the middle of the arc H'Z', which represents the distance between the horizon of the place and the zenith of the place

Fig. 32.1

arc representing the distance between the horizon H′ and the zenith of the place Z′, as shown in the drawing in Figure 32.1. In this figure, the two angles at the center, θ, correspond to the two halves of the arc delimited by the position of the star S′ according to visual observation.

This is perfectly possible by direct visual observation during a clear night, staring at a lone star, S′, located at a point *visually appreciated* as being halfway between the horizon H′ and the zenith of the place Z′. A special flashlight is fixed on a support, so that the direction of the beam of light emitted by the flashlight is in the direction of the star S′.

If the shape of the celestial vault, *under (direct) visual observation*, were a hemisphere with the center at the point of observation, O, using a theodolite, we would determine the angle between the line of the observer's horizon, H′H″, and the line passing through the observation point, O, to the point where we *visually located* the star S' (using the light beam emitted by the flashlight) in the lower half of the arc representing the distance between the horizon H′ and the zenith of the place Z, to be of 45°, as indicated by Figure 32.2.

However, performing this experiment, we find that the angle sought (i.e., the angle above the horizon that delimits, under direct

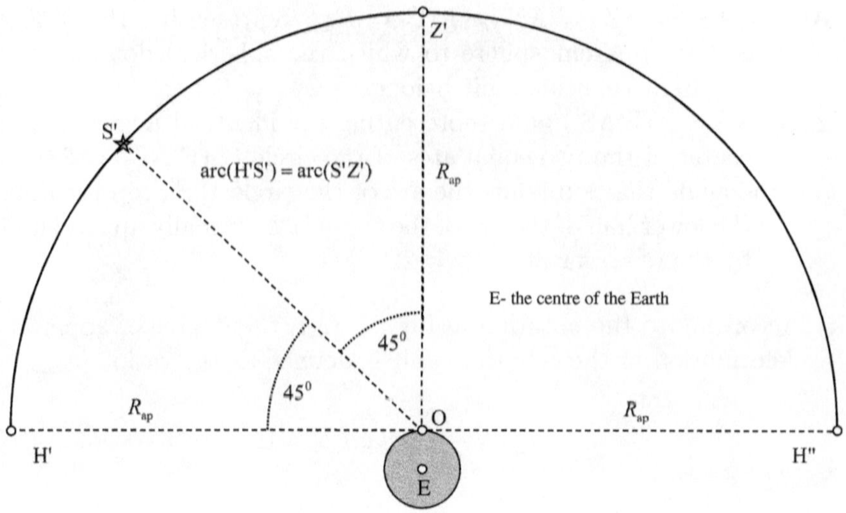

Fig. 32.2

visual observation, the lower half of the arc representing the distance between the horizon H′ and the zenith of the place Z′), *measured with the theodolite*, is not equal to 45°. Instead, its measured value is approximately 22°, as indicated in the drawing in Figure 32.1.

The conclusion is obvious: The apparent shape of the celestial vault is not a hemisphere, with the center at the observation point, O, as shown in the drawing in Figure 32.2, but is instead flattened at the zenith, as shown in the drawing in Figure 32.1.

We will agree that the apparent shape of the celestial vault, although not a hemisphere with its center at the point of observation, O, is still a spherical dome belonging to another sphere, different from the real celestial sphere, having its center at a point A. This is different from the center of the real celestial sphere, located below the observation point, O, on the same vertical, as shown in the drawing in Figure 32.1, with the following notation:

- H′H″ – the line of the horizon, passing through the observation point, O;
- H′Z′H″ – representing the intersection of the apparent celestial vault with the vertical plane passing through the apparent zenith, Z′;

AH′ = AS′ = AZ′ = AN = AH″ = R_{ap}, representing the radius of the apparent sphere to which the spherical dome of the apparent celestial vault belongs;

$\angle(\text{H}'\text{AS}') = \angle(\text{Z}'\text{AS}') = \theta$, representing the identical angles at the center of the two equal arcs of the circle ($\angle(\text{H}'\text{AZ}') = 2\theta$);

α – the angle that subtends the arc of the circle H′S′, representing the lower half of the arc of the circle H′Z′, visually appreciated by the observer at O ($\alpha = 22°$).

a) According to the notation in Figure 32.1, the degree of apparent deformation of the celestial vault is defined as the ratio

$$g_{\text{da}} = \frac{\text{OH}'}{\text{OZ}'} = \operatorname{ctg}\theta > 1.$$

Thus, *demonstrate* the existence of the relationship

$$\operatorname{ctg}^3\theta - \frac{3+\operatorname{tg}^2\alpha}{2\cdot\operatorname{tg}\alpha}\cdot\operatorname{ctg}^2\theta + \operatorname{ctg}\theta + \frac{1-\operatorname{tg}^2\alpha}{2\cdot\operatorname{tg}\alpha} = 0,$$

the equation that allows us to determine the degree of apparent deformation of the celestial vault, g_{da}. Then, *demonstrate* the existence of the relationship

$$\operatorname{ctg}\theta = 2\cdot\sqrt[3]{\begin{array}{l}-\frac{1}{2}\cdot\left(\frac{2b^3}{27}-\frac{b}{3}+d\right)\\ +\sqrt{\frac{1}{4}\cdot\left(\frac{2b^3}{27}-\frac{b}{3}+d\right)^2+\frac{1}{27}\cdot\left(1-\frac{b^2}{3}\right)^3}\end{array}} - \frac{b}{3},$$

representing the solution of the previous equation, where:

$$b = -\frac{3+\operatorname{tg}^2\alpha}{2\cdot\operatorname{tg}\alpha};\quad d = \frac{1-\operatorname{tg}^2\alpha}{2\cdot\operatorname{tg}\alpha}.$$

b) *Determine* the numerical value of the degree of apparent deformation of the celestial vault, g_{da}, knowing that:

$$\alpha = 22°;\quad \operatorname{tg}\alpha = 0.404026225;\quad \operatorname{tg}^2\alpha = 0.163237191.$$

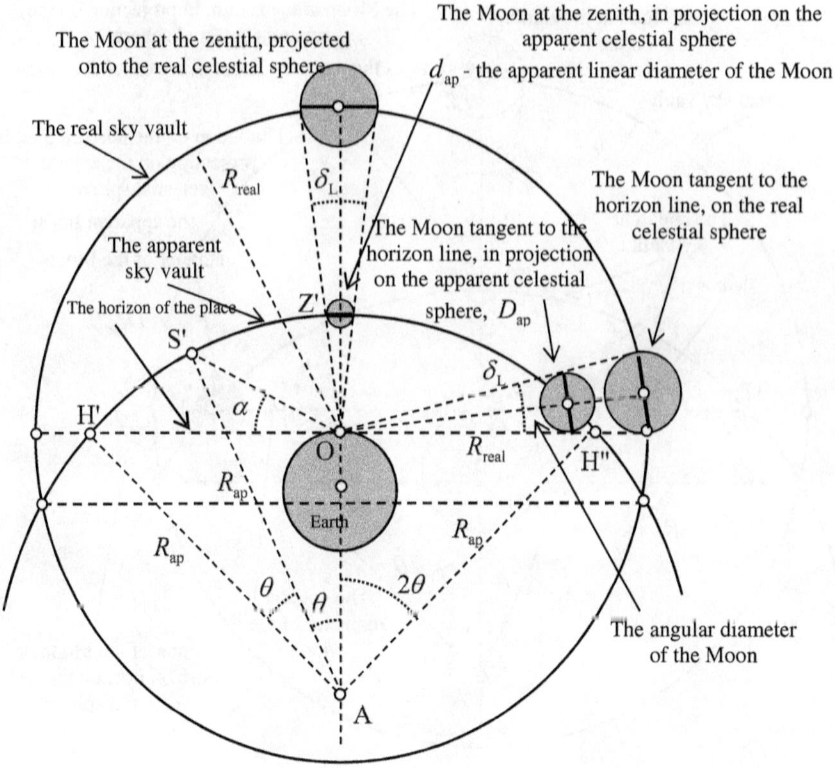

Fig. 32.3

c) *Determine*: **1)** the relationship between the radii of the apparent and real celestial spheres, R_{apparent} R_{real}, highlighted in Figures 32.3 and 32.4; **2)** the ratio between the apparent linear diameter of the Moon when the Moon's disk is tangent to the horizon line, D_{apparent}, and the apparent linear diameter of the Moon when its center is at the zenith of the observer, d_{apparent}, as indicated by Figures 32.3 and 32.4.

Solution

a) The problem statement specified that we will consider the apparent shape of the celestial vault to be, although not a hemisphere with its center at the point of observation, O, to nevertheless be a spherical

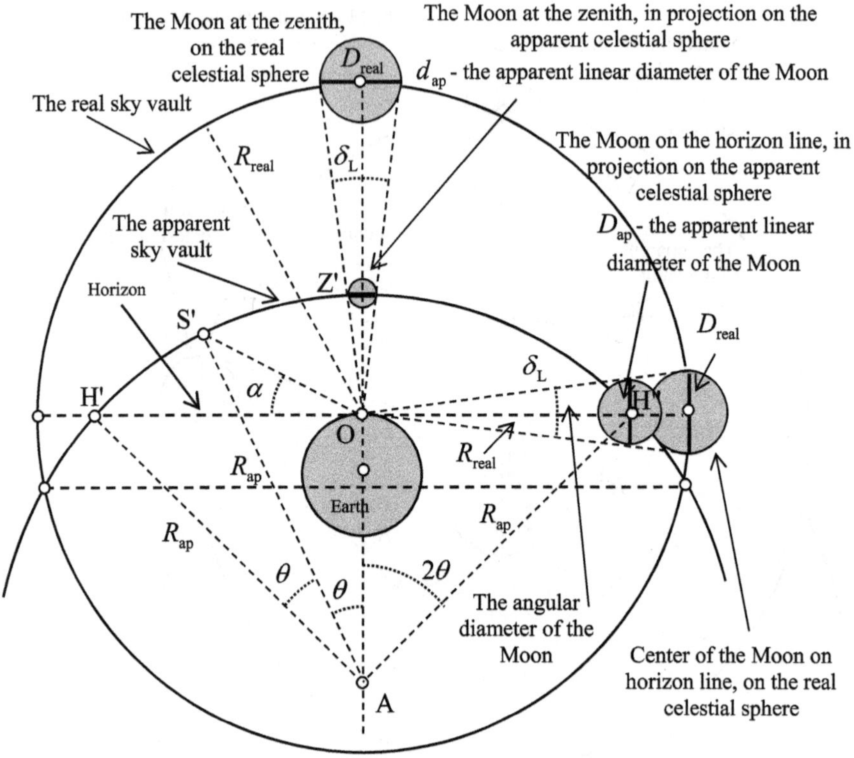

Fig. 32.4

dome belonging to a sphere that is different from the real celestial sphere, having its center at a point A, which different from the center of the real celestial sphere, located below the point of observation, O, on the same vertical as it. This is indicated by the drawing in Figure 32.3, where the notation is the same as in Figure 32.1.

Let us now divide the circular arc Z′H″ into n equal parts. On this arc, we note a point N, as indicated by the drawing in Figure 32.5, in such a way that the two angles at the center, ∠(NAH″) and ∠(Z′AN), are:

$$\angle(\mathrm{NAH''}) = \frac{2\theta}{n}\cdot k; \quad k < n;$$

$$\angle(\mathrm{Z'AN}) = 2\theta - \frac{2\theta}{n}\cdot k = 2\theta\left(1-\frac{k}{n}\right).$$

Under these conditions, from the triangle AMN, it follows that:

$$\sin\angle(\mathrm{Z'AN}) = \sin\left[2\theta\left(1-\frac{k}{n}\right)\right] = \frac{\mathrm{MN}}{\mathrm{AN}};$$

$$\mathrm{MN} = \mathrm{AN}\cdot\sin\left[2\theta\left(1-\frac{k}{n}\right)\right];$$

$$\mathrm{MN} = R_{\mathrm{ap}}\cdot\sin\left[2\theta\left(1-\frac{k}{n}\right)\right];$$

$$\cos\angle(\mathrm{Z'AN}) = \cos\left[2\theta\left(1-\frac{k}{n}\right)\right] = \frac{\mathrm{AM}}{\mathrm{AN}};$$

$$\mathrm{AM} = \mathrm{AN}\cdot\cos\left[2\theta\left(1-\frac{k}{n}\right)\right];$$

$$\mathrm{AM} = R_{\mathrm{ap}}\cdot\cos\left[2\theta\left(1-\frac{k}{n}\right)\right].$$

Similarly, from triangle AOH″, we obtain:

$$\cos\angle(\mathrm{OAH''}) = \cos(2\theta) = \frac{\mathrm{OA}}{\mathrm{AH''}};$$

$$\mathrm{OA} = \mathrm{AH''}\cdot\cos(2\theta);$$

$$\mathrm{OA} = R_{\mathrm{ap}}\cdot\cos(2\theta).$$

This Results in:

$$\mathrm{OM} = \mathrm{AM} - \mathrm{OA} = \mathrm{AM} = R_{\mathrm{ap}}\cdot\cos\left[2\theta\left(1-\frac{k}{n}\right)\right] - R_{\mathrm{ap}}\cdot\cos(2\theta);$$

$$\mathrm{tg}\,\beta = \frac{\mathrm{OM}}{\mathrm{MN}} = \frac{\cos\left[2\theta\left(1-\frac{k}{n}\right)\right]-\cos(2\theta)}{\sin\left[2\theta\left(1-\frac{k}{n}\right)\right]}.$$

For the particular case of the point where the star S′ is located, dividing the arc H′Z′ into two equal parts, arc(H′S′) = arc(S′Z′), when $k = n/2$, this becomes

$$\mathrm{tg}\,\alpha = \frac{\cos\theta - \cos 2\theta}{\sin\theta}.$$

172

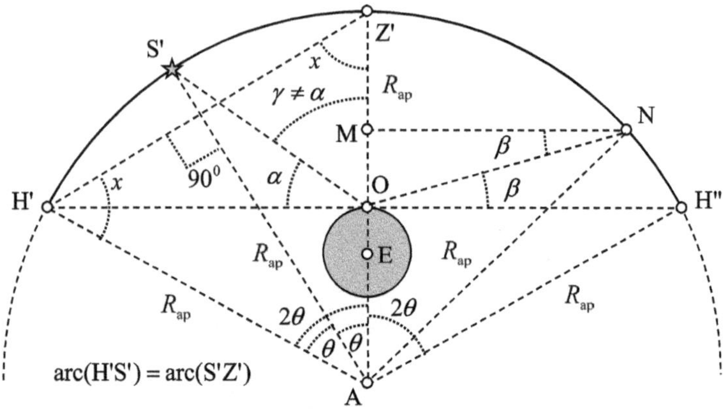

S' – the apparent position of the star, visually assessed as being in the middle of the arc H'Z', which represents the distance between the horizon of the place and the zenith of the place

Fig. 32.5

In the drawing in Figure 32.5, the triangle H′AZ′ is an isosceles triangle, so we get:

$$\angle \mathrm{H}' = \angle \mathrm{Z}' = x;$$

$$2x + 2\theta = 180^\circ; \quad x + \theta = 90^\circ; \quad x = 90^\circ - \theta;$$

$$\operatorname{tg} x = \frac{\mathrm{OH}'}{\mathrm{OZ}'} = \operatorname{tg}(90^\circ - \theta) = \operatorname{ctg}\theta,$$

the size of which,

$$\operatorname{ctg}\theta = \frac{\mathrm{OH}'}{\mathrm{OZ}'} = g_{\mathrm{da}},$$

defines the *degree of apparent deformation* of the celestial vault.

This results in:

$$\operatorname{tg}\alpha = \frac{\cos\theta - \cos 2\theta}{\sin\theta}; \quad \operatorname{tg}\alpha = \frac{\cos\theta}{\sin\theta} - \frac{\cos 2\theta}{\sin\theta};$$

$$\cos 2\theta = \frac{\operatorname{ctg}^2\theta - 1}{\operatorname{ctg}^2\theta + 1}; \quad \sin\theta = \frac{1}{\pm\sqrt{1 + \operatorname{ctg}^2\theta}};$$

$$\operatorname{tg}\alpha = \operatorname{ctg}\theta - \frac{\frac{\operatorname{ctg}^2\theta-1}{\operatorname{ctg}^2\theta+1}}{\frac{1}{\pm\sqrt{1+\operatorname{ctg}^2\theta}}} = \operatorname{ctg}\theta - \frac{\operatorname{ctg}^2\theta - 1}{\operatorname{ctg}^2\theta + 1} \cdot \frac{\pm\sqrt{\operatorname{ctg}^2\theta + 1}}{1};$$

$$\operatorname{tg}\alpha = \operatorname{ctg}\theta - \frac{\operatorname{ctg}^2\theta - 1}{\operatorname{ctg}^2\theta + 1} \cdot \left(\pm\sqrt{\operatorname{ctg}^2\theta + 1}\right);$$

$$\operatorname{tg}\alpha = \operatorname{ctg}\theta - \frac{\operatorname{ctg}^2\theta - 1}{\pm\sqrt{\operatorname{ctg}^2\theta + 1}}; \quad \operatorname{tg}\alpha - \operatorname{ctg}\theta = -\frac{\operatorname{ctg}^2\theta - 1}{\pm\sqrt{\operatorname{ctg}^2\theta + 1}};$$

$$\operatorname{tg}\alpha - \operatorname{ctg}\theta = \frac{1 - \operatorname{ctg}^2\theta}{\pm\sqrt{\operatorname{ctg}^2\theta + 1}}; \quad (\operatorname{tg}\alpha - \operatorname{ctg}\theta)^2 = \frac{(1 - \operatorname{ctg}^2\theta)^2}{\operatorname{ctg}^2\theta + 1};$$

$$(\operatorname{tg}\alpha - \operatorname{ctg}\theta)^2 \cdot (1 + \operatorname{ctg}^2\theta) = (1 - \operatorname{ctg}^2\theta)^2;$$

$$(\operatorname{tg}^2\alpha + \operatorname{ctg}^2\theta - 2 \cdot \operatorname{tg}\alpha \cdot \operatorname{ctg}\theta) \cdot (1 + \operatorname{ctg}^2\theta) = (1 + \operatorname{ctg}^4\theta - 2\operatorname{ctg}^2\theta);$$

$$\operatorname{tg}^2\alpha + \operatorname{tg}^2\alpha \cdot \operatorname{ctg}^2\theta + \operatorname{ctg}^2\theta + \operatorname{ctg}^4\theta - 2 \cdot \operatorname{tg}\alpha \cdot \operatorname{ctg}\theta - 2 \cdot \operatorname{tg}\alpha \cdot \operatorname{ctg}^3\theta$$
$$= 1 + \operatorname{ctg}^4\theta - 2 \cdot \operatorname{ctg}^2\theta;$$

$$\operatorname{tg}^2\alpha + (3 + \operatorname{tg}^2\alpha) \cdot \operatorname{ctg}^2\theta - 2 \cdot \operatorname{tg}\alpha \cdot \operatorname{ctg}\theta - 2 \cdot \operatorname{tg}\alpha \cdot \operatorname{ctg}^3\theta = 1;$$

$$2 \cdot \operatorname{tg}\alpha \cdot \operatorname{ctg}^3\theta - (3 + \operatorname{tg}^2\alpha) \cdot \operatorname{ctg}^2\theta + 2 \cdot \operatorname{tg}\alpha \cdot \operatorname{ctg}\theta + 1 - \operatorname{tg}^2\alpha = 0;$$

$$\operatorname{ctg}^3\theta - \frac{3 + \operatorname{tg}^2\alpha}{2 \cdot \operatorname{tg}\alpha} \cdot \operatorname{ctg}^2\theta + \operatorname{ctg}\theta + \frac{1 - \operatorname{tg}^2\alpha}{2 \cdot \operatorname{tg}\alpha} = 0.$$

This represents the equation that allows the determination of the degree of apparent deformation of the celestial vault:

$$\operatorname{ctg}\theta = \frac{\mathrm{OH}'}{\mathrm{OZ}'} = g_{\mathrm{da}}.$$

To solve the previous equation, we first make the substitution

$$\operatorname{ctg}\theta = y,$$

resulting in:

$$y^3 - \frac{3 + \operatorname{tg}^2\alpha}{2 \cdot \operatorname{tg}\alpha} \cdot y^2 + y + \frac{1 - \operatorname{tg}^2\alpha}{2 \cdot \operatorname{tg}\alpha} = 0;$$
$$a \cdot y^3 + b \cdot y^2 + c \cdot y + d = 0;$$

$$a = 1; \quad b = -\frac{3 + \operatorname{tg}^2\alpha}{2 \cdot \operatorname{tg}\alpha}; \quad c = 1; \quad d = \frac{1 - \operatorname{tg}^2\alpha}{2 \cdot \operatorname{tg}\alpha};$$

$$y^3 + b \cdot y^2 + y + d = 0.$$

This represents the normal form of the cubic equation, so we make a second substitution.

$$y = z - \frac{b}{3},$$

resulting in:

$$\left(z - \frac{b}{3}\right)^3 + b \cdot \left(z - \frac{b}{3}\right)^2 + \left(z - \frac{b}{3}\right) + d = 0;$$

$$z^3 - 3z^2 \cdot \frac{b}{3} + 3z \cdot \frac{b^2}{9} - \frac{b^3}{27} + b \cdot z^2 - \frac{2}{3}b^2 z + \frac{b^3}{9} + z - \frac{b}{3} + d = 0;$$

$$z^3 + 3z \cdot \frac{b^2}{9} - \frac{b^3}{27} - \frac{2}{3}b^2 z + \frac{b^3}{9} + z - \frac{b}{3} + d = 0;$$

$$z^3 + \left(1 - \frac{b^2}{3}\right) \cdot z + \frac{2b^3}{27} - \frac{b}{3} + d = 0;$$

$$p = 1 - \frac{b^2}{3}; \quad q = \frac{2b^3}{27} - \frac{b}{3} + d;$$

$$z^3 + p \cdot z + q = 0,$$

representing the *reduced form of the cubic equation.*

Instead of a single unknown, z, we now introduce two unknowns, u and w, by the relation

$$z = u + w,$$

so, we get:

$$(u + w)^3 + p \cdot (u + w) + q = 0;$$

$$u^3 + 3u^2 w + 3uw^2 + w^3 + p \cdot (u + w) + q = 0;$$

$$u^3 + 3uw \cdot (u + w) + w^3 + p \cdot (u + w) + q = 0;$$

$$u^3 + w^3 + q + (u + w) \cdot (3uw + p) = 0.$$

The two unknowns introduced, u and w, meet the condition

$$3uw + p = 0,$$

so, we get:

$$u^3 + w^3 = -q; \quad uw = -\frac{p}{3}.$$

These relations allow the expression of the two unknowns introduced, as follows:

$$(u^3 + w^3)^2 = q^2; \quad u^6 + 2u^3w^3 + w^6 = q^2;$$

$$(uw)^3 = -\left(\frac{p}{3}\right)^3;$$

$$4 \cdot (uw)^3 = -4 \cdot \left(\frac{p}{3}\right)^3; \quad 4u^3w^3 = -4 \cdot \left(\frac{p}{3}\right)^3;$$

$$u^6 + 2u^3w^3 + w^6 - 4u^3w^3 = q^2 + 4 \cdot \left(\frac{p}{3}\right)^3;$$

$$u^6 - 2u^3w^3 + w^6 = q^2 + 4 \cdot \left(\frac{p}{3}\right)^3;$$

$$(u^3 - w^3)^2 = q^2 + 4 \cdot \left(\frac{p}{3}\right)^3;$$

$$u^3 - w^3 = \pm\sqrt{q^2 + 4 \cdot \left(\frac{p}{3}\right)^3}; \quad u^3 + w^3 = -q;$$

$$2u^3 = -q \pm \sqrt{q^2 + 4 \cdot \left(\frac{p}{3}\right)^3}; \quad u^3 = -\frac{q}{2} \pm \sqrt{\left(\frac{q}{2}\right)^2 + \left(\frac{p}{3}\right)^3};$$

$$u_1 = \sqrt[3]{-\frac{q}{2} + \sqrt{\left(\frac{q}{2}\right)^2 + \left(\frac{p}{3}\right)^3}},$$

where u_1 has a real, positive value,

$$u_1 > 0,$$

because

$$\sqrt{\left(\frac{q}{2}\right)^2 + \left(\frac{p}{3}\right)^2} > \frac{q}{2};$$

$$-\frac{q}{2}+\sqrt{\left(\frac{q}{2}\right)^2+\left(\frac{p}{3}\right)^2}>0.$$

Similarly, we get:

$$u^3-w^3=\pm\sqrt{q^2+4\cdot\left(\frac{p}{3}\right)^3}; \quad u^3+w^3=-q;$$

$$w^3-u^3=\mp\sqrt{q^2+4\cdot\left(\frac{p}{3}\right)^3}; \quad u^3+w^3=-q;$$

$$2w^3=-q\mp\sqrt{q^2+4\cdot\left(\frac{p}{3}\right)^3};$$

$$w^3=-\frac{q}{2}\mp\sqrt{\left(\frac{q}{2}\right)^2+\left(\frac{p}{3}\right)^3};$$

$$w_1=\sqrt[3]{-\frac{q}{2}+\sqrt{\left(\frac{q}{2}\right)^2+\left(\frac{p}{3}\right)^3}}.$$

Thus:

$$z=u+w;$$

$$z_1=u_1+w_1;$$

$$u_1=\sqrt[3]{-\frac{q}{2}+\sqrt{\left(\frac{q}{2}\right)^2+\left(\frac{p}{3}\right)^3}}; \quad w_1=\sqrt[3]{-\frac{q}{2}+\sqrt{\left(\frac{q}{2}\right)^2+\left(\frac{p}{3}\right)^3}};$$

$$z_1=\sqrt[3]{-\frac{q}{2}+\sqrt{\left(\frac{q}{2}\right)^2+\left(\frac{p}{3}\right)^3}}+\sqrt[3]{-\frac{q}{2}+\sqrt{\left(\frac{q}{2}\right)^2+\left(\frac{p}{3}\right)^3}};$$

$$z_1=2\cdot\sqrt[3]{-\frac{q}{2}+\sqrt{\left(\frac{q}{2}\right)^2+\left(\frac{p}{3}\right)^3}},$$

representing the real solution of the equation in reduced form;

$$y=z-\frac{b}{3}; \quad y_1=z_1-\frac{b}{3};$$

$$y_1=2\cdot\sqrt[3]{-\frac{q}{2}+\sqrt{\left(\frac{q}{2}\right)^2+\left(\frac{p}{3}\right)^3}}-\frac{b}{3},$$

representing the real solution of the cubic equation in normal form;

$$p = 1 - \frac{b^2}{3}; \quad q = \frac{2b^3}{27} - \frac{b}{3} + d;$$

$$y_1 = 2 \cdot \sqrt[3]{\begin{array}{l} -\frac{1}{2} \cdot \left(\frac{2b^3}{27} - \frac{b}{3} + d\right) \\ + \sqrt{\frac{1}{4} \cdot \left(\frac{2b^3}{27} - \frac{b}{3} + d\right)^2 + \frac{1}{27} \cdot \left(1 - \frac{b^2}{3}\right)^3} \end{array}} - \frac{b}{3};$$

$$b = -\frac{3 + \operatorname{tg}^2\alpha}{2 \cdot \operatorname{tg}\alpha}; \quad c = 1; \quad d = \frac{1 - \operatorname{tg}^2\alpha}{2 \cdot \operatorname{tg}\alpha}; \quad \operatorname{ctg}\theta = y; \quad \operatorname{ctg}\theta_1 = y_1;$$

$$\operatorname{ctg}\theta = \frac{\mathrm{OH}'}{\mathrm{OZ}'},$$

which defines the degree of apparent deformation of the celestial vault.

b) We know that:

$$\operatorname{ctg}\theta = 2 \cdot \sqrt[3]{\begin{array}{l} -\frac{1}{2} \cdot \left(\frac{2b^3}{27} - \frac{b}{3} + d\right) \\ + \sqrt{\frac{1}{4} \cdot \left(\frac{2b^3}{27} - \frac{b}{3} + d\right)^2 + \frac{1}{27} \cdot \left(1 - \frac{b^2}{3}\right)^3} \end{array}} - \frac{b}{3};$$

$$\alpha = 22^\circ; \quad \operatorname{tg}\alpha = 0.404026225; \quad \operatorname{tg}^2\alpha = 0.163237191.$$

This results in:

$$b = -\frac{3 + \operatorname{tg}^2\alpha}{2 \cdot \operatorname{tg}\alpha}; \quad b = -3.914643401;$$

$$d = \frac{1 - \operatorname{tg}^2\alpha}{2 \cdot \operatorname{tg}\alpha}; \quad d = 1.035530316;$$

$$-\frac{1}{2} \cdot \left(\frac{2b^3}{27} - \frac{b}{3} + d\right) = -\frac{1}{2} \cdot (-17.39542559 + 1.304881134$$

$$+ 1.0335530316) = 7.528495713;$$

$$\frac{1}{4}\cdot\left(\frac{2b^3}{27}-\frac{b}{3}+d\right)^2=\frac{1}{4}\cdot(-15.055699143)^2=56.67824769;$$

$$\frac{1}{27}\cdot\left(1-\frac{b^2}{3}\right)^3=\frac{1}{27}\cdot(1-5.108144319)^3=\frac{1}{27}\cdot(-4.108144319)^3$$
$$=\frac{1}{27}\cdot(-69.33253441)=-2.567871645;$$

$$\frac{1}{4}\cdot\left(\frac{2b^3}{27}-\frac{b}{3}+d\right)^2+\frac{1}{27}\cdot\left(1-\frac{b^2}{3}\right)^3$$
$$=56.67824769-2.567871645=54.11037605;$$

$$\sqrt{\frac{1}{4}\cdot\left(\frac{2b^3}{27}-\frac{b}{3}+d\right)^2+\frac{1}{27}\cdot\left(1-\frac{b^2}{3}\right)^3}=\sqrt{54.11037605}$$
$$=7.355975533;$$

$$-\frac{b}{3}=-\frac{-3.914643401}{3}=1.304881134;$$

$$\operatorname{ctg}\theta=2\cdot\sqrt[3]{-\frac{1}{2}\cdot\left(\frac{2b^3}{27}-\frac{b}{3}+d\right)+\sqrt{\frac{1}{4}\cdot\left(\frac{2b^3}{27}-\frac{b}{3}+d\right)^2+\frac{1}{27}\cdot\left(1-\frac{b^2}{3}\right)^3}}-\frac{b}{3};$$

$$\operatorname{ctg}\theta=2\cdot\sqrt[3]{7.528495713+7.355975533}+1.304881134$$
$$=2\cdot\sqrt[3]{14.88447125}+1.304881134;$$

$$\sqrt[3]{14.88447125}=(14.88447125)^{1/3}=s;\ \log(14.88447125)^{1/3}=\log s;$$

$$\frac{1}{3}\cdot\log(14.88447125)=\log s;\quad \frac{1}{3}\cdot\log(14.88447125)^{1/3}=\log s;$$

$$\frac{1}{3}\cdot 1.172733412=\log s;\quad 0.390911137=\log s;\quad s\approx 2.5;$$

$$\operatorname{ctg}\theta=2\cdot\sqrt[3]{14.88447125}+1.304881134=2\cdot 2.5+1.304881134$$
$$=6.1304881134;$$

$$\text{ctg}\theta \approx 6 = \frac{1}{\text{tg}\theta}; \quad \text{tg}\theta = \frac{1}{6} = 0.1666666666 \approx 0.17; \quad \theta \approx 10°;$$

$$\text{ctg}\,\theta = \frac{\text{OH}'}{\text{OZ}'} = g_{\text{da}} = 6; \quad \text{OH}' = 6 \cdot \text{OZ}'.$$

c) On clear evenings, when the Moon can be observed in the sky, its apparent linear diameter, at the time of its appearance above the horizon, is larger than its apparent linear diameter when the Moon is at its zenith. Then, as the Moon approaches the horizon again, its apparent linear diameter becomes larger. The decrease in the apparent linear diameter of a star, from the horizon to the zenith, is a consequence of the apparent shape of the celestial vault.

If we consider the fact that the Moon's orbit is a circle around the Earth and that when the Moon is on the horizon, it is at a greater distance from the observer than when the Moon is at its zenith, we expect the observed phenomenon to be exactly the opposite, that is, the apparent linear diameter of the Moon at the horizon should be smaller than when the Moon was at its zenith.

Another reason can be proposed. Since the radius of the Earth is much smaller than the distance between the Earth and the Moon, for simplicity of explanation, suppose that the Moon follows its trajectory on the celestial vault, which is shaped like a hemisphere with its center at the point of observation. As a result, as long as the Moon is above the horizon, it is always at the same distance from the observer, and so its apparent linear diameter should always be the same.

Neither of the two situations presented is evident to the observer. He observes a star whose apparent linear diameter increases from its zenith to the horizon, this being the effect of the apparent flattening of the celestial vault.

So far, we have established that the actual shape of the sky is a hemisphere with its center at the point of observation, and the apparent shape of the sky is a spherical dome belonging to another sphere, with the center below the point of observation, on the same vertical. It is easy to see now that the position and apparent linear dimension of a star are given by the projection of the star from the actual sky onto the apparent sky.

In order to present these conclusions in a drawing, we should specify:

1) *the place where the two spherical surfaces intersect (above the horizon or below the horizon)*;
2) *the relation between the radii of the two spheres,* R_{real} and R_{apparent}.

To clarify the answer to this question, we must use the conclusions of another set of determinations.

By determining through other methods the linear diameter of a star and comparing it with the visually determined one, it is established that an observer overestimates the linear dimensions of a star or other celestial body when it is near the horizon on the celestial vault, and when the body is near the zenith, its linear dimensions are underestimated.

The drawing in Figure 32.6, in which we represent the positions of the two celestial vaults (the real one and the apparent one), must meet several conditions:

1) The actual sky must be a hemisphere, with its center at the point of observation and the hemisphere above the horizon.
2) The apparent celestial vault must be a spherical dome comprising part of another sphere whose center is below the point of observation on the same vertical.

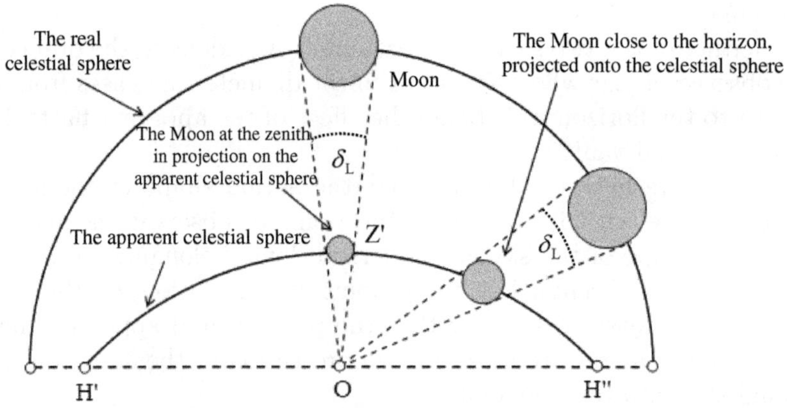

Fig. 32.6

3) The apparent positions and sizes of stars should be determined by projecting their locations from the real celestial hemisphere onto the apparent vault.

4) Near the horizon, the apparent linear dimensions must be larger than near the zenith.

Knowing the effects of the apparent shape of the celestial vault is important in practice when visually determining the positions of different bodies or in assessing the distance or degree of cloud cover. For example, when clouds are close to the horizon, the degree of cloudiness is overestimated, and when they are closer to the zenith, the cloud cover is underestimated.

For the observer at point O on the Earth's surface, when the Moon is evolving on the real celestial spherical vault, centered at point O, the distance between the center of the Moon and the observer is constant, so the angular diameter of the Moon, δ_{L}, is constant.

The apparent positions of the Moon, in its evolution from the zenith to the horizon, for the observer at O, meaning the projections of the Moon from the real celestial hemisphere (vault) onto the apparent spherical dome, move farther and farther from the observer. Its angular dimensions are the same, δ_{L}, but its linear dimensions (diameter) become larger because the Moon is farther from the observer, as shown in the drawing in Figure 32.3.

This is the result of the degree of deformation of the apparent celestial vault compared to the real celestial vault. The two celestial spheres intersect below the line of the observer's horizon.

Under these conditions, according to the notation in Figures 32.3 and 32.7, the positions of the center of the apparent Moon relative to the horizon line, are approximately the same, following the evolution of the Moon from the zenith to the horizon line, on the real celestial sphere. The evolution of the projections of the Moon from the real celestial sphere to the apparent celestial sphere results in:

$$R_{\mathrm{ap}} \cdot \cos(90^\circ - 2\theta) = R_{\mathrm{real}}; \quad \theta \approx 10^\circ;$$

$$R_{\mathrm{ap}} \cdot \sin(2\theta) = R_{\mathrm{real}} = 0.34 \cdot R_{\mathrm{ap}};$$

$$R_{\mathrm{real}} < R_{\mathrm{ap}};$$

$$d_{\mathrm{ap}} = \delta_{\mathrm{L}} \cdot (R_{\mathrm{ap}} - R_{\mathrm{ap}} \cdot \cos(2\theta)) = \delta_{\mathrm{L}} \cdot R_{\mathrm{ap}} \cdot (1 - \cos(2\theta));$$

$$D_{\mathrm{ap}} = \delta_{\mathrm{L}} \cdot R_{\mathrm{ap}} \cdot \cos(90^\circ - 2\theta) = \delta_{\mathrm{L}} \cdot R_{\mathrm{ap}} \cdot \sin(2\theta);$$

$$\frac{D_{\mathrm{ap}}}{d_{\mathrm{ap}}} = \frac{\sin(2\theta)}{1 - \cos(2\theta)} = \frac{2 \cdot \sin\theta \cdot \cos\theta}{\sin^2\theta + \cos^2\theta - (\cos^2\theta - \sin^2\theta)}$$

$$= \frac{2 \cdot \sin\theta \cdot \cos\theta}{\sin^2\theta + \cos^2\theta - \cos^2\theta + \sin^2\theta};$$

$$\frac{D_{\mathrm{ap}}}{d_{\mathrm{ap}}} = \frac{2 \cdot \sin\theta \cdot \cos\theta}{2 \cdot \sin^2\theta} = \frac{\cos\theta}{\sin\theta} = \operatorname{ctg}\theta = \frac{1}{\operatorname{tg}\theta} = \frac{1}{0.17632698} \approx 5.67.$$

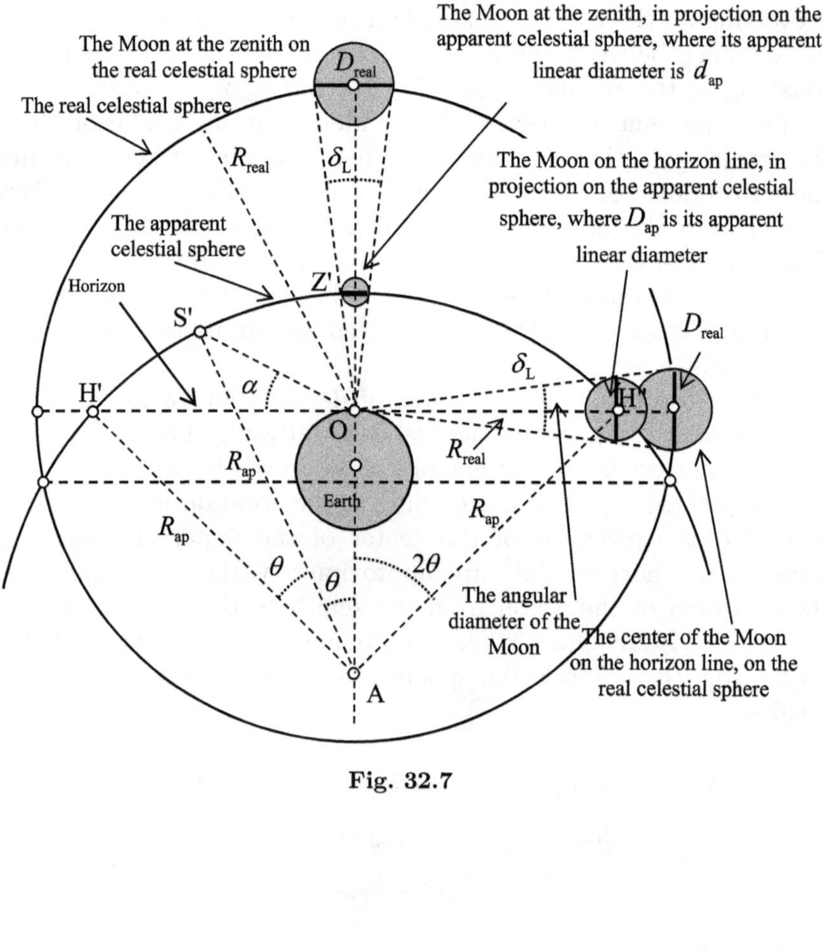

Fig. 32.7

Problem 33

The Sun as Seen from Saturn

It is known that: the distance from Saturn to the Sun is 9.54 times greater than the distance from the Earth to the Sun; the angular diameter of the Sun's disk, seen from Earth, is $32'$.

Determine the angular diameter of the Sun seen from Saturn.

Solution

Since the distance between Saturn and the Sun is 9.54 times greater than the distance from Earth to the Sun, it turns out that the angular diameter of the Sun's disk, observed from Saturn, is 9.54 times smaller than the angular diameter of the Sun observed from Earth. Therefore, using the drawing in Figure 33.1, it results that:

$$d_{\text{Sun}} = r_{\text{Saturn–Sun}} \alpha_{\text{Sun–Saturn}};$$

$$d_{\text{Sun}} = r_{\text{Earth–Sun}} \alpha_{\text{Sun–Earth}};$$

$$r_{\text{Saturn–Sun}} = 9.54 r_{\text{Earth–Sun}};$$

$$\alpha_{\text{Sun–Saturn}} = \frac{1}{9.54} \alpha_{\text{Sun–Earth}} = \frac{32'}{9.54} \approx 3.35' \approx 0.00097 \,\text{radians};$$

$$\alpha_{\text{Sun–Saturn}} \approx 0.001 \,\text{radians},$$

representing the angular diameter of the Sun seen from Saturn.

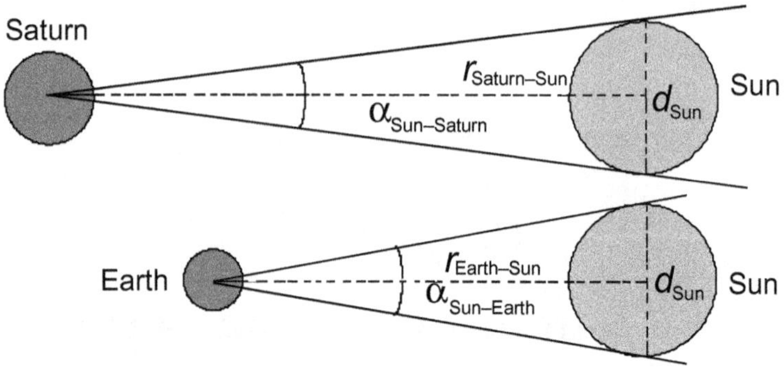
Saturn
$r_{\text{Saturn–Sun}}$
Sun
d_{Sun}
$\alpha_{\text{Sun–Saturn}}$
Earth
$r_{\text{Earth–Sun}}$
d_{Sun}
Sun
$\alpha_{\text{Sun–Earth}}$

Fig. 33.1

Problem 34

The Gauss Problem

On January 1, 1801, the Italian astronomer Giuseppe Piazzi observed Ceres for the first time for about one month until the asteroid disappeared in the Sun's light. One year later, on January 1, 1802, the German mathematician Carl Friedrich Gauss rediscovered the same asteroid when it reappeared from behind the Sun at the exact point foreseen by his precise calculus.

Gauss's method, known as *Gauss's problem*, assumes that it is possible to determine the shape and parameters of an asteroid/ satellite/projectile's orbit relative to the Sun/Earth.

Gauss succeeded in determining the shape of Ceres, orbit knowing the spherical equatorial coordinates (right ascension α; declination δ) corresponding to three moments in time.

Gauss's method is further simplified if the following are considered to be known: two position vectors of the asteroid/satellite/projectile relative to the Sun/Earth, the asteroid/satellite/projectile's flight period between two locations and the asteroid/satellite/projectile's movement direction. The transfer of the asteroid/satellite/projectile between these two known locations during a known time interval, always in the same direction, can occur on a conic orbit: ellipse, parabola or hyperbola.

In modern astrodynamics terms, solving *Gauss's problem* means solving the problems of cosmic interceptions, cosmic encounters and interplanetary flights.

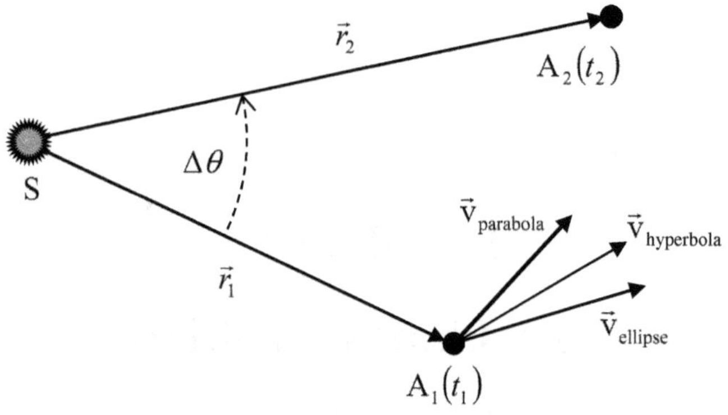

Fig. 34.1

For this, the following elements are *given*: the position vectors, $\vec{r}_1$ and $\vec{r}_2$, of an asteroid/satellite/projectile relative to the Sun/the Earth, as presented in Figure 34.1, corresponding to the moments t_1 and t_2, respectively; the duration of the asteroid/satellite/ projectile's flight between the two positions $\Delta t = t_2 - t_1$; the movement direction of the asteroid/satellite/projectile.

To be determined:

a) The elements of the asteroid/satellite/projectile's velocity so its transfer can occur on an ellipse, $\vec{v}_1 = \vec{v}_{\text{ellipse}}$;

b) The elements of the asteroid/satellite/projectile's velocity vector so its transfer can occur on a parabola/hyperbola, $\vec{v}_2 = \vec{v}_{\text{parabola/hyperbola}}$.

Solution

Depending on the characteristics of the asteroid/satellite/projectile's velocity (vector direction and magnitude) at point $A_1(\vec{r}_1, t_1)$, the shape of the transfer orbit can be calculated exactly (ellipse, parabola, or hyperbola), as well as its elements. Thus, after the same time interval, Δt, the asteroid/satellite/projectile to passes through the point $A_2(\vec{r}_2, t_2)$, knowing that each time, $\angle(\vec{r}_1, \vec{r}_2) = \Delta\theta$, as presented in Figures 34.1 and 34.2.

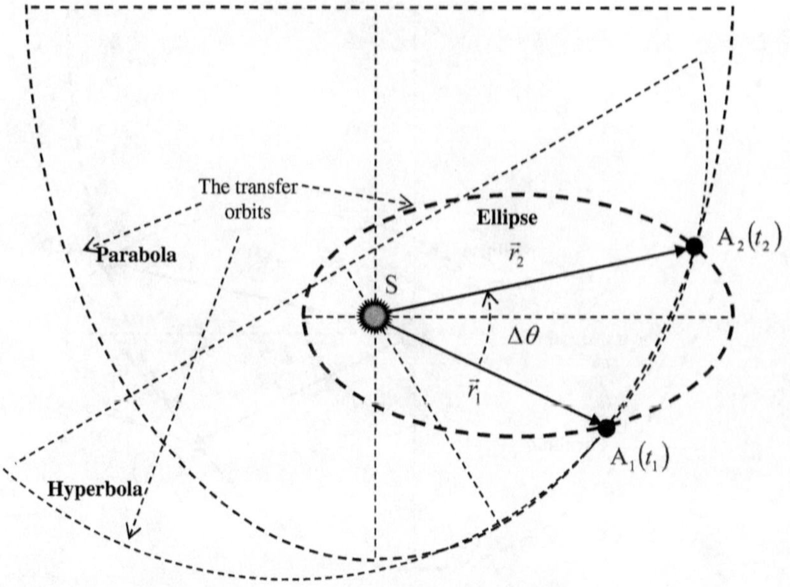

Fig. 34.2

Figure 34.3 presents the sectors corresponding to the asteroid/satellite/projectile's transfer orbits from point A_1 to point A_2. These sectors pertain to an ellipse or a parabola and are determined by the values and orientations of the asteroid/satellite/projectile's velocities at the point $A_1(\vec{r}_1, t_1)$, $\vec{v}_{\text{ellipse}}$ and $\vec{v}_{\text{parabola}}$, respectively.

a)

1) The transfer on a single ellipse

If the projectile M is launched from satellite S and meets the asteroid at point A, it means that points A and S must be on the same ellipse with its focus F_1 in the center of the Earth, E, as presented in Figures 34.4 and 34.5.

Since the problem has a single solution, it means that the second focus of the ellipse, F_2, is on the SA sector, at the tangency point on the circles with the centers in points S and A, and of radii x and y, respectively, as presented in Figures 34.6 and 34.7. Using the

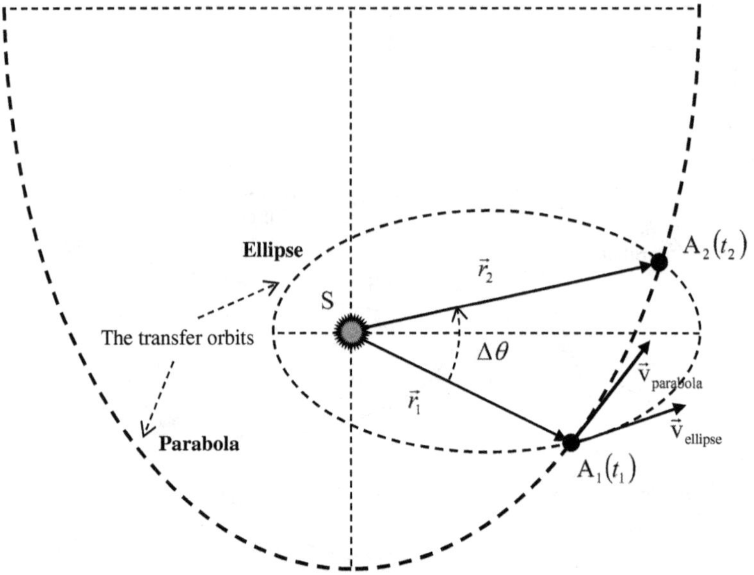

Fig. 34.3

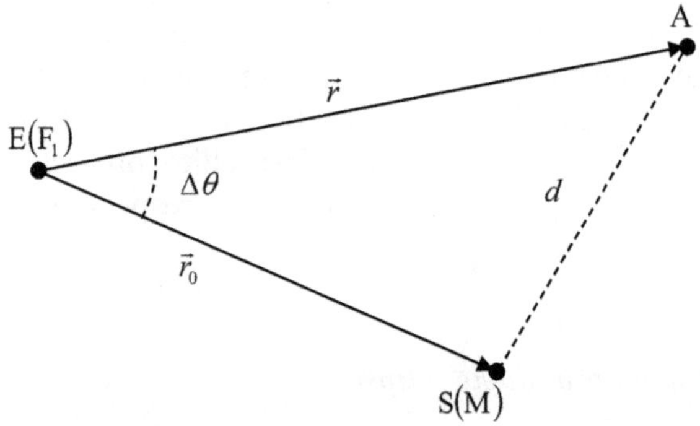

Fig. 34.4

ellipse relations to find the focus F_2 on the SA segment, it results that:

$$ES + SF_2 = 2a; \quad EA + AF_2 = 2a;$$

$$r_0 + x = 2a; \quad r + y = 2a;$$

$$r_0 + x = r + y;$$

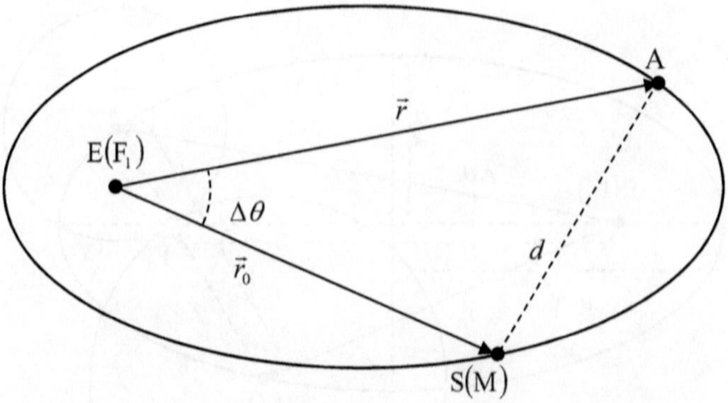

Fig. 34.5

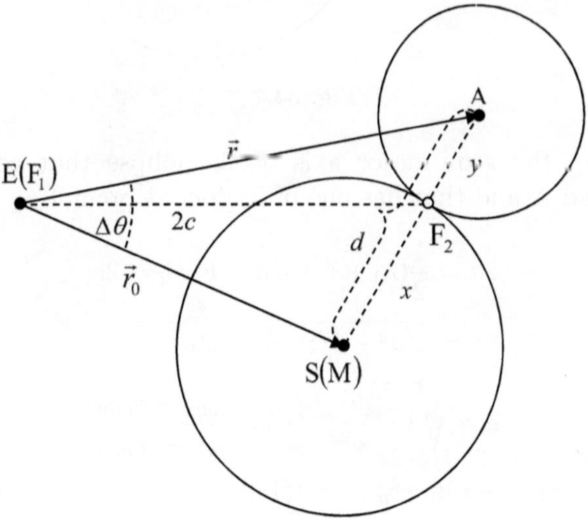

Fig. 34.6

$$x - y = r - r_0 = \Delta r; \quad x + y = d;$$

$$x = \frac{1}{2}(d + \Delta r); \quad y = \frac{1}{2}(d - \Delta r);$$

$$a = \frac{1}{2}(r_0 + x) = \frac{1}{2}\left[r_0 + \frac{1}{2}(d + \Delta r)\right] = \frac{1}{2}\left[r_0 + \frac{1}{2}(d + r - r_0)\right];$$

$$a = \frac{1}{2}\left[r_0 + \frac{1}{2}(d + r) - \frac{r_0}{2}\right] = \frac{1}{4}(r_0 + r + d),$$

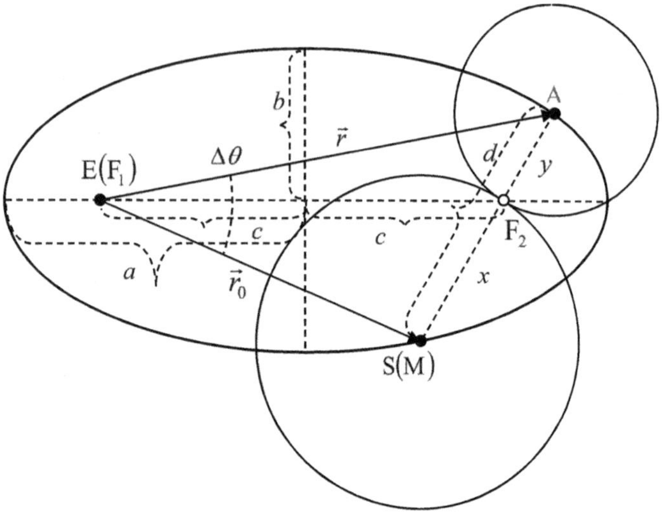

Fig. 34.7

representing the semi-major axis of the ellipse that contains the points S and A and that has one of its foci at point E;

$$2a = \frac{1}{2}(r_0 + r + d); \quad F_1F_2 = 2c;$$

$$c = \sqrt{a^2 - b^2}; \quad b = \sqrt{a^2 - c^2};$$

$$e = \sqrt{1 - \frac{b^2}{a^2}}; \quad e = \frac{r_{\max} - r_{\min}}{r_{\max} + r_{\min}};$$

$$r_{\min} = a(1 - e); \quad r_{\max} = a(1 + e); \quad r_{\min} + r_{\max} = 2a.$$

All of these are geometric elements that allow the representation of the elliptical transfer orbit according to the known rules.

To determine the direction of the initial velocity $\vec{v}_0$, establish the direction of the tangent to the ellipse at point S. This can be done using the ellipse's optical properties, according to which the light rays emitted from the focus of a concave ellipsoidal mirror will all pass, after reflection, through the other focus of the mirror, as presented in Figure 34.8.

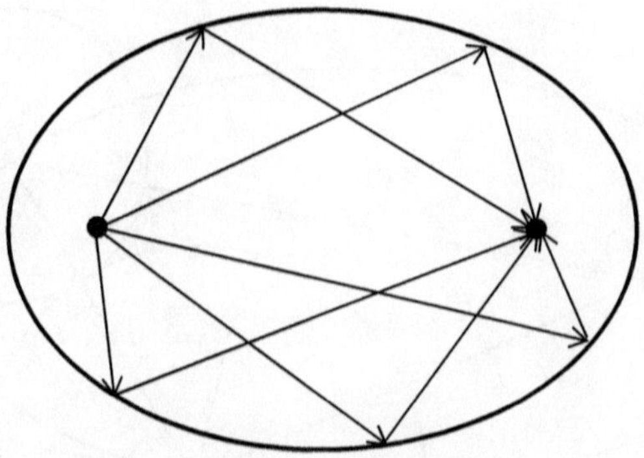

Fig. 34.8

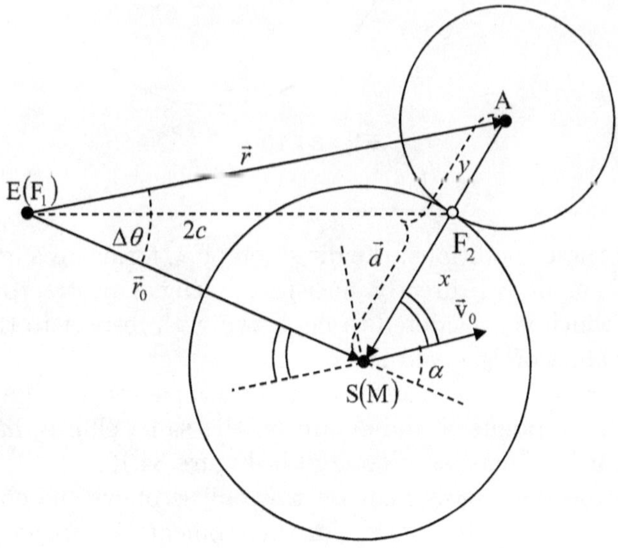

Fig. 34.9

As a consequence, the tangent to the ellipse will form equal angles with the focal chords S′. A focal chord is a line that passes through the foci, as presented in Figures 34.9, 34.10 and 34.11. The tangent to the ellipse at point S is perpendicular to the bisector of angle ASE.

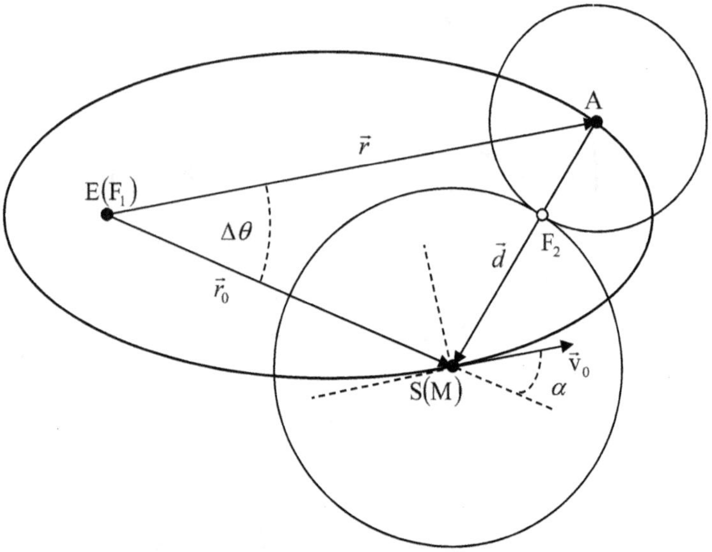

Fig. 34.10

Under these conditions, the direction of $\vec{v}_0$ forms an angle α with the direction of $\vec{r}_0$, directly measured using a protractor in Figure 34.9, which is considered to be drawn with respect to the known numeric values of r_0, r and $\Delta\theta$.

If projectile M launched from satellite S passes through point A, it means that points S and A are on the same ellipse, having the Earth, E, in its focus, as presented in Figure 34.12.

The projectile's movement on this ellipse obeys the energy and momentum conservation laws. For the moments when the projectile passes through one of the points B_1 or B_2 (the ellipse's apogee and perigee, respectively), the $E = E_0$ and $L = L_0$ equations become:

$$\frac{v_{1,2}^2}{2} - K\frac{M}{r_{1,2}} = \frac{v_0^2}{2} - K\frac{M}{r_0};$$

$$v_{1,2}r_{1,2} = v_0 r_0 \sin\alpha;$$

$$\left(\frac{v_0^2}{2} - K\frac{M}{r_0}\right) r_{1,2}^2 + KMr_{1,2} - \frac{1}{2}v_0^2 r_0^2 \sin^2\alpha = 0.$$

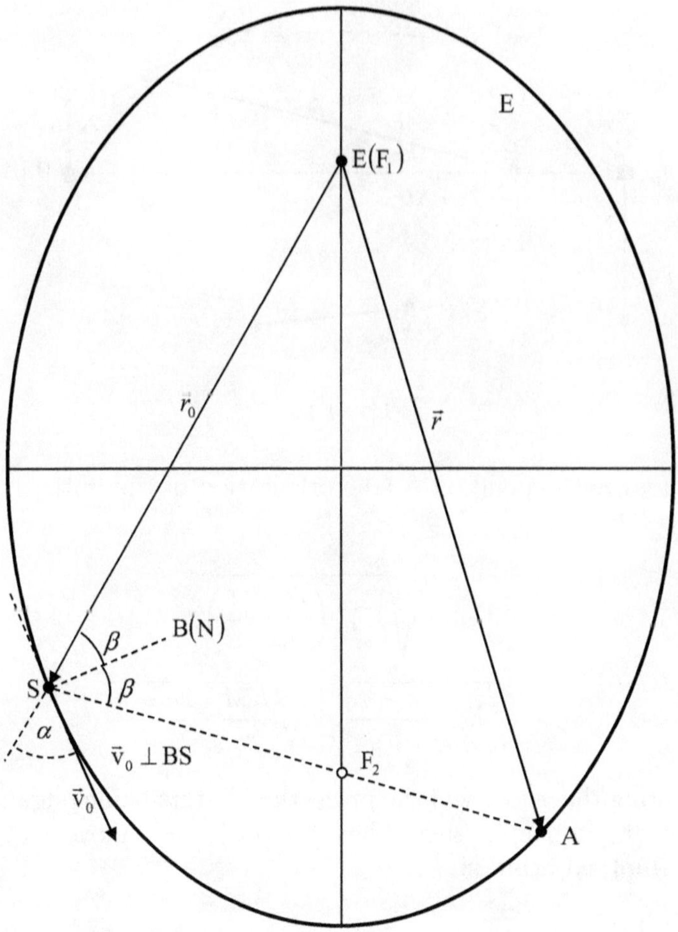

Fig. 34.11

The two roots of this equation, r_1 and r_2, represent the $r_{1,2}$ values at apogee and perigee ($r_{\min}, r_{\max}$); thus,

$$a = \frac{1}{2}(r_1 + r_2) = \frac{1}{2}(r_{\min} + r_{\max}),$$

and from Viète's formula, the result is:

$$a = \frac{KMr_0}{2KM - r_0 v_0^2},$$

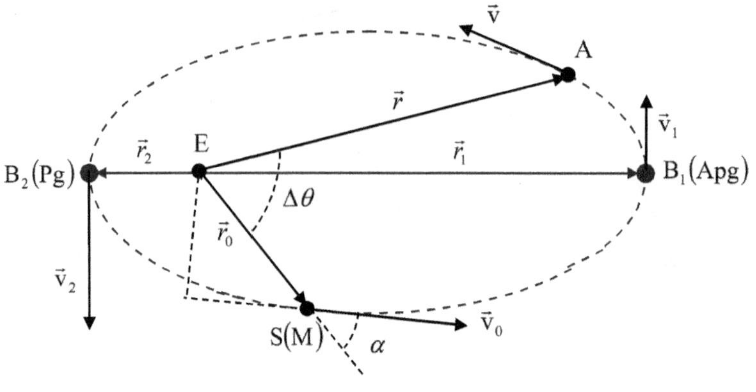

Fig. 34.12

which does not depend on α (the orientation of the initial injection velocity $\vec{v}_0$);

$$v_0 = \sqrt{KM\left(\frac{2}{r_0} - \frac{1}{a}\right)};$$

$$v_0 = \sqrt{KM\frac{2a - r_0}{ar_0}} = \sqrt{\frac{KM}{r_0} \cdot \frac{2a - r_0}{a}},$$

representing the velocity of the projectile M at its launching moment.

According to Kepler's third law, the projectile's movement period on the elliptical orbit is:

$$T^2 = \frac{4\pi^2}{K(M+m)} \cdot a^3; \quad m \ll M; \quad T^2 = \frac{4\pi^2}{KM} \cdot a^3;$$

$$T = 2\pi\sqrt{\frac{a^3}{KM}}.$$

A model of the transfer ellipse (tr.ell.) presented in Figure 34.10 is constructed using homogeneous plasticine. There are two identical, plane uniform disks of Δh thickness. From one of the two elliptical plasticine disks, the SAF_1 sector is cut off, and then the gravitational balance of the system presented in Figure 34.13 is realized. For this

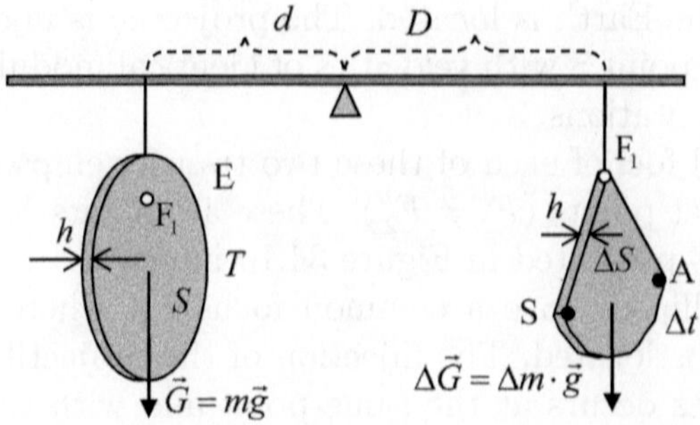

Fig. 34.13

system:

$$G \cdot d = \Delta G \cdot D; \quad mg \cdot d = \Delta m \cdot g \cdot D;$$
$$m \cdot d = \Delta m \cdot D; \quad \rho \cdot V \cdot d = \rho \cdot \Delta V \cdot D;$$
$$V \cdot d = \Delta V \cdot D; \quad S \cdot h \cdot d = \Delta S \cdot h \cdot D;$$
$$S \cdot d = \Delta S \cdot D; \quad \frac{\Delta S}{S} = \frac{d}{D}.$$

According to Kepler's second law (the Law of Equal Areas), the result is:

$$T \ldots\ldots\ldots\ldots\ldots\ldots S;$$
$$\Delta t \ldots\ldots\ldots\ldots\ldots\ldots \Delta S;$$
$$\Delta t = \frac{\Delta S}{S} \cdot T = \frac{d}{D} \cdot T,$$

representing the time taken by the projectile to meet with the asteroid.

2) Transfer on two different ellipses

Since the problem accepts two solutions for projectile M's flight between two points, S and A, it means that the projectile can move from S to A on either of the two ellipses that cross each other at points S and A and that have a common focus ($F_1 \equiv F_2$) at the point where the Earth is located. The projectile is injected onto the two ellipses at point S with velocities of identical moduli but different (opposite) orientations.

The second foci of each of these two transfer ellipses from S to A are two distinct points ($F_2' \neq F_2''$). These are points X and Y, where the two circles presented in Figure 34.10 intersect.

The two ellipses have a common focus F_1, where the center of the Earth, E, is located. The injection of the projectile onto each of the two ellipses occurs at the same point and with the same initial velocity as presented in Figure 34.14. Thus, the semi-major axes of

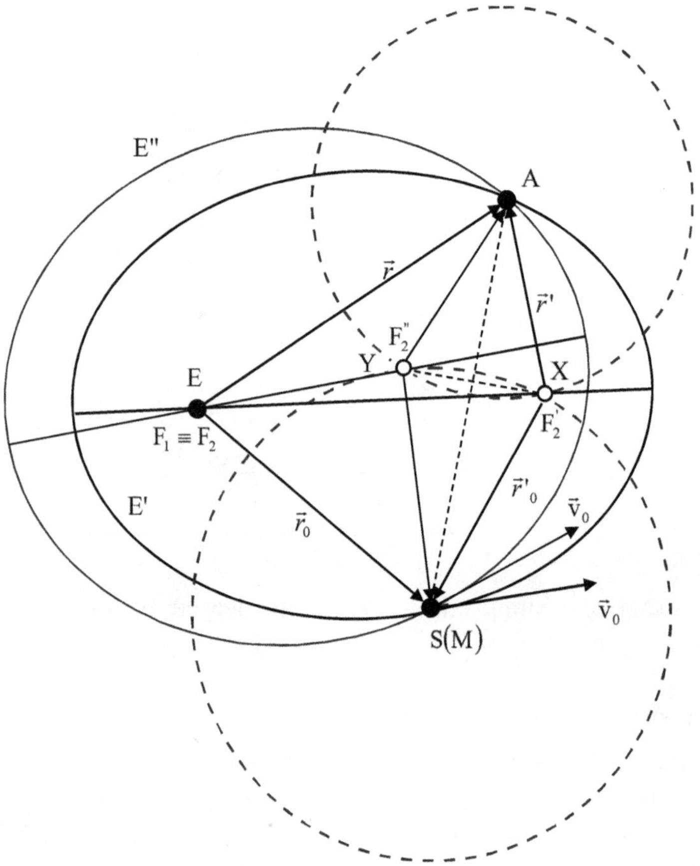

Fig. 34.14

the two ellipses are identical:

$$a = \frac{KMr_0}{2KM - r_0 v_0^2}.$$

The two ellipses differ in the distances between the foci, $2c_1 \neq 2c_2$; the minor axis, $2b_1 \neq 2b_2$; and the eccentricities, $e_1 \neq e_2$.

Measuring the distances on the initial drawing (Figure 34.1), it results that

$$\mathrm{ES} + \mathrm{SX} = \mathrm{EA} + \mathrm{AX},$$

which implies that point X is the second focus of one of the two ellipses, $\mathrm{X} \equiv F_2'$.

After locating the foci of the two ellipses, the directions of the ellipses' axes can be drawn. For each of the two ellipses whose foci have been located, the following relations can be written:

$$r_0 + r_0' = 2a; \quad r + r' = 2a.$$

Measuring r_0 and r_0', or r and r', we can calculate

$$a = \frac{1}{2}(r_0 + r_0') = \frac{1}{2}(r + r').$$

Measuring the distances between the two ellipses' foci, it results that:

$$\mathrm{F_1F_2'} = 2c_1; \quad b_1 = \sqrt{a^2 - c_1^2};$$

$$\mathrm{F_1F_2''} = 2c_2; \quad b_2 = \sqrt{a^2 - c_2^2}.$$

The graphs of the two ellipses are drawn using the method presented in Figures 34.15–34.19.

According to the optical properties (characteristics) of the ellipse, the angles α' and α'' can be measured.

The velocity of the projectile launched on either of the two ellipses is the same:

$$\mathrm{v}_0 = \sqrt{KM\frac{2a - r_0}{ar_0}} = \sqrt{\frac{KM}{r_0} \cdot \frac{2a - r_0}{a}}.$$

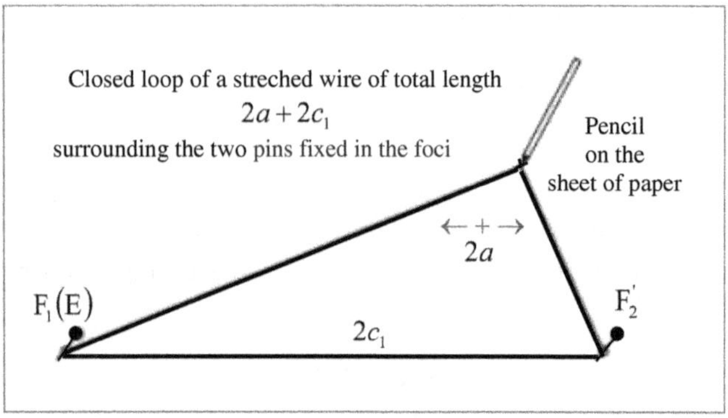

Fig. 34.15

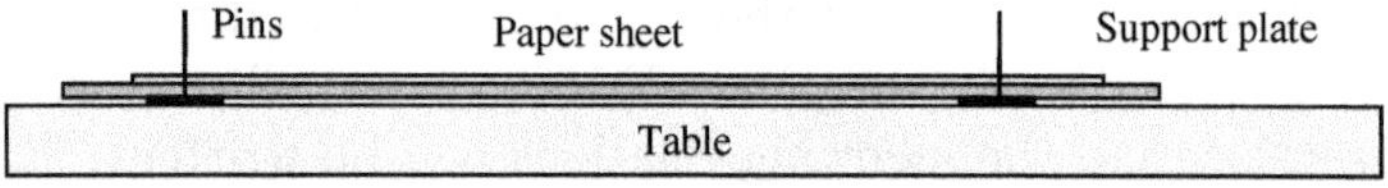

Fig. 34.16

The movement period of the projectile on either of the two ellipses is the same:

$$T^2 = \frac{4\pi^2}{K(M+m)} \cdot a^3; \quad m \ll M; \quad T^2 = \frac{4\pi^2}{KM} \cdot a^3;$$

$$T = 2\pi\sqrt{\frac{a^3}{KM}}.$$

Each of the two ellipses, E′ and E″, presented in Figures 34.17 and 34.18, is duplicated in homogeneous plasticine. The ellipses are in the shape of two identical, plane uniform disks of the same thickness Δh. Each model has the SAF_1 sector cut off, and the gravitational balance of the system presented in Figure 34.20 is realized. For this system:

$$G \cdot d = \Delta G \cdot D; \quad mg \cdot d = \Delta m \cdot g \cdot D;$$

$$m \cdot d = \Delta m \cdot D; \quad \rho \cdot V \cdot d = \rho \cdot \Delta V \cdot D;$$

$$V \cdot d = \Delta V \cdot D; \quad S \cdot h \cdot d = \Delta S \cdot h \cdot D;$$

$$S \cdot d = \Delta S \cdot D; \quad \frac{\Delta S}{S} = \frac{d}{D}.$$

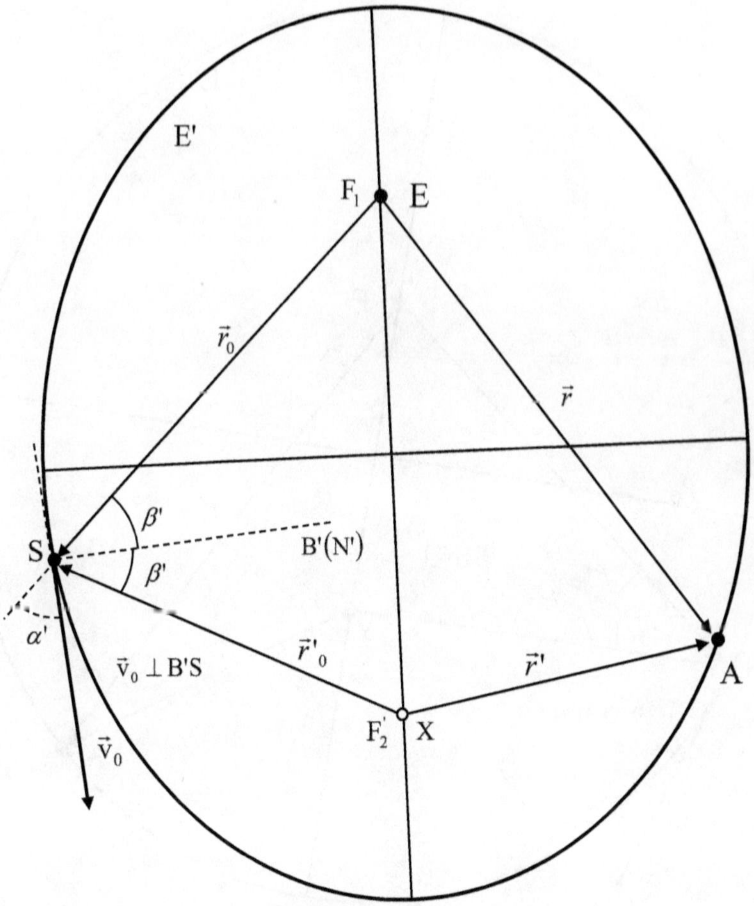

Fig. 34.17

According to Kepler's second law (the Law of Equal Areas), it results that:

$$T \ldots\ldots\ldots\ldots\ldots\ldots\ldots\ldots\ldots\ldots S;$$

$$\Delta t \ldots\ldots\ldots\ldots\ldots\ldots\ldots\ldots \Delta S;$$

$$\Delta t = \frac{\Delta S}{S} \cdot T = \frac{d}{D} \cdot T,$$

representing the time taken by the projectile to travel from S to A.

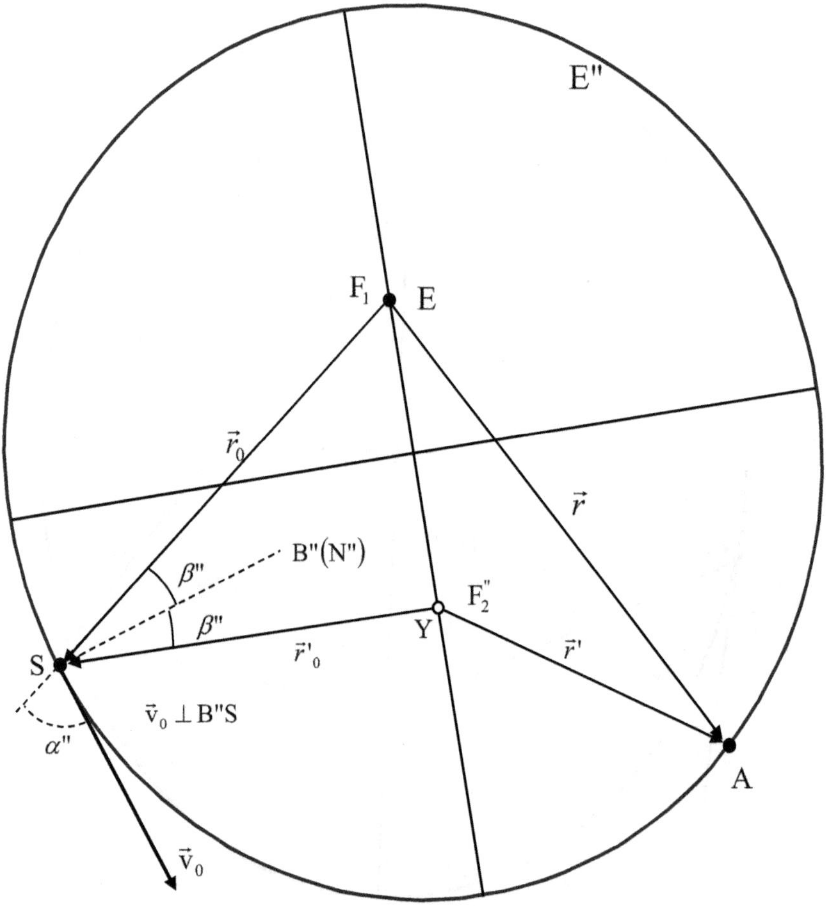

Fig. 34.18

Through measurements, $\Delta t'$ is determined by the projectile moving on the ellipse E′ and $\Delta t'' \neq \Delta t'$ (the ellipse interval) is determined by the projectile moving on E″.

b)

1) Transfer on a parabola

A parabola is the locus of a point moving so that it is equidistant from a fixed point called the focus and a fixed line called a directrix.

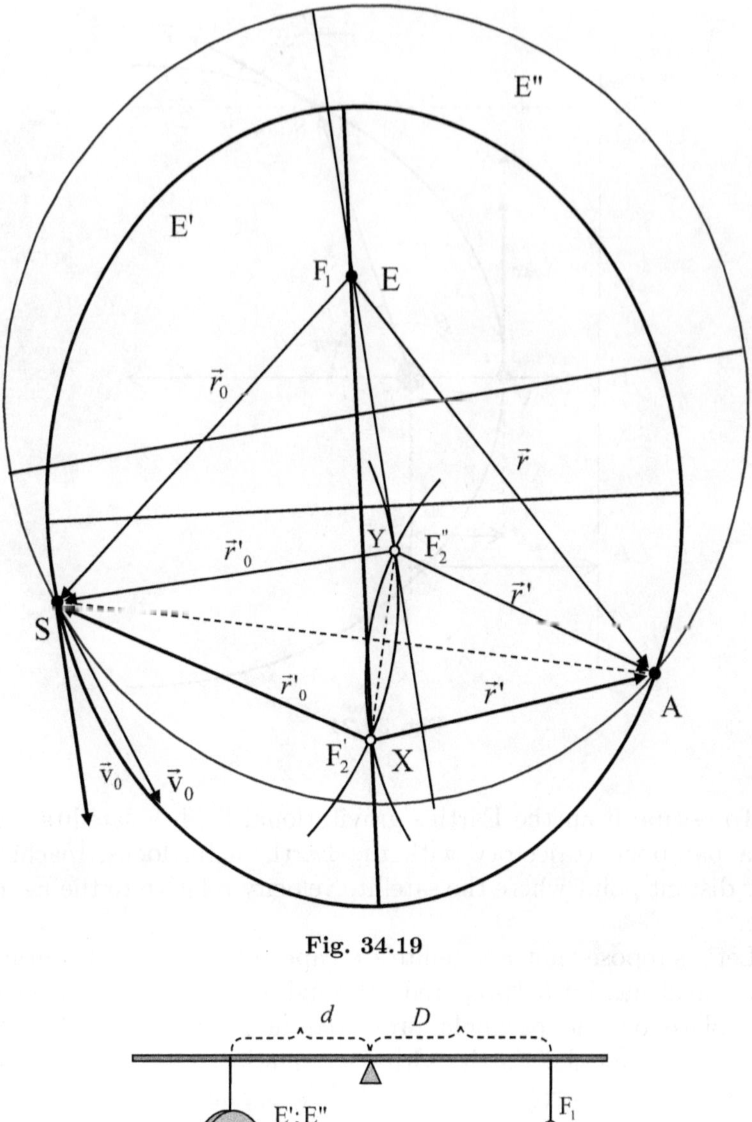

Fig. 34.19

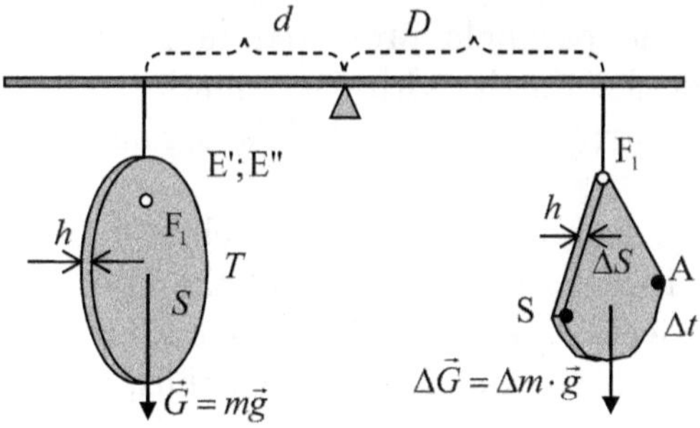

Fig. 34.20

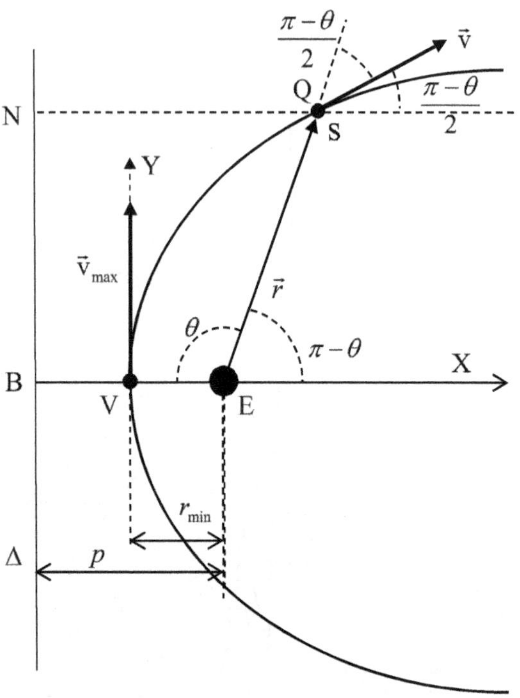

Fig. 34.21

To escape from the Earth's gravitational field, a satellite moves on a parabolic trajectory with the Earth in its focus, reaching a very distant point where the satellite velocity relative to the Earth is zero.

Let's suppose that a satellite's escape from the Earth's gravitational field has been "prepared" through calculus. The escape will take place on the parabola presented in Figure 34.21, in whose focus is located the Earth, with the equation $y^2 = 2px$ in Cartesian coordinates $(x;\, y)$. The parabola's parameter is known to be $p = 2r_{\text{min}}$.

To succeed in such an escape, the satellite is first moved upward using a carrier rocket to the altitude of the injection point (Q). At this level, the injection velocity $\vec{v}$ is transferred to the satellite in the direction of the tangent to the parabola; thus, the total mechanical energy of the satellite–Earth system is:

$$E = \frac{m\mathrm{v}^2}{2} - K\frac{mM}{r} = 0.$$

When the escape succeeds and the satellite is located very far from the Earth ($r \to \infty$), it is considered to be stationary relative to the Earth ($\mathrm{v}_\infty = 0$).

The *optical properties of the parabola* can be demonstrated as follows: All light rays emitted by a source placed in the focus of a concave parabolic mirror will be parallel to the main optical axis of the paraboloid after reflection, and reciprocally incident rays parallel to the main optical axis (of the paraboloid) will be reflected through the focus.

Consequently, the tangent to the parabola at point Q is the bisector of the EQN angle.

Considering the definition of the parabola, it results that:

$$\mathrm{QE} = \mathrm{QN};$$

$$r = \mathrm{EB} + \mathrm{QE}\cos(\pi - \theta);$$

$$r = 2r_{\min} + r(-\cos\theta);$$

$$2r_{\min} = r(1 + \cos\theta);$$

$$r_{\min} = r\,\cos^2\frac{\theta}{2}.$$

If the satellite injection takes place at point V, representing the parabola vertex where $r = r_{\min}$, the injection velocity should be $\mathrm{v} = \mathrm{v}_{\max}$; thus:

$$E = \frac{m\mathrm{v}_{\max}^2}{2} - K\frac{mM}{r_{\min}} = 0;$$

$$\mathrm{v}_{\max} = \sqrt{2\frac{KM}{r_{\min}}};$$

$$r_{\min}\mathrm{v}_{\max} = r\mathrm{v}\sin\left(\frac{\pi}{2} - \frac{\theta}{2}\right) = r\mathrm{v}\cos\frac{\theta}{2};$$

$$r\cos^2\frac{\theta}{2}\mathrm{v}_{\max} = r\mathrm{v}\cos\frac{\theta}{2}; \quad \mathrm{v}_{\max} = \frac{\mathrm{v}}{\cos\frac{\theta}{2}}.$$

The polar coordinates of the satellite on the parabola

From the equation of the trajectory (the equation of the conic) written in polar coordinates, for $e = 1$ (parabola), it results that:

$$r = \frac{p}{1+\cos\theta};$$

$$r = \frac{p}{2\cos^2\frac{\theta}{2}} = \frac{p}{2}\left(1 + \mathrm{tg}^2\frac{\theta}{2}\right).$$

Since the movement on the parabola takes place under the action of the central force of gravitational attraction, we have:

$$r^2\dot{\theta} = C = \sqrt{pKM}; \quad r^2 \mathrm{d}\theta = C\mathrm{d}t;$$

$$\frac{p^2}{4}\left(1 + \mathrm{tg}^2\frac{\theta}{2}\right)^2 \mathrm{d}\theta = C\mathrm{d}t;$$

$$\mathrm{tg}\frac{\theta}{2} = u; \quad \frac{1}{2}\frac{\mathrm{d}\theta}{2\cos^2\frac{\theta}{2}} = \mathrm{d}u;$$

$$\mathrm{d}\theta = 2\cos^2\frac{\theta}{2}\mathrm{d}u = \frac{2}{1+\mathrm{tg}^2\frac{\theta}{2}}\mathrm{d}u; \quad \mathrm{d}\theta = \frac{2}{1+u^2}\mathrm{d}u;$$

$$\frac{p^2}{2C}(1+u^2)\mathrm{d}u = \mathrm{d}t;$$

$$t - t_0 = \frac{p^2}{2c}\int_0^u (1+u^2)\mathrm{d}u,$$

where t_0 is the moment when the satellite passes through the point corresponding to $r_{\min}$, for which $\theta = 0$ and $u = 0$, and t is the moment when the satellite's polar coordinates are r and θ;

$$t - t_0 = \frac{p^2}{2C}\left(u + \frac{u^3}{3}\right);$$

$$\mathrm{tg}\frac{\theta}{2} + \frac{1}{3}\mathrm{tg}^3\frac{\theta}{2} = \frac{2\sqrt{pKM}}{p^2}(t - t_0).$$

By solving the above equation, we have $\theta = f(t)$.

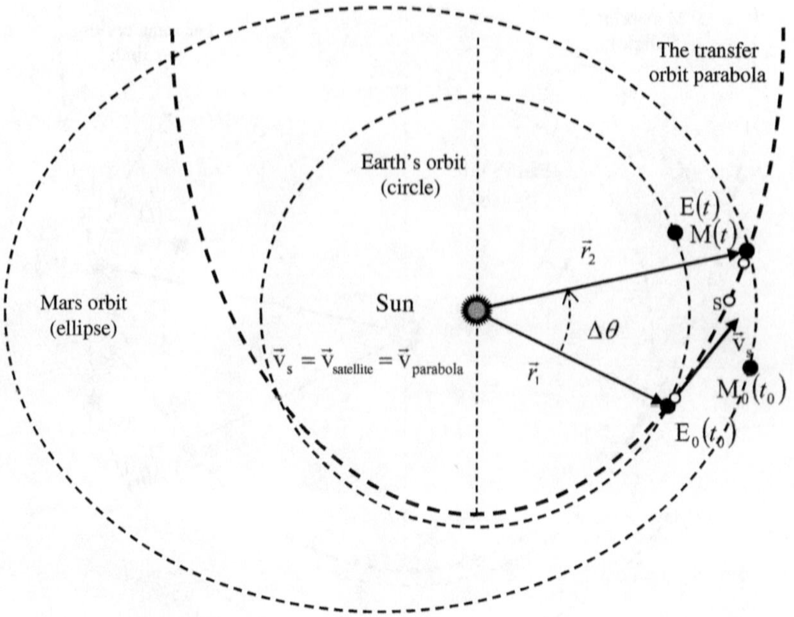

Fig. 34.22

Then, from the parabola equation

$$r = \frac{p}{2\cos^2\frac{\theta}{2}},$$

we deduce the $r = f(t)$ dependence.

In Figure 34.22, the parabolic orbit for the satellite's transfer from the Earth to Mars is presented, admitting that Mars' orbit is elliptical and the Earth's orbit is circular. Mars' elliptical orbit has the largest eccentricity among the planets of our solar system. We also consider that the two orbits are in the ecliptic.

Notation: $E_0(t_0)$ and $M_0(t_0)$ – the Earth's and Mars' positions at the moment when the satellite s is on the parabolic orbit towards Mars; $E(t)$ and $M(t)$ – the Earth's and Mars' positions when the satellite s has reached Mars. To solve this problem, it is important to know that during the satellite's transfer, Mars' angular velocity is variable.

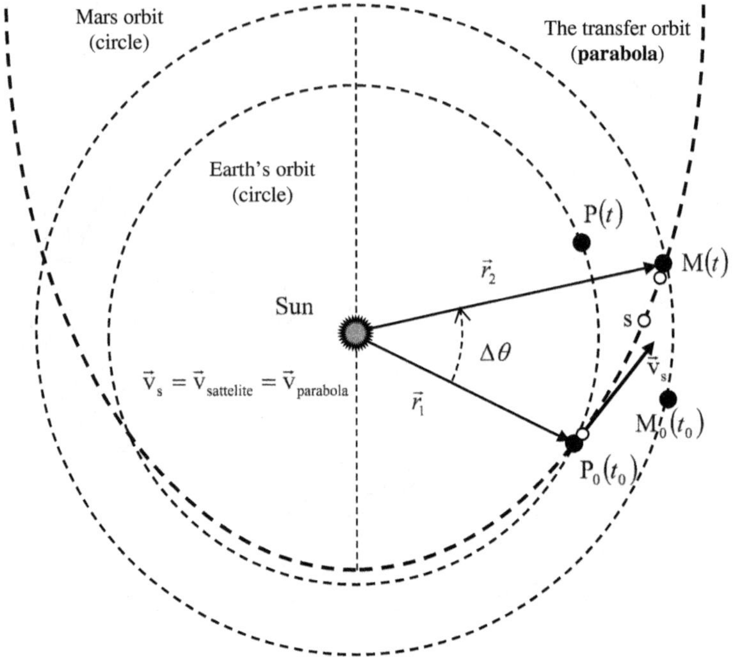

Fig. 34.23

Figure 34.23 presents the same problem, in a simplified version, where we consider that Mars' orbit is also circular and the satellite's transfer occurs on a parabolic orbit, too.

2) Transfer on a hyperbola

A hyperbola is the geometric locus of all points in a plane for which the difference between the distances to two fixed points (foci) in the plane is constant.

A satellite is moving along a hyperbolic trajectory when it has to escape the Earth's gravitational field and reach a very distant point in space while still having a certain velocity relative to the Earth, which is located in one of the hyperbola's foci.

Let's suppose that a satellite's escape has been "prepared" through calculus. The satellite will escape from the Earth's gravitational field, having the Earth in the closest focus, as presented in

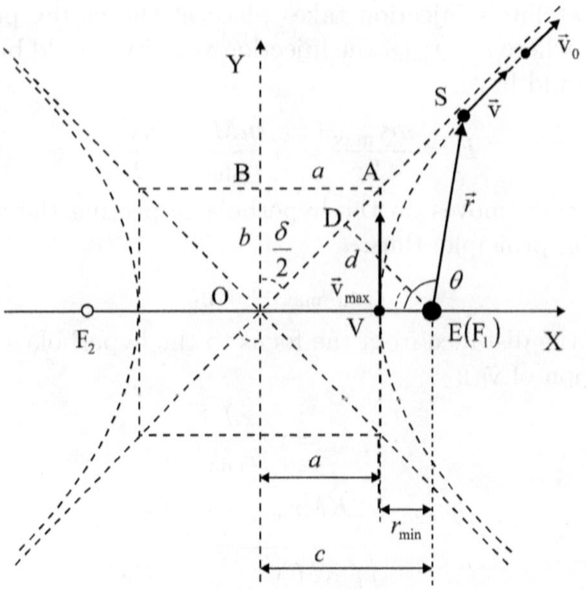

Fig. 34.24

Figure 34.24. The equation of the trajectory, in Cartesian coordinates $(x; y)$ is

$$\frac{x^2}{a^2} - \frac{y^2}{b^2} = 1,$$

for which the a and b parameters are known.

To succeed in such an escape, the satellite is first moved upwards using a carrier rocket to the altitude of the chosen injection point. At this level, a certain injection velocity $\vec{v}$ is transferred to the satellite in the direction of the tangent to the hyperbola; thus, the total mechanical energy of the satellite–Earth system is

$$E = \frac{m\mathrm{v}^2}{2} - K\frac{mM}{r} > 0.$$

Under these conditions, when the "escape" has succeeded and the satellite has traveled very far from the Earth, we have:

$$r \to \infty;$$

$$E = \frac{m\mathrm{v}_0^2}{2}; \quad \mathrm{v}_0 = \sqrt{\frac{2E}{m}}.$$

If the satellite's injection takes place at the vertex point of the hyperbola, where $r = r_{\min}$, the injection velocity should be $\mathrm{v} = \mathrm{v}_{\max}$; thus, we would have

$$E = \frac{m\mathrm{v}_{\max}^2}{2} - K\frac{mM}{r_{\min}} = \frac{m\mathrm{v}_0^2}{2}.$$

The satellite moves on the hyperbola respecting the momentum conservation principle; thus:

$$r_{\min}v_{\max} = d\mathrm{v}_0,$$

where d is the distance from the focus to the hyperbola's asymptote (the direction of $\vec{\mathrm{v}}_0$);

$$\mathrm{v}_{\max} = \mathrm{v}_0\frac{d}{r_{\min}};$$

$$\mathrm{v}_0^2 r_{\min}^2 + 2KMr_{\min} - \mathrm{v}_0^2 d^2 = 0;$$

$$r_{\min} = \sqrt{\left(\frac{KM}{\mathrm{v}_0^2}\right)^2 + d^2} - \frac{KM}{\mathrm{v}_0^2};$$

$$r_{\min} = c - a; \quad a = \frac{KM}{\mathrm{v}_0^2}; \quad \mathrm{v}_0 = \sqrt{\frac{KM}{a}};$$

$$c = \sqrt{a^2 + d^2},$$

where c represents the hyperbola's eccentricity.

From the right triangle OED, where OE $= c$ and ED $= d$, the result is:

$$\mathrm{OE}^2 = \mathrm{OD}^2 + \mathrm{ED}^2;$$

$$c^2 = x^2 + d^2; \quad x = \mathrm{OD} = a;$$

$$\mathrm{tg}\,\frac{\delta}{2} = \frac{\mathrm{OD}}{\mathrm{ED}} = \frac{a}{d}.$$

From the right triangle OAB, the result is:

$$\mathrm{tg}\,\frac{\delta}{2} = \frac{\mathrm{AB}}{\mathrm{OB}} = \frac{a}{d}; \quad d = b;$$

$$r_{\min} = \sqrt{a^2 + b^2} - a.$$

Any hyperbola is characterized by a and b (already marked) and by the following parameters: e – numerical eccentricity, and p – the hyperbola's parameter. The relation among them is:

$$e = \frac{\sqrt{a^2+b^2}}{c} = \frac{c}{a}; \quad p = \frac{b^2}{a}; \quad e > 1.$$

Under these conditions, the result is:

$$r_{\min} = VF_1;$$
$$r_{\min} = a(e-1); \quad r_{\max} = VF_2;$$
$$r_{\max} - r_{\min} = 2a;$$
$$r_{\max} = a(e+1).$$

The satellite's polar coordinates on the hyperbola

From the satellite's trajectory equation in polar coordinates,

$$r = \frac{p}{1 + e\cos\theta},$$

where $e > 1$ (the hyperbolic trajectory). Corresponding to the minimal distance between the satellite and the Earth, we have:

$$\theta = 0; \quad r_{\min} = \frac{p}{1+e};$$
$$r_{\min} = a(e-1);$$
$$p = a(e^2-1).$$

Differentiating the hyperbola equation yields:

$$\frac{1}{r} = \frac{1}{p} + \frac{e}{p}\cos\theta;$$
$$\frac{p\mathrm{d}r}{r^2} = e\sin\theta\mathrm{d}\theta;$$
$$\cos\theta = \frac{1}{e}\left(\frac{p}{r} - 1\right); \quad \sin\theta = \sqrt{1 - \frac{1}{e^2}\left(\frac{p}{r}-1\right)^2};$$
$$\frac{p\mathrm{d}r}{r^2} = \sqrt{e^2 - \left(\frac{p}{r}-1\right)^2}\mathrm{d}\theta;$$
$$r^2\mathrm{d}\theta = \frac{p\mathrm{d}r}{\sqrt{e^2 - \left(\frac{p}{r}-1\right)^2}}.$$

Under the action of the central force of gravitational attraction, the movement leads to

$$r^2\dot{\theta} = C = \sqrt{pKM}.$$

This results in:

$$r^2\frac{\mathrm{d}\theta}{\mathrm{d}t} = C; \quad \mathrm{d}t = \frac{r^2\mathrm{d}\theta}{C};$$

$$\mathrm{d}t = \frac{p}{C}\frac{r\mathrm{d}r}{\sqrt{e^2r^2 - (p-r)^2}};$$

$$p = a(e^2 - 1);$$

$$\begin{aligned}
e^2r^2 - (p-r)^2 &= e^2r^2 - [a(e^2-1) - r]^2 \\
&= e^2r^2 - a^2(e^2-1)^2 + 2a(e^2-1)r - r^2 \\
&= r^2(e^2-1) - a^2(e^2-1)^2 + 2a(e^2-1)r \\
&= (e^2-1)[r^2 - a^2(e^2-1) + 2ar] \\
&= (e^2-1)[(a^2 + 2ar + r^2) - a^2e^2] \\
&= (e^2-1)[(a+r)^2 - a^2e^2];
\end{aligned}$$

$$\mathrm{d}t = \frac{p}{C}\frac{r\mathrm{d}r}{\sqrt{e^2-1}\sqrt{(a+r)^2 - a^2e^2}}; \quad C = \sqrt{pKM}; \quad p = a(e^2-1);$$

$$\mathrm{d}t = \sqrt{\frac{a}{KM}}\frac{r\mathrm{d}r}{\sqrt{(a+r)^2 - a^2e^2}};$$

$$a + r = ae\mathrm{ch}u; \quad \mathrm{ch}u = \frac{e^u + e^{-u}}{2}; \quad \mathrm{sh}u = \frac{e^u - e^{-u}}{2};$$

$$r = a(e\mathrm{ch}u - 1); \quad \mathrm{d}r = ae\mathrm{sh}u\mathrm{d}u;$$

$$(a+r)^2 - a^2e^2 = a^2e^2\mathrm{sh}^2u;$$

$$\mathrm{d}t = a\sqrt{\frac{a}{KM}}(e\mathrm{ch}u - 1)\mathrm{d}u;$$

$$t - t_0 = a\sqrt{\frac{a}{KM}}\int_{u_0}^{u}(e\mathrm{ch}u - 1)\mathrm{d}u,$$

where t_0 is the time at which the satellite is at a minimum distance from the Earth. Corresponding to this moment, we have:

$$a + r_{\min} = ae\text{ch}u_0;$$

$$a + a(e - 1) = ae\text{ch}u_0;$$

$$chu_0 = 1.$$

t is the time at which the satellite is at a distance r from the Earth, where we have:

$$a + r = ae\text{ch}u;$$

$$t - t_0 = a\sqrt{\frac{a}{KM}}(e\text{sh}u - u)|_{u_0}^{u};$$

$$t - t_0 = a\sqrt{\frac{a}{KM}}[(e\text{sh}u - u) - (e\text{sh}u_0 - u_0)];$$

$$e\text{sh}u - u = \frac{1}{a}\sqrt{\frac{KM}{a}}(t - t_0) + (e\text{sh}u_0 - u_0).$$

Solving this equation using special methods allows us to establish the $u = f(t)$ dependency. Afterwards, the time dependency can be established:

$$r = a(e\text{ch}u - 1) = f(t).$$

To establish the time dependency for the other polar coordinate θ, we proceed as follows:

$$r = \frac{p}{1 + e\cos\theta};$$

$$e\cos\theta = \frac{p}{r} - 1;$$

$$p = a(e^2 - 1); \quad r = a(e\text{ch}u - 1);$$

$$e\cos\theta = \frac{e^2 - 1}{e\text{ch}u - 1} - 1;$$

$$\cos\theta = \frac{e - \text{ch}u}{e\text{ch}u - 1};$$

$$1 - \cos\theta = \frac{(e + 1)(\text{ch}u - 1)}{e\text{ch}u - 1};$$

$$1 + \cos\theta = \frac{(e-1)(\text{ch}u + 1)}{e\text{ch}u - 1};$$

$$\frac{1-\cos\theta}{1+\cos\theta} = \frac{e+1}{e-1}\frac{\text{ch}u - 1}{\text{ch}u + 1};$$

$$\text{tg}\frac{\theta}{2} = \sqrt{\frac{e+1}{e-1}}\text{th}\frac{u}{2}.$$

In Figure 34.25, the sectors of the asteroid/satellite/projectile's transfer orbit from point A_1 to point A_2 belonging to an ellipse and to a hyperbola are indicated.

They are determined by the values and orientations of the asteroid/satellite/projectile's velocities $\vec{v}_{\text{ellipse}}$ and $\vec{v}_{\text{hyperbola}}$, respectively, at the moment when the asteroid/satellite/projectile passes through point $A_1(\vec{r}_1, t_1)$.

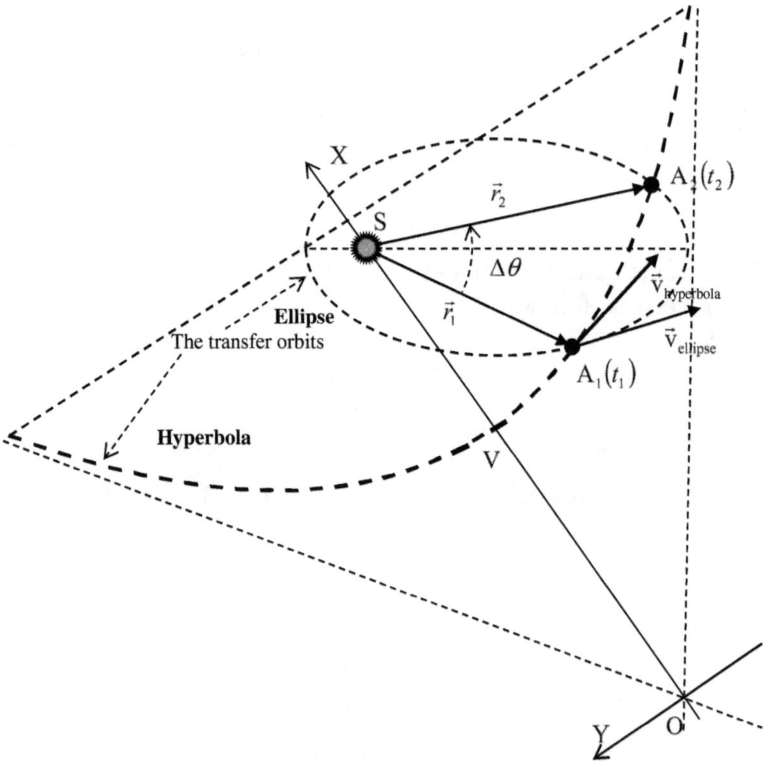

Fig. 34.25

Let's consider that *Gauss's problem* must be solved when an asteroid dangerously approaching the Earth is intercepted.

From a circumterrestrial satellite, at a certain point in its orbit ($\vec{r}_1$) and at a certain moment (t_1), a projectile is launched aiming to intercept and to destroy the asteroid at a point ($\vec{r}_2$) and at the moment (t_2), far from the Earth. We know that $\angle(\vec{r}_1; \vec{r}_2) = \Delta\theta$, as well as the direction of the projectile's movement. For this purpose, the projectile's velocity $\vec{v}$ at the launching moment must be established. It will determine the shape and the parameters of the projectile's transfer orbit from point ($\vec{r}_1$) to point ($\vec{r}_2$).

Knowing the projectile's movement direction, we can specify that the center angle $\Delta\theta$ swept by the projectile's vector radius in the time interval $\Delta t = t_2 - t_1$ is either $\Delta\theta < \pi$, if the projectile travels on a short path, or $\Delta\theta > \pi$, if the projectile travels on a longer trajectory, as indicated in Figure 34.26.

Generally, the movement from the initial to the final position implies the existence of an infinite number of orbits that pass through

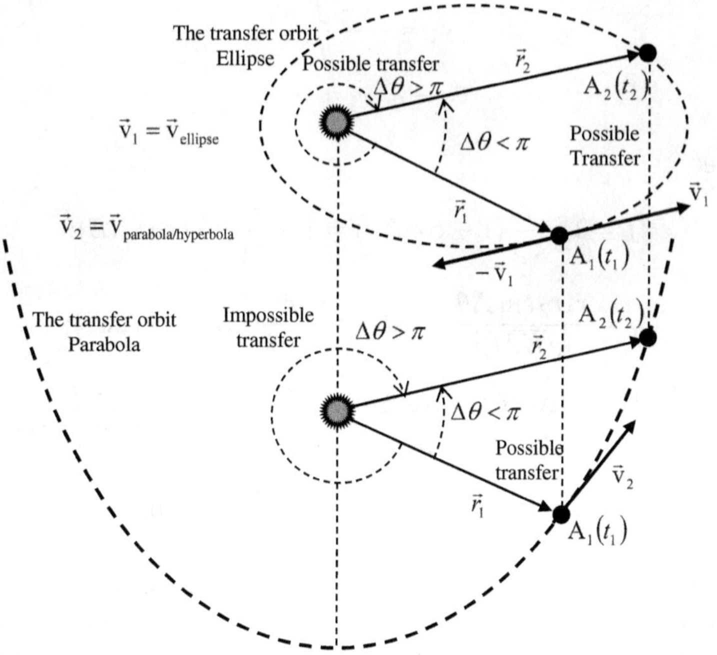

Fig. 34.26

the two positions. However, if the condition of flight duration is imposed, there are only two transfer orbits that represent solutions to Gauss's problem, one for each possible direction in which the projectile can fly between the two positions.

The vectors of the two positions, $\vec{r}_1$ and $\vec{r}_2$, are uniquely defined in the plane of the projectile's transfer orbits. If $\Delta\theta = \pi$, it means that the two position vectors are collinear, but they have opposite directions. Thus, the plane of the transfer orbit is not determined. In such cases, there is no unique solution for the $\vec{v}_1$ and $\vec{v}_2$ vectors. If $\Delta\theta = 0$ or $\Delta\theta = 2\pi$, it means that the two position vectors are collinear and have the same direction; thus, thus their transfer orbit is a degenerate conic. In such a case, it is possible to obtain a unique solution for the $\vec{v}_1$ and $\vec{v}_2$ vectors.

To solve the problem, Gauss proposed that the relations between the four vectors ($\vec{r}_1, \vec{r}_2, \vec{v}_1$ and $\vec{v}_2$) be established by the scalar functions f, g, q and u, corresponding to the following expressions:

$$\vec{r}_2 = f \cdot \vec{r}_1 + g \cdot \vec{v}_1;$$

$$\vec{v}_1 = \frac{\vec{r}_2 - f \cdot \vec{r}_1}{g};$$

$$\vec{v}_2 = q \cdot \vec{r}_1 + u \cdot \vec{v}_1,$$

where:

$$f = 1 - \frac{r_2}{p}(1 - \cos\Delta\theta) = 1 - \frac{a}{r_1}(1 - \cos\Delta E);$$

$$g = \frac{r_1 r_2 \sin\Delta\theta}{\sqrt{KMp}} = \Delta t - \sqrt{\frac{a^3}{KM}}(\Delta E - \sin\Delta E);$$

$$q = \sqrt{\frac{KM}{p}}\tan\frac{\Delta\theta}{2}\left(\frac{1-\cos\Delta\theta}{p} - \frac{1}{r_1} - \frac{1}{r_2}\right) = -\frac{\sqrt{KMa}}{r_1 r_2}\sin\Delta E;$$

$$u = 1 - \frac{r_1}{p}(1 - \cos\Delta\theta) = 1 - \frac{a}{r_2}(1 - \cos\Delta E).$$

Here, $\Delta\theta$ is the variation of the projectile's true anomaly, corresponding to its evolution from point $\vec{r}_1$ to point $\vec{r}_2$; ΔE is the variation of the projectile's eccentric anomaly corresponding to its movement from point $\vec{r}_1$ to point $\vec{r}_2$; p is the parameter of the transfer conic (*semi-latus rectum*); and a is the geometric constant of the transfer conic.

In the four previous expressions, there are seven physical quantities: $\vec{r}_1, \vec{r}_2, \Delta\theta, \Delta t, p, a$ and ΔE. Among these, the first four are known, i.e., $\vec{r}_1, \vec{r}_2, \Delta\theta$ and Δt. Thus, to determine the other three unknown parameters (elements of the conic on which the projectile's transfer will take place, p, a and ΔE), we need to solve transcendental equations, which requires numerical methods.

To establish the relation between the true and the eccentric anomaly $(\theta; E)$, as well as the relation between the variations of the two anomalies $(\Delta\theta; \Delta E)$, as a consequence of the projectile's transfer between the two points, we use the images in Figures 34.27, 34.28 and 34.29.

Let's suppose that the transfer of the projectile M must take place on an elliptical orbit. From the equation of the projectile's elliptical orbit presented in Figure 34.27, the differentiation results in:

$$r = \frac{p}{1 + e\cos\theta},$$

where θ is the projectile's true anomaly;

$$e = \sqrt{1 - \frac{b^2}{a^2}} < 1; \quad p = \frac{b^2}{a}; \quad p = a(1 - e^2);$$

$$p = b\sqrt{1 - e^2}; \quad p = \frac{b^2}{a};$$

$$-\frac{1}{r^2}\mathrm{d}r = -\frac{e}{p}\sin\theta \mathrm{d}\theta;$$

$$\cos\theta = \frac{p}{e}\left(\frac{1}{r} - \frac{1}{p}\right); \quad \sin\theta = \sqrt{1 - \frac{p^2}{e^2}\left(\frac{1}{r} - \frac{1}{p}\right)^2};$$

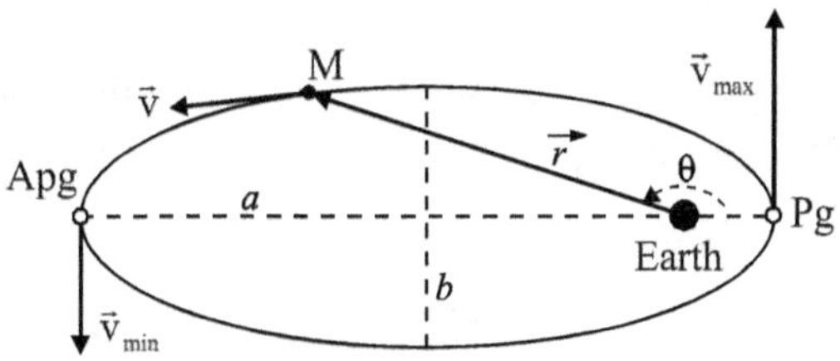

Fig. 34.27

$$\frac{p}{r^2}\mathrm{d}r = \sqrt{e^2 - \left(\frac{p}{r} - 1\right)^2}\,\mathrm{d}\theta;$$

$$r^2\mathrm{d}\theta = \frac{p\mathrm{d}r}{\sqrt{e^2 - \left(\frac{p}{r} - 1\right)^2}}.$$

On the other hand, knowing that $r^2\dot{\theta} = C$, it results in:

$$r^2\frac{\mathrm{d}\theta}{\mathrm{d}t} = C; \quad \mathrm{d}t = \frac{r^2\mathrm{d}\theta}{C};$$

$$\mathrm{d}t = \frac{p}{C}\frac{r\mathrm{d}r}{\sqrt{e^2r^2 - (p-r)^2}};$$

$$p = a(1-e^2);$$

$$\begin{aligned}
e^2\,r^2 - (p-r)^2 &= e^2\,r^2 - [a(1-e^2) - -r]^2 \\
&= e^2\,r^2 - a^2(1-e^2)^2 + 2a(1-e^2)r - r^2 \\
&= -r^2(1-e^2) - a^2(1-e^2)^2 + 2a(1-e^2)r \\
&= (1-e^2)[2ar - r^2 - a^2(1-e^2)] \\
&= (1-e^2)[a^2e^2 - (a^2 - 2ar + r^2)] \\
&= (1-e^2)[a^2e^2 - (a-r)^2];
\end{aligned}$$

$$\mathrm{d}t = \frac{p}{C}\frac{r\mathrm{d}r}{\sqrt{1-e^2}\sqrt{a^2e^2 - (a-r)^2}};$$

$$C = \sqrt{pKM}; \quad p = a(1-e^2);$$

$$\mathrm{d}t = \sqrt{\frac{a}{KM}}\frac{r\mathrm{d}r}{\sqrt{a^2e^2 - (a-r)^2}}.$$

Noting that:

$$a - r = ae\cos E,$$

where E is the *eccentrical anomaly* whose signification is presented in Figure 34.28, the result is:

$$r = a(1 - e\,\cos E);$$

$$dr = ae\,\sin E\mathrm{d}E;$$

$$\mathrm{d}t = a\sqrt{\frac{a}{KM}}(1 - e\,\cos E)\,\mathrm{d}E.$$

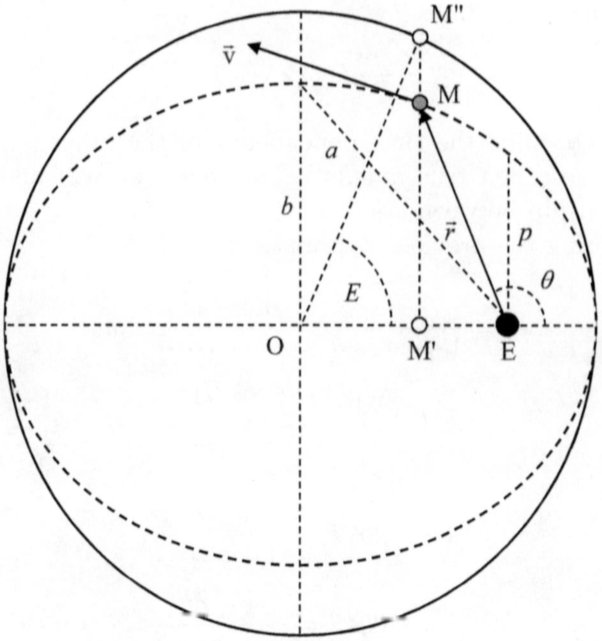

Fig. 34.28

By integration, from the $E = 0$ value corresponding to the projectile's passing through the perigee, where $r_{\min} = a(1 - e)$, we obtain:

$$t - t_0 = a\sqrt{\frac{a}{KM}} \int_0^E (1 - e \cos E)\, \mathrm{d}E,$$

where t_0 is the moment of the projectile's passage through perigee;

$$E - e \sin E = \frac{1}{a}\sqrt{\frac{KM}{a}}(t - t_0);$$

$$n = \frac{1}{a}\sqrt{\frac{KM}{a}}; \quad n = \frac{2\pi}{T},$$

where n represents the *medium movement* of the projectile;

$$E - e \sin E = n(t - t_0),$$

i.e., Kepler's equation, where $n(t - t_0)$ is the average anomaly.

Solving the previous equations, we determine the eccentric anomaly E of the projectile at moment t, and we then determine

r's dependency on time:

$$r = a(1 - e\cos E).$$

Let's determine the time dependency of the other polar coordinates, the so-called true anomaly. For these, we will first establish the relationship between θ and E.

Comparing the previous expressions,

$$r = \frac{p}{1 + e\cos\theta} = \frac{a(1-e^2)}{1+e\cos\theta} \quad \text{and}$$
$$r = a(1 - e\cos E),$$

results in:

$$1 - e\cos E = \frac{1-e^2}{1+e\cos\theta};$$

$$1 + e\cos\theta = \frac{1-e^2}{1-e\cos E};$$

$$\cos\theta = \frac{\cos E - e}{1 - e\cos E};$$

$$1 - \cos\theta = \frac{(1+e)(1-\cos E)}{1-e\cos E};$$

$$1 - e\cos E = \frac{r}{a}; \quad 1 - \cos\theta = 2\sin^2\frac{\theta}{2}; \quad 1 - \cos E = 2\sin^2\frac{E}{2};$$

$$\sin^2\frac{\theta}{2} = \frac{a(1+e)\sin^2\frac{E}{2}}{r};$$

$$1 + \cos\theta = \frac{(1-e)(1+\cos E)}{1-e\cos E};$$

$$1 - e\cos E = \frac{r}{a}; \quad 1 + \cos\theta = 2\cos^2\frac{\theta}{2}; \quad 1 + \cos E = 2\cos^2\frac{E}{2};$$

$$\cos^2\frac{\theta}{2} = \frac{a(1-e)\cos^2\frac{E}{2}}{r};$$

$$\sqrt{r}\sin\frac{\theta}{2} = \sqrt{a(1+e)}\sin\frac{E}{2}; \quad \sqrt{r}\cos\frac{\theta}{2} = \sqrt{a(1-e)}\cos\frac{E}{2};$$

$$\mathrm{tg}\frac{\theta}{2} = \sqrt{\frac{1+e}{1-e}}\,\mathrm{tg}\frac{E}{2},$$

representing the relation between the true and the eccentric anomaly of the projectile moving on an ellipse with the Earth in one of its foci.

To establish the geometrical interpretation of the eccentric anomaly E using Figure 34.28, where, besides the ellipse on which the projectile is moving around the Earth, we have also presented the circle with its center in the ellipse's center and its radius equal to the ellipse's semi-major axis (the eccentric circle), we write:

$$\mathrm{OM''} = a; \quad \mathrm{OM'} = \mathrm{OE} - \mathrm{EM'} = a \cdot \cos(\angle \mathrm{M'OM''});$$

$$\mathrm{OS} = a - r_{\min} = a - \frac{p}{1+e} = a - \frac{a(1-e^2)}{1+e} = ae = c;$$

$$\mathrm{EM'} = -r\cos\theta; \theta > 90^\circ; \quad \cos\theta < 0; \quad EM' > 0;$$

$$ae + r\cos\theta = a\cos(\angle \mathrm{M'OM''}).$$

Using the equation of the ellipse drawn in polar coordinates, the result is:

$$r = \frac{p}{1+e\cos\theta} = \frac{a(1-e^2)}{1+e\cos\theta};$$

$$1+e\cos\theta = \frac{a(1-e^2)}{r}; \quad e\cos\theta = \frac{a(1-e^2)}{r} - 1;$$

$$\cos\theta = \frac{a(1-e^2)}{er} - \frac{1}{e} = \frac{1}{r}\left[\frac{a(1-e^2)}{e} - \frac{r}{e}\right];$$

$$\cos\theta = \frac{1}{r}\left(\frac{a}{e} - ae - \frac{r}{e}\right);$$

$$r\cos\theta + ae = \frac{a-r}{e};$$

$$ae - r\cos\theta = a\cos(\angle \mathrm{M'OM''});$$

$$a - r = ae \cdot \cos(\angle \mathrm{M'OM''});$$

$$a - r = aecosE;$$

$$E = \angle(\mathrm{M'OM''}).$$

The projectile's movement on the elliptical orbit described by the position vector $\vec{r}$ is

$$v = \sqrt{KM\left(\frac{2}{r} - \frac{1}{a}\right)}.$$

Figure 34.29 presents the two anomalies.

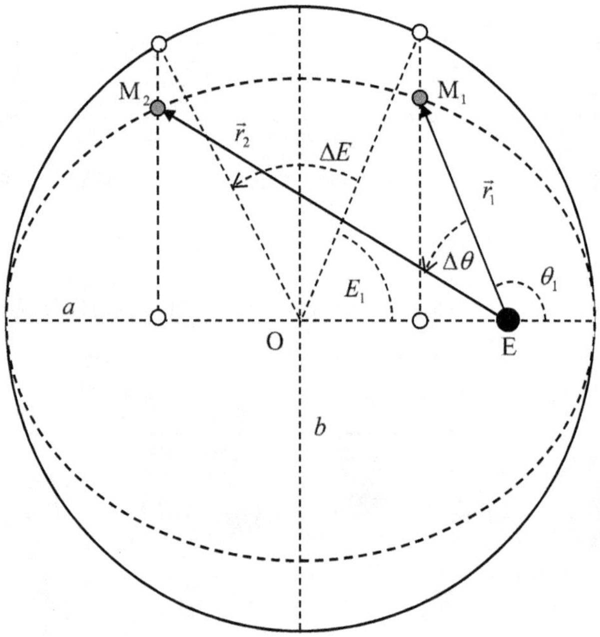

Fig. 34.29

Problem 35

104 Tauri Star

For the 104 Tauri star in the constellation Taurus in 2019, we know: the distance to Earth, $R = 50.3$ ly; the declination, $\delta = 18°38'.42$; the apparent magnitude, $m = 4.92$; the proper motion in right ascension, $\mu_\alpha = 534.73 \cdot 10^{-3}\,''/\text{year}$; the proper motion in declination, $\mu_\delta = 17.432 \cdot 10^{-3}\,''/\text{year}$; and the wavelength of $H\alpha$ radiation recorded on Earth, $\lambda = 656.045$ nm.

Under laboratory conditions, the same radiation has the wavelength $\lambda_0 = 656$ nm.

a) *Determine* the speed of the star in relation to the observer, $\vec{v}$.

b) *Establish* if the star has already passed through the position at which it is at a minimum distance from the Earth, or if it will pass through this position in the future.

Depending on the variant identified, *determine* the minimum distance, R_{min}, and *determine* when the star passed/will pass will pass through the position corresponding to the minimum distance from the Earth, Δt.

c) *Determine* the angle η between the direction of the tangent to the star's displacement and the tangent to the celestial meridian of the star.

d) *Determine* the apparent magnitude of the 104 Tauri star when it was, or will be, at a minimum distance from the Earth, m_0.

We know: the speed of light in vacuum, $c = 300\,000$ km/s; 1 ly $= 9.5 \cdot 10^{12}$ km; 1 year $= 31.536 \cdot 10^6$ s.

Solution

a) By comparing observations of the same stars made at very long time intervals, astronomers discovered important differences between the equatorial coordinates of the same star, thus highlighting that stars are not "fixed" in the celestial sphere.

As the drawing in Figure 35.1 indicates, in a time interval Δt, the displacement of a star σ on the celestial sphere is represented by the large circular arc $\sigma\sigma'$, corresponding to the two positions,

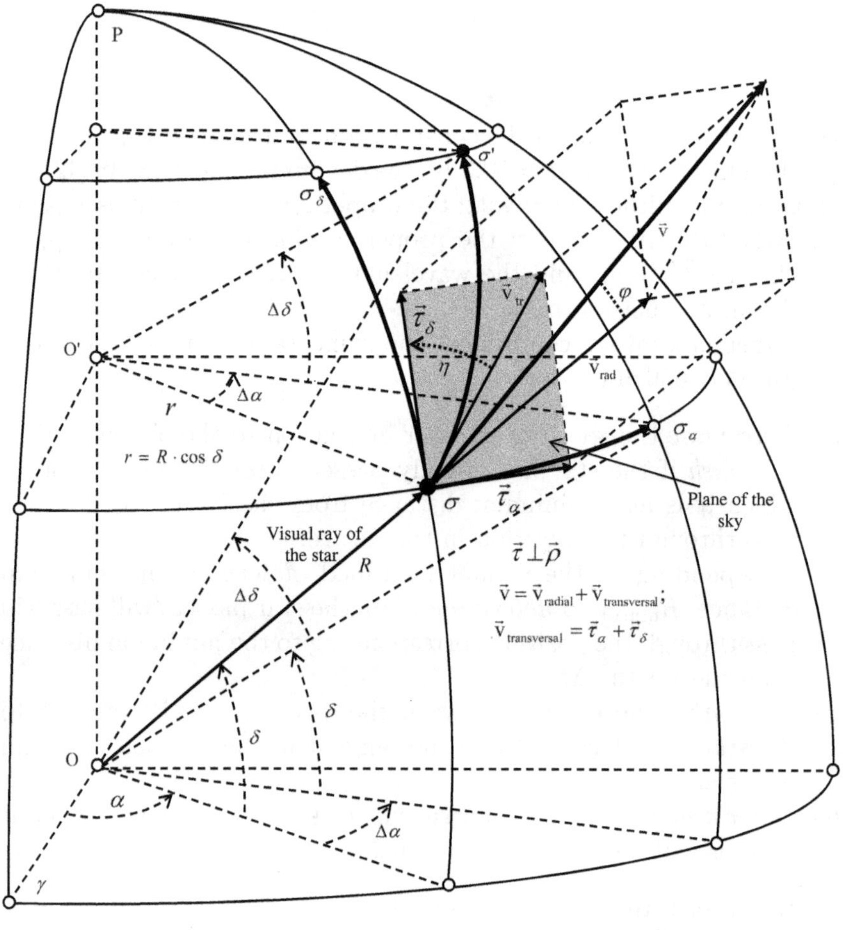

Fig. 35.1

and the equatorial coordinates of the star σ are:

$$\sigma(\alpha, \delta); \quad \sigma'(\alpha + \Delta\alpha, \delta + \Delta\delta).$$

Therefore, the total displacement $\sigma\sigma'$ is the result of the composition of the displacement in the right ascension (small circular arc $\sigma\sigma_\alpha$, with center at O$'$ and radius $R \cdot \cos\delta$) and the declination displacement (large circular arc $\sigma\sigma_\delta$).

In order to compare the displacements of the stars on the celestial sphere, the so-called *proper motion* of a star, μ, is defined as a size equal numerically to the angle under which, from the center of the heliocentric celestial sphere, the large circular arc is represented, indicating the displacement of the star within a year, and having as a unit of measure (arcseconds)/year, that is:

$$\mu = \frac{\angle(\sigma \mathrm{O} \sigma')}{\Delta t},$$

from which we recognize that "proper motion" refers to an angular velocity;

$$\mathrm{arc}(\sigma\sigma') = R \cdot \angle(\sigma\mathrm{O}\sigma\prime); \quad \angle(\sigma\mathrm{O}\sigma') = \frac{\mathrm{arc}(\sigma\sigma')}{R};$$

$$\mu = \frac{\mathrm{arc}(\sigma\sigma')}{R \cdot \Delta t};$$

$$\mathrm{arc}(\sigma\sigma') = \mu \cdot R \cdot \Delta t.$$

Corresponding to the movement of the star in the right ascension, the "proper ascension movement" of the star is defined as

$$\mu_\alpha = \frac{\angle(\sigma\mathrm{O}'\sigma_\alpha)}{\Delta t} = \frac{\Delta\alpha}{\Delta t},$$

where $\Delta\alpha$ (the variation in the right ascension of a star, expressed in seconds of time, because the right ascension of a star is expressed in units of time, $x^{\mathrm{h}}y^{\mathrm{m}}z^{\mathrm{s}}$) must be expressed in arcsec, based on the relation

$$1\mathrm{s}\ (\text{second of time}) = 15''\ (\text{seconds of arc}).$$

Thus, μ_α is expressed in (arcseconds)/year:

$$\mu_{\alpha(\text{arcsec/year})} = \frac{(\Delta\alpha)_{\text{arcsec}}}{(\Delta t)_{\text{years}}};$$

$$\text{arc}(\sigma\sigma_\alpha) = r \cdot \angle(\sigma\text{O}'\sigma_\alpha) = R\cos\delta \cdot \angle(\sigma\text{O}'\sigma_\alpha);$$

$$\Delta\alpha = \angle(\sigma\text{O}'\sigma_\alpha) = \frac{\text{arc}(\sigma\sigma_\alpha)}{R\cdot\cos\delta};$$

$$\mu_\alpha = \frac{\text{arc}(\sigma\sigma_\alpha)}{R\cdot\cos\delta\cdot\Delta t};$$

$$\text{arc}(\sigma\sigma_\alpha) = \mu_\alpha \cdot R \cdot \cos\delta \cdot \Delta t.$$

Corresponding to the movement of the star in declination, the "proper movement in declination" of the star is defined as:

$$\mu_\delta = \frac{\angle(\sigma\text{O}\sigma_\delta)}{\Delta t} = \frac{\Delta\delta}{\Delta t},$$

where $\Delta\delta$ (the variation in the star's declination) can be expressed in (arcseconds)/year as μ_δ is expressed in arcseconds/year;

$$\mu_\delta = \frac{(\Delta\delta)_{\text{arcsec}}}{(\Delta t)_{\text{years}}};$$

$$\text{arc}(\sigma\sigma_\delta) = R\cdot\angle(\sigma\text{O}\sigma_\delta); \quad \Delta\delta = \angle(\sigma\text{O}\sigma_\delta) = \frac{\text{arc}(\sigma\sigma_\delta)}{R};$$

$$\mu_\delta = \frac{\text{arc}(\sigma\sigma_\delta)}{R\cdot\Delta t};$$

$$\text{arc}(\sigma\sigma_\delta) = \mu_{\delta}. \cdot R \cdot \Delta t.$$

From Figure 35.1, we write the relation between the total displacement and the ascending and declining displacements of the star as follows:

$$\text{arc}(\sigma\sigma') = \sqrt{(\text{arc}(\sigma\sigma_\alpha))^2 + (\text{arc}(\sigma\sigma_\delta))^2};$$

$$\text{arc}(\sigma\sigma') = \mu R\Delta t; \quad \text{arc}(\sigma\sigma_\alpha) = \mu_\alpha R\cos\delta\Delta t;$$

$$\text{arc}(\sigma\sigma_\delta) = \mu_\delta\, R\,\Delta t;$$

$$\mu = \sqrt{(\mu_\alpha\cos\delta)^2 + \mu_\delta^2},$$

representing the relation between the proper motion and the proper motion in ascension and declination of the star, where μ_α and μ_δ are expressed in the same unit of measure, for example (arcsec/year), as in the problem statement.

The speed of the star σ, corresponding to the year 2019, is represented in the drawing in Figure 35.1 by the vector $\vec{v}$, tangent to the large circular arc $(\sigma\sigma')$, such that:

$$\vec{v} = \vec{v}_{\text{radial}} + \vec{v}_{\text{transversal}}; \quad v = \sqrt{v_{\text{rad}}^2 + v_{\text{tr}}^2};$$

$$\vec{v}_{\text{transversal}} = \vec{\tau}_\alpha + \vec{\tau}_\delta; \quad v_{\text{tr}} = \sqrt{\tau_\alpha^2 + \tau_\delta^2};$$

$$v = \sqrt{v_{\text{rad}}^2 + \tau_\alpha^2 + \tau_\delta^2}.$$

Because proper motion, μ, proper motion in ascention, μ_α, and proper motion in declination, μ_δ, refer to angular velocities, the result is:

$$\text{arc}(\sigma\sigma') = v_{\text{tr}} \cdot \Delta t = \mu \cdot R \cdot \Delta t;$$

$$\text{arc}(\sigma\sigma_\alpha) = \tau_\alpha \cdot \Delta t = \mu_\alpha \cdot r \cdot \Delta t = \mu_\alpha \cdot R \cdot \cos\delta \cdot \Delta t;$$

$$\text{arc}(\sigma\sigma_\delta) = \tau_\delta \cdot \Delta t = \mu_\delta \cdot R \cdot \Delta t.$$

Considering that the total displacement of the star is very small, from Figure 35.1, the result is:

$$\text{arc}\sigma\sigma' = \sqrt{(\text{arc}\sigma\sigma_\alpha)^2 + (\text{arc}\sigma\sigma_\delta)^2};$$

$$v_{\text{tr}} \cdot \Delta t = \sqrt{R^2 \cdot (\mu_\alpha \cdot \cos\delta)^2 \cdot (\Delta t)^2 + R^2 \cdot \mu_\delta^2 \cdot (\Delta t)^2};$$

$$v_{\text{tr}} \cdot \Delta t = R \cdot (\Delta t) \cdot \sqrt{(\mu_\alpha \cdot \cos\delta)^2 + \mu_\delta^2};$$

$$v_{\text{tr}} = R \cdot \sqrt{(\mu_\alpha \cdot \cos\delta)^2 + \mu_\delta^2};$$

$$\mu = \sqrt{(\mu_\alpha \cdot \cos\delta)^2 + \mu_\delta^2};$$

$$v_{\text{tr}} = R \cdot \mu;$$

$$\mu_\alpha = 534.73 \cdot 10^{-3}\,''/\text{year}; \quad \mu_\delta = 17.43 \cdot 10^{-3}\,''/\text{year};$$

$$\delta = 18^\circ 38'.42 = 18^\circ.64; \quad \cos\delta = 0.947;$$

$$\mu = \sqrt{(534.73 \cdot 0.947)^2 + (17.43)^2} \cdot 10^{-3}\,''/\text{year} = 506 \cdot 10^{-3}\,''/\text{year};$$

$$1\,\text{rad} = 206265\,\text{arcsec} = 206265''; \quad 1'' = \frac{1}{206265}\,\text{rad} = \frac{1}{206265};$$

$$\mu = 506 \cdot 10^{-3} \cdot \frac{1}{206265}\,\frac{1}{\text{year}};$$

$$R = 50.3\ \text{ly} = 50.3 \cdot 9.5 \cdot 10^{12}\ \text{km} = 477.85 \cdot 10^{12}\ \text{km};$$

$$\text{v}_{\text{tr}} = R \cdot \mu = 477.85 \cdot 10^{12} \cdot 506 \cdot 10^{-3} \cdot \frac{1}{206265}\,\frac{\text{km}}{\text{year}};$$

$$1\ \text{year} = 31.536 \cdot 10^6\ \text{s};$$

$$\text{v}_{\text{tr}} = 477.85 \cdot 10^{12} \cdot 506 \cdot 10^{-3} \cdot \frac{1}{206265} \cdot \frac{1}{31.536 \cdot 10^6}\,\frac{\text{km}}{\text{s}};$$

$$\text{v}_{\text{tr}} = \frac{477.85 \cdot 506}{206265 \cdot 31.536} \cdot 10^3\,\frac{\text{km}}{\text{s}};$$

$$\text{v}_{\text{tr}} = 37.171\ \frac{\text{km}}{\text{s}}.$$

According to the longitudinal Doppler effect:

$$\text{v}_{\text{rad}} = c \cdot \frac{\Delta\lambda}{\lambda_0} = c \cdot \frac{\lambda - \lambda_0}{\lambda_0};$$

$$c = 300000\ \frac{\text{km}}{\text{s}}; \quad \lambda = 656.045\,\text{nm}; \quad \lambda_0 = 656\,\text{nm};$$

$$\lambda > \lambda_0,$$

which indicates the radial movement of the star away from the observer;

$$\text{v}_{\text{rad}} = 300000 \cdot \frac{656.045 - 656}{656}\,\frac{\text{km}}{\text{s}} = 300000 \cdot \frac{0.045}{656}\,\frac{\text{km}}{\text{s}};$$

$$\text{v}_{\text{rad}} = 20.579\ \frac{\text{km}}{\text{s}}; \quad \text{v}_{\text{tr}} = 37.171\ \frac{\text{km}}{\text{s}};$$

$$\text{v} = \sqrt{\text{v}_{\text{rad}}^2 + \text{v}_{\text{tr}}^2} = \sqrt{(20.579)^2 + (31.171)^2}\ \frac{\text{km}}{\text{s}};$$

$$\text{v} = 37.351\ \frac{\text{km}}{\text{s}}.$$

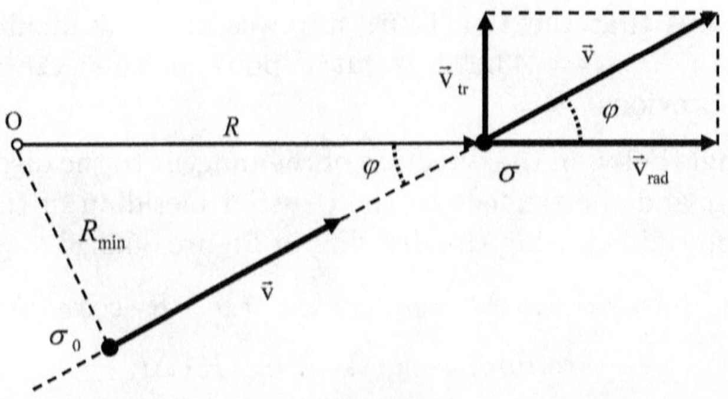

Fig. 35.2

b) Under these conditions, according to the drawing in Figure 35.2, the angle φ between the direction of the star's movement and the direction of the star's observation line is determined as follows:

$$\tan\varphi = \frac{\mathrm{v_{tr}}}{\mathrm{v_{rad}}};$$

$$\mathrm{v_{tr}} = 37.171\ \frac{\mathrm{km}}{\mathrm{s}}; \qquad \mathrm{v_{rad}} = 20.579\ \frac{\mathrm{km}}{\mathrm{s}},$$

$$\tan\varphi = \frac{37.171}{20.579} = 1.806;$$

$$\varphi = 61^\circ;$$

$$R_{\mathrm{min}} = R \cdot \sin\varphi = 50.3\ \mathrm{ly} \cdot \sin 61^\circ;$$

$$R_{\mathrm{min}} = 43.993\ \mathrm{ly};$$

$$\Delta t = \frac{\sigma_0\sigma}{\mathrm{v}} = \frac{R\cdot\cos\varphi}{\mathrm{v}} = \frac{50.3\cdot 9.5\cdot 10^{12}\ \mathrm{km}}{37.351\ \frac{\mathrm{km}}{\mathrm{s}}} = \frac{50.3\cdot 9.5\cdot 10^{12}}{37.351}\ \mathrm{s};$$

$$\Delta t = \frac{50.3\cdot 9.5\cdot 10^{12}}{37.351}\ \mathrm{s};$$

$$1\ \mathrm{year} = 31.536\cdot 10^6\ \mathrm{s};$$

$$\Delta t = \frac{50.3\cdot 9.5\cdot 10^{12}}{37.351}\cdot\frac{1}{31.536\cdot 10^6}\ \mathrm{years}$$

$$= \frac{50.3\cdot 9.5}{37.351\cdot 31.536}\cdot 10^6\ \mathrm{years};$$

$$\Delta t = 0.405\cdot 10^6\ \mathrm{years},$$

which proves that the 104 Tauri star was at a minimum distance from Earth, $R_{\text{min}} = 43.993$ ly, at a point in time $\Delta t = 0.405 \cdot 10^6$ years previous.

c) The angle between the direction of the tangent to the displacement of the star and the tangent to the celestial meridian of the star, η, can be determined using the drawing in Figure 35.3:

$$\text{arc}(\sigma\sigma_\alpha) = \tau_\alpha \cdot \Delta t = \mu_\alpha \cdot r \cdot \Delta t = \mu_\alpha \cdot R \cdot \cos\delta \cdot \Delta t;$$

$$\text{arc}(\sigma\sigma_\delta) = \tau_\delta \cdot \Delta t = \mu_\delta \cdot R \cdot \Delta t;$$

$$\frac{\tau_\alpha \cdot \Delta t}{\tau_\delta \cdot \Delta t} = \frac{\mu_\alpha \cdot R \cdot \cos\delta \cdot \Delta t}{\mu_\delta \cdot R \cdot \Delta t};$$

$$\frac{\tau_\alpha}{\tau_\delta} = \frac{\mu_\alpha \cdot \cos\delta}{\mu_\delta};$$

$$\tan\eta = \frac{\tau_\alpha}{\tau_\delta} = \frac{\mu_\alpha \cdot \cos\delta}{\mu_\delta};$$

$$\mu_\alpha = 534.73 \cdot 10^{-3}\,''/\text{year}; \quad \mu_\delta = 17.43 \cdot 10^{-3}\,''/\text{year};$$

$$\cos\delta = 0.947;$$

$$\tan\eta = \frac{534.73 \cdot 10^{-3} \cdot 0.947}{17.43 \cdot 10^{-3}} = 29.052;$$

$$\eta = 88^\circ.03.$$

d) According to Pogson's formula, it follows that:

$$\log\frac{E_0}{E} = -0.4(m_0 - m) = \log\frac{\frac{L}{4\pi R_{\text{min}}^2}}{\frac{L}{4\pi R^2}} = \log\left(\frac{R}{R_{\text{min}}}\right)^2$$

$$= 2 \cdot \log\frac{R}{R_{\text{min}}};$$

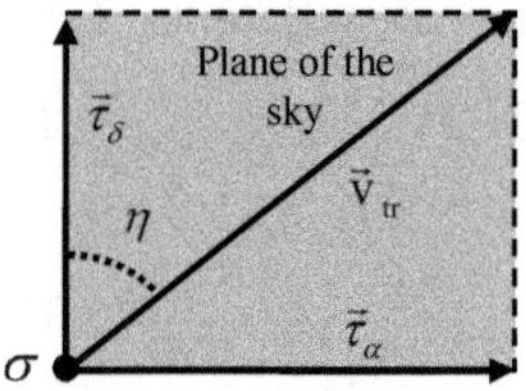

Fig. 35.3

$$m_0 = m - 5 \cdot \log \frac{R}{R_{\min}};$$

$$m = 4.92; \quad R = 50.3 \text{ ly}; \quad R_{\min} = 43.993 \text{ ly};$$

$$m_0 = 4.92 - 5 \cdot \log \frac{50.3}{43.993} = 4.92 - 0.05;$$

$$m_0 = 4.87 < m.$$

Problem 36

Saturn's Mass

When Saturn was observed with the help of a 2.5 m telescope (the Nordic Optical Telescope in Las Palmas, Canary Islands), a spectroscopic slit was placed over the image of the planet, as indicated by the drawing in Figure 36.1, where we notice that the width of the shadow projected by Saturn over its rings is very small.

Figure 36.2 indicates the positions of the following spectral lines of the solar radiation reaching the Earth observer after its reflection on the surface of the planet Saturn: 1) λ_{H} and λ_{Q}, which correspond to the radiation reflected by Saturn at the extreme points H (W) and Q (E), respectively, diametrically opposite on its equator; 2) λ_U and λ_V, which correspond to the radiation reflected by the inner ring of Saturn at the points U (W) and Z (E), respectively, diametrically opposite on the inner ring. In addition, two spectral lines from transitions D_1 and D_2 of Na I (neutral sodium) are indicated, having the wavelengths $\lambda_{01} = 589.00\,\mathrm{nm}$ and $\lambda_{02} = 589.60\,\mathrm{nm}$, respectively.

a) Saturn's ring is flat and in the plane of Saturn's equator. *Determine* the angle α between the observer's direction of view and Saturn's equatorial plane.

b) *Determine* the longitudinal scale of the spectrum, $S_{\text{longitudinal spectrum}}$.

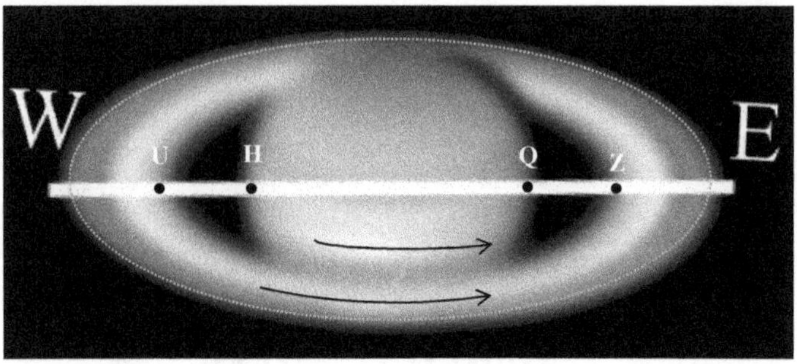

Fig. 36.1

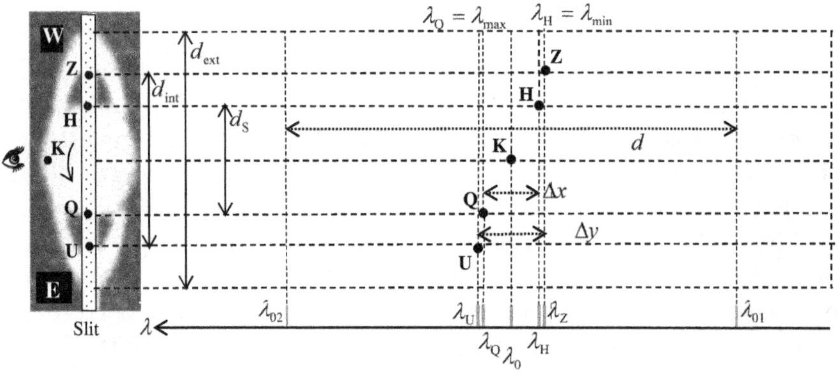

Fig. 36.2

c) The velocity scale, S_{speed}, is the ratio between the velocity of a reflecting point on Saturn's equator and the distance between the spectral lines, measured with a ruler, which correspond to two diametrically opposite reflecting points on Saturn's equator.

From the image of the given spectrum, *determine* the speed scale, S_{speed}. *Specify* the significance of the spectral line with the wavelength λ_0 and *determine* its value.

d) *Estimate* the length of Saturn's diameter, if the period of Saturn's rotation is $T_{\text{proper rotation, Saturn}} = 10.66\,\text{h}$.

e) *Determine* the transverse scale of the spectrum, $S_{\text{transversal spectrum}}$, the radius of the inner ring of Saturn,

$R_{\text{interior ring, Saturn}}$, and the radius of the outer ring of Saturn, $R_{\text{exterior ring, Saturn}}$.

f) *Estimate* the mass of the planet Saturn.

It is known that:

$$d_{\text{Earth–Sun}} = a_P = 1\text{ AU} = 1.496 \cdot 10^6\text{ km}; \quad M_{\text{Sun}} = 1.99 \cdot 10^{30}\text{ kg};$$

$$T_{\text{Earth}} = 1\text{ year} = 3.16 \cdot 10^7\text{ s}; \quad c = 3 \cdot 10^5\text{ km/}s.$$

Solution

a) Analyzing the drawing in Figure 36.1, from the statement of the problem, as well as the drawings in Figures 36.3 and 36.4, it is obvious that the observer's line of sight is not in the plane of Saturn's equator and, therefore, not in the plane of Saturn's rings. The angle between the direction of the observer's line of sight and the plane of Saturn's equator (Saturn's ring plane) is α. From the statement of the problem, the plane of Saturn's image and its rings (BCEF plane) is the projection of Saturn's equator plane (ABCD plane).

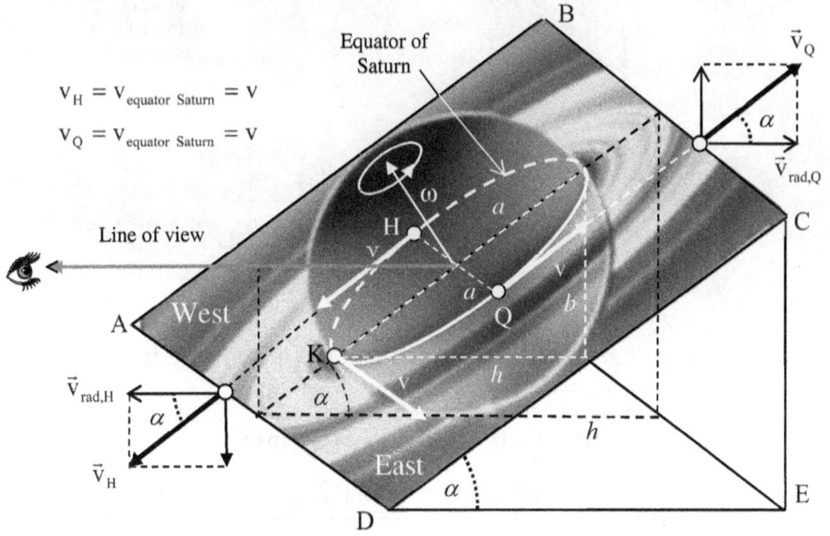

Fig. 36.3

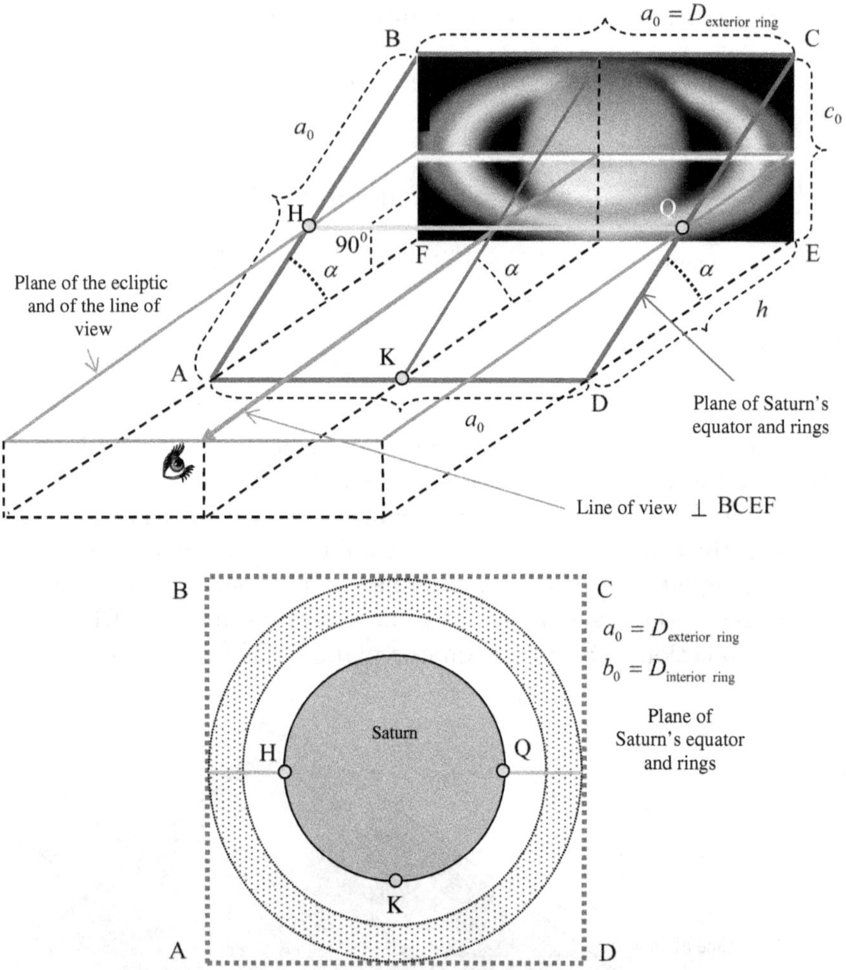

Fig. 36.4

After taking measurements on the image of the ring in Figure 36.1 (a_0 — large axis; c_0 — small axis), we find that:

$$a_0 = 116\ \text{mm}; \quad c_0 = 51\ \text{mm};$$

$$\cos\alpha = \frac{h}{a_0} = \frac{\sqrt{a_0^2 - c_0^2}}{a_0} = \frac{\sqrt{13456\ \text{mm}^2 - 2601\ \text{mm}^2}}{116\ \text{mm}}$$

$$= \frac{104\ \text{mm}}{116\ \text{mm}} \approx 0.90,$$

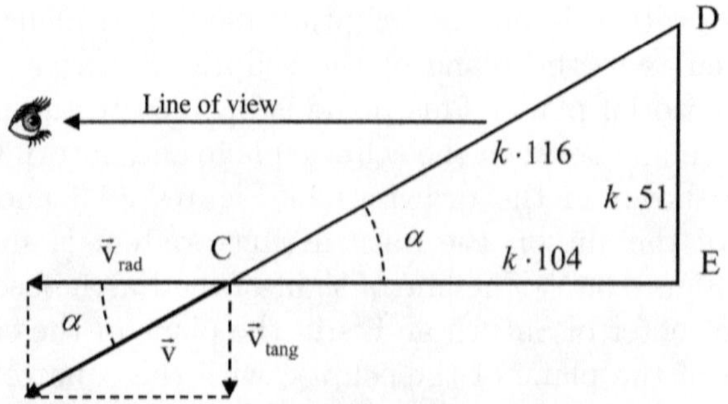

Fig. 36.5

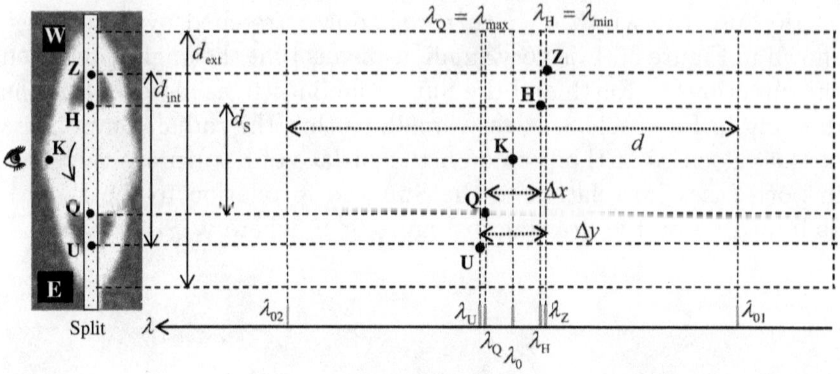

Fig. 36.6

so that the sides of the rectangular triangle CDE, regardless of their physical significance, have values directly proportional to the numbers noted on the drawing in Figure 36.5.

b) Now, we determine the longitudinal scale of the spectrum. The distance between the two spectral lines λ_{02} and λ_{01} measured with a ruler on the spectrum image in Figures 36.2 or 36.6 is $d = 85$ mm.

The difference in the wavelengths of the two spectral lines is

$$\Delta\lambda = \lambda_{02} - \lambda_{01} = 589.60 \text{ nm} - 589.00 \text{ nm} = 0.60 \text{ nm}.$$

It follows that the longitudinal scale of the spectrum is

$$S_{\text{longitudinal spectrum}} = \frac{\Delta\lambda}{d} = \frac{0.60 \text{ nm}}{85 \text{ mm}} \approx 0.00694 \, \frac{\text{nm}}{\text{mm}}.$$

c) The observatory is on the ecliptic plane. The plane of the slit is perpendicular to the plane of the ecliptic. Saturn's rings are in Saturn's equatorial plane. This plane is inclined to the plane of the ecliptic. The angle between the ecliptic plane and Saturn's equatorial plane is as shown in the drawings in Figures 36.3 and 36.4. The projection of the slit on the Saturn–rings system is shown in the drawing in Figure 36.7. The line of sight of the Earth observer passes through the center of Saturn and is in the plane of the ecliptic. The intersection of the plane of the ecliptic with the equatorial plane of Saturn is along a large circle on the surface of Saturn. As a result, half of each of Saturn's rings will be above the ecliptic, and the other half will be below the ecliptic plane.

Because the width of Saturn's shadow, stretched over its rings, shown in Figure 36.1, is very small, it means that the angle γ between the directions to Earth and the Sun, from Saturn, as indicated by the drawing in Figure 36.8, is very small, so that the radial components of the velocities of the points on the surface of Saturn are the same in both cases (in relation to the Sun and in relation to the Earth), as indicated by the drawing in Figure 36.9, where $v_{rad,Q} = v_{rad,H}$.

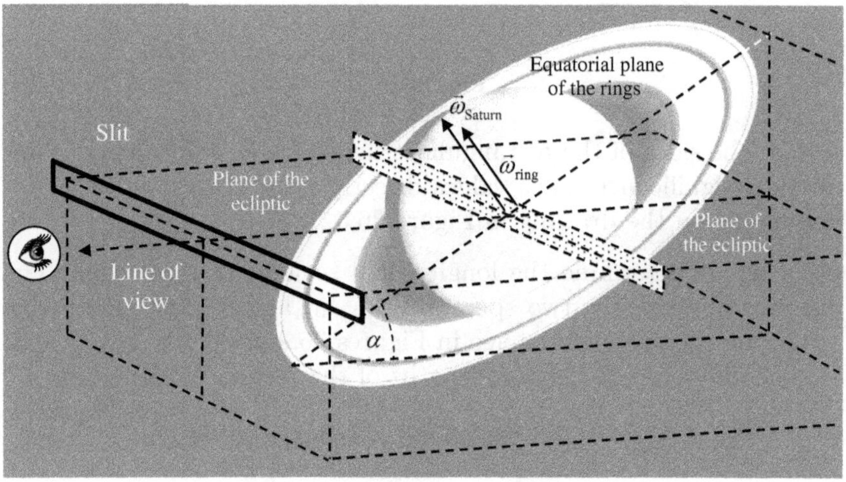

Fig. 36.7

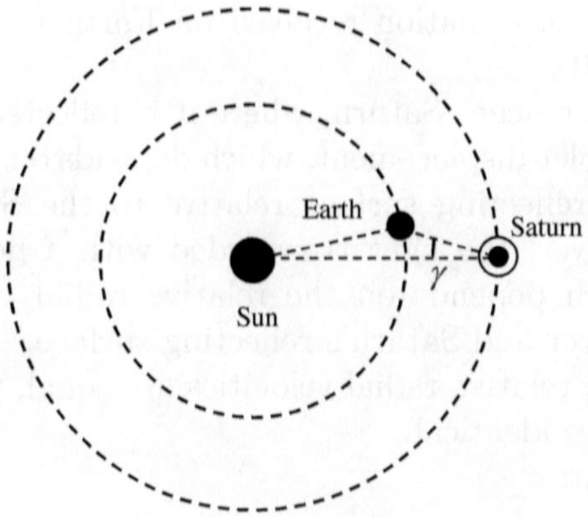

Fig. 36.8

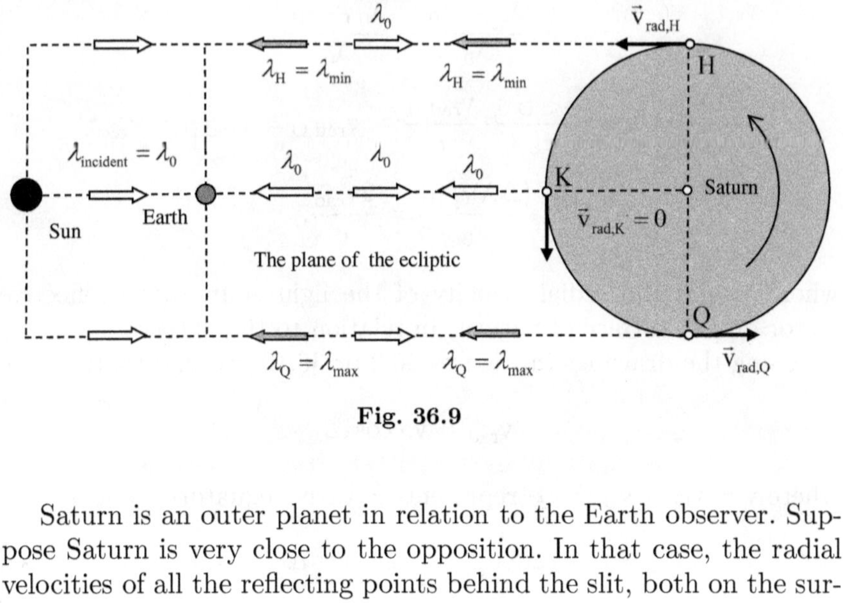

Fig. 36.9

Saturn is an outer planet in relation to the Earth observer. Suppose Saturn is very close to the opposition. In that case, the radial velocities of all the reflecting points behind the slit, both on the surface of Saturn and on the surface of each ring, in relation to the Sun and to the Earth, are equal, as the drawing in Figure 36.8 shows.

Saturn's orbital velocity around the Sun has no radial component.

Behind the slit, Saturn and its rings rotate, reflecting the light from the Sun. Since Saturn and its rings are rotating toward the Sun and the Earth, the radiation received on Earth will be affected by two Doppler shifts.

Sunlight first reaches Saturn, where it is reflected on its surface, with a first Doppler displacement, which depends on the radial velocity of Saturn's reflecting surface relative to the Sun. Arriving at the Earth observer, the light is recorded with a new Doppler displacement, which depends on the relative radial velocity between the Earth observer and Saturn's reflecting surface.

Since the two relative radial velocities are equal, the two Doppler displacements are identical.

It follows that:

$$\frac{(\Delta\lambda)_{\mathrm{Q-H}}}{\lambda_0} = \frac{(\Delta\lambda)_{\mathrm{W-E}}}{\lambda_0} = \frac{\lambda_{\mathrm{Q}} - \lambda_{\mathrm{H}}}{\lambda_0} = \frac{\lambda_{\mathrm{Q}} - \lambda_K + \lambda_K - \lambda_{\mathrm{H}}}{\lambda_0}$$

$$= \frac{\lambda_{\mathrm{Q}} - \lambda_0}{\lambda_0} + \frac{\lambda_0 - \lambda_{\mathrm{H}}}{\lambda_0};$$

$$\frac{(\Delta\lambda)_{\mathrm{Q-H}}}{\lambda_0} = \frac{\mathrm{v_{rad,Q}}}{c} + \frac{\mathrm{v_{rad,H}}}{c}; \quad \mathrm{v_{rad,Q}} = \mathrm{v_{rad,H}} = \mathrm{v_{rad}};$$

$$\frac{(\Delta\lambda)_{\mathrm{Q-H}}}{\lambda_0} = \frac{2\mathrm{v_{rad}}}{c},$$

where $\mathrm{v_{rad}}$ is the radial velocity of the light source (the reflective sector on the surface of Saturn) in relation to the observer.

Using the drawings in Figures 36.2 or 36.10, we find that:

$$\mathrm{v_{rad}} = \mathrm{v} \cdot \cos\alpha,$$

where $\mathrm{v} = \mathrm{v_Q} = \mathrm{v_H} = \mathrm{v}_K$ represents Saturn's equatorial velocity;

$$\frac{(\Delta\lambda)_{\mathrm{Q-H}}}{\lambda_0} = \frac{2\mathrm{v} \cdot \cos\alpha}{c};$$

$$\Delta x = d_{\mathrm{Q-H}} = 10 \text{ mm};$$

$$\frac{(\Delta\lambda)_{\mathrm{Q-H}}}{\lambda_0} = \frac{S_{\text{longitudinal spectrum}} \cdot d_{\mathrm{Q-H}}}{\lambda_0} = \frac{S_{\text{longitudinal spectrum}} \cdot \Delta x}{\lambda_0};$$

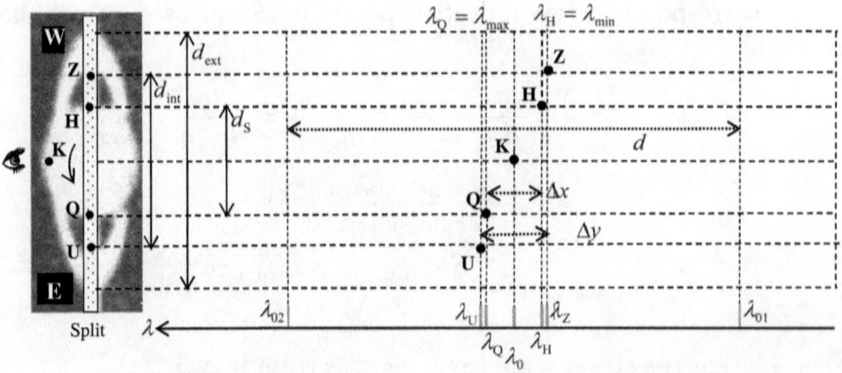

Fig. 36.10

$$S_{\text{longitudinal spectrum}} = 0.00694\,\frac{\text{nm}}{\text{mm}};$$

$$\frac{(\Delta\lambda)_{\text{Q-H}}}{\lambda_0} = \frac{2\text{v}\cdot\cos\alpha}{c};$$

$$\frac{2\text{v}_{\text{rad}}}{c} = \frac{2\text{v}\cdot\cos\alpha}{c} = \frac{S_{\text{longitudinal spectrum}}\cdot d_{\text{Q-H}}}{\lambda_0}$$

$$= \frac{S_{\text{longitudinal spectrum}}\cdot\Delta x}{\lambda_0};$$

$$\lambda_0 = \lambda_{0\text{m}} = \frac{\lambda_{01}+\lambda_{02}}{2} = 589.30\ \text{nm};$$

$$\lambda_0 = \frac{\lambda_{\text{max}}+\lambda_{\text{min}}}{2} = 589.3\ \text{nm},$$

representing the wavelength of the radiation reflected at point K on Saturn's equator, unaffected by the Doppler effect, λ_0, as well as the wavelength of the spectral line $\lambda_{0\text{m}}$, with respect to which the spectral lines λ_{01} and λ_{02} are symmetrical.

Defining the speed scale, S_{speed}, as the ratio between the equatorial velocity of the reflective points on Saturn's equator and the distance between the spectral lines measured with a ruler, which correspond to two reflective points on Saturn's equator, the result is:

$$S_{\text{speed(Q-H)}} = \frac{v_{\text{equator, Saturn}}}{d_{Q-H}} = \frac{v_Q}{d_{Q-H}} = \frac{v_H}{d_{Q-H}} = \frac{v}{\Delta x}$$
$$= \frac{S_{\text{longitudinal spectrum}} \cdot c}{2 \cdot \lambda_0 \cdot \cos\alpha};$$
$$S_{\text{speed(Q-H)}} = \frac{S_{\text{longitudinal spectrum}} \cdot c}{2 \cdot \lambda_0 \cdot \cos\alpha},$$

representing the speed scale from the spectrum image;

$$S_{\text{speed(Q-H)}} = \frac{0.00694\,\frac{\text{nm}}{\text{mm}} \cdot 3 \cdot 10^5\,\frac{\text{km}}{\text{s}}}{2 \cdot 589.3\ \text{nm} \cdot 0.90} = 1.96\,\frac{\text{km/s}}{\text{mm}} \approx 2\,\frac{\text{km/s}}{\text{mm}}.$$

d) According to the drawing in Figure 36.11, the difference in speed between the diametrically opposite points, H (W) and Q (E), on Saturn's equator is:

$$\Delta\vec{v}_{\text{Q-H,Saturn}} = \vec{v}_Q - \vec{v}_H;$$
$$\Delta v_{\text{Saturn,Q-H}} = v_Q + v_H = 2v_{\text{equator, Saturn}},$$

where $v_{\text{equator, Saturn}}$ is the velocity of a point on Saturn's equator.

Using the drawing in Figure 36.10, measuring the distance Δx between the spectral lines of the radiation reflected at the points

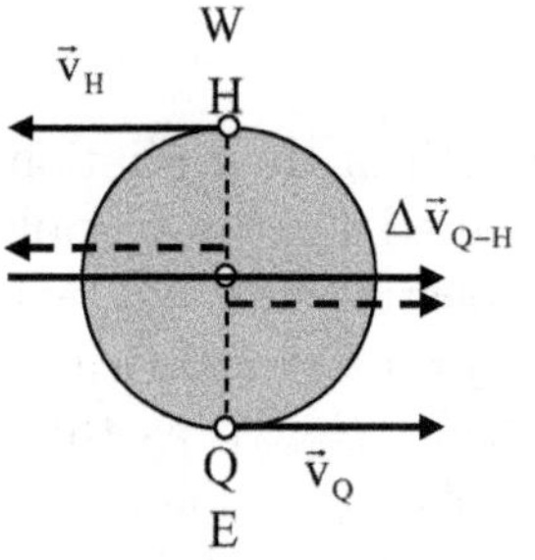

Fig. 36.11

E (H) and W (Q) on the surface of Saturn, having the wavelengths $\lambda_H = \lambda_{min}$ and $\lambda_Q = \lambda_{max}$, respectively, we find that:

$$\Delta x = 10 \text{ mm};$$

$$\Delta v_{\text{Saturn,Q-H}} = S_{\text{speed}} \cdot \Delta x = 1.96 \frac{\text{km/s}}{\text{mm}} \cdot 10 \text{ mm} = 19.6 \frac{\text{km}}{\text{s}};$$

$$v_{\text{equator, Saturn}} = \frac{1}{2} \cdot \Delta v_{\text{Saturn}} = 0.5 \cdot 19.6 \frac{\text{km}}{\text{s}};$$

$$v_{\text{equator, Saturn}} = 9.8 \frac{\text{km}}{\text{s}},$$

representing the velocity of a point on Saturn's equator;

$$T_{\text{proper rotation, Saturn}} = 10.66 \text{ h} = 38376 \text{ s};$$

$$L_{\text{equator circumference, Saturn}} = v_{\text{equator, Saturn}} \cdot T_{\text{proper rotation, Saturn}};$$

$$L_{\text{equator circumference, Saturn}} = 9.8 \frac{\text{km}}{\text{s}} \cdot 38376 \text{ s} = 376084.8 \text{ km}$$

$$= \pi D_{\text{Saturn}},$$

representing Saturn's equatorial circumference;

$$D_{\text{Saturn}} = \frac{L_{\text{equator circumference, Saturn}}}{\pi}$$

$$= \frac{376084.8}{3.14} \text{ km} \approx 119772.22 \text{ km},$$

representing the diameter of Saturn;

$$R_{\text{Saturn}} = \frac{D_{\text{Saturn}}}{2} \approx 59886 \text{ km},$$

representing the radius of Saturn.

e) In the image of the spectrum in the problem statement, Figure 36.2, or in the drawings in Figure 36.6 or 36.10, the diameter of Saturn, measured with a ruler, is $d_{\text{Saturn}} = 20$ mm, such that:

$$D_{\text{Saturn}} = 119772.22 \text{ km};$$

$$S_{\text{transversal spectrum}} = \frac{D_{\text{Saturn}}}{d_{\text{Saturn}}} = \frac{119772.22 \text{ km}}{20 \text{ mm}} \approx 5988.6 \ \frac{\text{km}}{\text{mm}},$$

representing the cross-sectional scale of the spectrum image from the problem statement.

In the same figures, measuring with a ruler the diameter of the inner ring, we find that

$$d_{\text{interior ring}} = 32 \text{ mm},$$

so the diameter of the inner circle of Saturn's ring is

$$D_{\text{interior ring}} = d_{\text{interior ring}} \cdot S_{\text{transversal spectrum}}$$
$$= 32 \text{ mm} \cdot 5988.6 \ \frac{\text{km}}{\text{mm}} = 191635.2 \text{ km},$$

and the radius of Saturn's inner ring is

$$R_{\text{interior ring}} = a_{\text{interior ring}} = \frac{D_{\text{interior ring}}}{2} = 95817.6 \text{ km}.$$

Similarly, for Saturn's outer ring, the result is:

$$d_{\text{exterior ring}} = 47 \text{ mm};$$

$$D_{\text{exterior ring}} = d_{\text{exterior ring}} \cdot S_{\text{transversal spectrum}}$$
$$= 47 \text{ mm} \cdot 5988.6 \ \frac{\text{km}}{\text{mm}} = 281464.2 \text{ km};$$

$$R_{\text{exterior ring}} = a_{\text{exterior ring}} = \frac{D_{\text{exterior ring}}}{2} = 140732.1 \text{ km}.$$

f) We have defined the speed scale, $S_{\text{speed(Q−H)}}$, as the ratio between the equatorial velocity of the reflective points H and Q on Saturn's equator and the distance between the spectral lines λ_Q and λ_H, measured with a ruler, which correspond to the two reflective points on Saturn's equator. This results in:

$$S_{\text{speed(Q-H)}} = \frac{\text{v}_{\text{equatorial, Saturn}}}{d_{\text{Q-H}}} = \frac{\text{v}_{\text{Q}}}{d_{\text{Q-H}}} = \frac{\text{v}_{\text{H}}}{d_{\text{Q-H}}} = \frac{\text{v}}{\Delta x}$$

$$= \frac{S_{\text{longitudinal spectrum}} \cdot c}{2 \cdot \lambda_0 \cdot \cos\alpha};$$

$$S_{\text{speed(Q-H)}} = \frac{S_{\text{longitudinal spectrum}} \cdot c}{2 \cdot \lambda_0 \cdot \cos\alpha},$$

representing the speed scale from the spectrum image;

$$S_{\text{speed(Q-H)}} = \frac{0.00694\ \frac{\text{nm}}{\text{mm}} \cdot 3 \cdot 10^5\ \frac{\text{km}}{\text{s}}}{2 \cdot 589.3\ \text{nm} \cdot 0.90} = 1.96\ \frac{\text{km/s}}{\text{mm}} \approx 2\ \frac{\text{km/s}}{\text{mm}}.$$

Using the spectrum image from Figure 36.2, 36.6, or 36.10, where we measure with a ruler the distance between the spectral lines λ_U and λ_Z, we find that

$$d_{\text{U-Z}} = \Delta y = 12\ \text{mm}.$$

Then, according to the drawing in Figure 36.12, the difference in speed between the diametrically opposite points, U (W) and Z (E), on the inner ring of Saturn is:

$$\Delta\vec{\text{v}}_{\text{U-Z,interior ring}} = \vec{\text{v}}_{\text{interior ring,U}} - \vec{\text{v}}_{\text{interior ring,Z}};$$

$$\Delta\text{v}_{\text{U-Z,interior ring}} = \text{v}_{\text{interior ring,U}} + \text{v}_{\text{interior ring,Z}} = 2\text{v}_{\text{interior ring}},$$

where $\text{v}_{\text{interior ring}}$ is the speed of a point on Saturn's inner ring.

The light from the Sun also reaches Saturn's inner ring, where it is reflected on its surface with a first Doppler displacement, which depends on the relative radial velocity of the inner ring's reflecting surface in relation to the Sun. Arriving at the Earth observer, the light is recorded with a new Doppler displacement, which depends on the relative radial velocity between the Earth observer and the reflecting surface of Saturn's inner ring.

As a result, the two Doppler shifts are identical.

It follows that:

$$\frac{(\Delta\lambda)_{U-Z}}{\lambda_0} = \frac{\lambda_U - \lambda_Z}{\lambda_0} = \frac{\lambda_U - \lambda_K + \lambda_K - \lambda_Z}{\lambda_0} = \frac{\lambda_U - \lambda_0}{\lambda_0} + \frac{\lambda_0 - \lambda_Z}{\lambda_0};$$

$$\frac{(\Delta\lambda)_{U-Z}}{\lambda_0} = \frac{v_{rad,U}}{c} + \frac{v_{rad,Z}}{c};$$

$$v_{rad,U} = v_{rad,Z} = v_{rad,\ interior\ ring} = v_U \cos\alpha = v_V \cos\alpha$$

$$= v_{interior\ ring} \cos\alpha;$$

$$\frac{(\Delta\lambda)_{U-Z}}{\lambda_0} = \frac{2v_{rad,\ interior\ ring}}{c} = \frac{2 \cdot v_{interior\ ring} \cdot \cos\alpha}{c},$$

where $v_{rad,\ interior\ ring}$ is the radial velocity of the light source (the reflective sector on the surface of Saturn's inner ring) relative to the observer.

Using the drawings in Figures 36.3 or 36.11, we find that:

$$v_{rad,\ interior\ ring} = v_{interior\ ring} \cdot \cos\alpha,$$

where $v_{interior\ ring} = v_U = v_Z$ represents the equatorial velocity of a point on Saturn's inner ring;

$$\frac{(\Delta\lambda)_{U-Z}}{\lambda_0} = \frac{2v_{interior\ ring} \cdot \cos\alpha}{c};$$

$$\Delta y = d_{U-Z} = 12\ \text{mm};$$

$$\frac{(\Delta\lambda)_{U-Z}}{\lambda_0} = \frac{S_{longitudinal\ spectrum} \cdot d_{U-Z}}{\lambda_0} = \frac{S_{longitudinal\ spectrum} \cdot \Delta y}{\lambda_0};$$

$$S_{longitudinal\ spectrum} = 0.00694\ \frac{\text{nm}}{\text{mm}};$$

$$\frac{(\Delta\lambda)_{U-Z}}{\lambda_0} = \frac{2v_{interior\ ring} \cdot \cos\alpha}{c}.$$

We now define the speed scale, $S_{speed(U-Z)}$, as the ratio between the equatorial velocity of the reflecting points U and Z on Saturn's inner ring and the distance between the spectral lines λ_U and λ_Z, measured with a ruler, which correspond to the two reflecting points on Saturn's inner ring. This results in:

$$S_{speed(U-Z)} = \frac{v_{interior\ ring}}{d_{U-Z}} = \frac{v_U}{d_{U-Z}} = \frac{v_Z}{d_{U-Z}} = \frac{v_{interior\ ring}}{\Delta y},$$

where $d_{U-Z} = \Delta y$, representing the distance between the spectral lines λ_U and λ_Z, measured in the drawings in Figure 36.3 or 36.11;

$$(\Delta\lambda)_{U-Z} = \lambda_U - \lambda_Z = S_{\text{longitudinal spectrum}} \cdot \Delta y;$$

$$\Delta y = \frac{(\Delta\lambda)_{U-Z}}{S_{\text{longitudinal spectrum}}};$$

$$\frac{(\Delta\lambda)_{U-Z}}{\lambda_0} = \frac{2v_{\text{interior ring}} \cdot \cos\alpha}{c};$$

$$S_{\text{speed(U-Z)}} = \frac{v_{\text{interior ring}}}{\Delta y} = \frac{v_{\text{interior ring}}}{\frac{(\Delta\lambda)_{U-Z}}{S_{\text{longitudinal spectrum}}}}$$

$$= \frac{v_{\text{interior ring}}}{(\Delta\lambda)_{U-Z}} \cdot S_{\text{longitudinal spectrum}};$$

$$\frac{(\Delta\lambda)_{U-Z}}{\lambda_0} = \frac{2v_{\text{interior ring}} \cdot \cos\alpha}{c};$$

$$(\Delta\lambda)_{U-Z} = \frac{2v_{\text{interior ring}} \cdot \cos\alpha \cdot \lambda_0}{c};$$

$$S_{\text{speed(U-Z)}} = \frac{v_{\text{interior ring}}}{\frac{2v_{\text{interior ring}} \cdot \cos\alpha \cdot \lambda_0}{c}} \cdot S_{\text{longitudinal spectrum}};$$

$$S_{\text{speed(U-Z)}} = \frac{c \cdot S_{\text{longitudinal spectrum}}}{2\lambda_0 \cos\alpha};$$

$$S_{\text{speed(H-Q)}} = \frac{S_{\text{longitudinal spectrum}} \cdot c}{2 \cdot \lambda_0 \cdot \cos\alpha};$$

$$S_{\text{speed(U-Z)}} = S_{\text{speed(H-Q)}};$$

$$S_{\text{speed(U-Z)}} = S_{\text{speed(H-Q)}} = 1.96\ \frac{\text{km/s}}{\text{mm}} \approx 2\ \frac{\text{km/s}}{\text{mm}}.$$

According to the drawing in Figure 36.12, the difference in speed between the diametrically opposite points, U and Z, on the inner ring of Saturn is:

$$\Delta\vec{v}_{U-Z,\ \text{interior ring}} = \vec{v}_U - \vec{v}_Z;$$

$$\Delta v_{U-Z} = v_U + v_Z = 2v_{\text{interior ring}},$$

where $v_{\text{interior ring}}$ is the speed of a point on Saturn's inner ring.

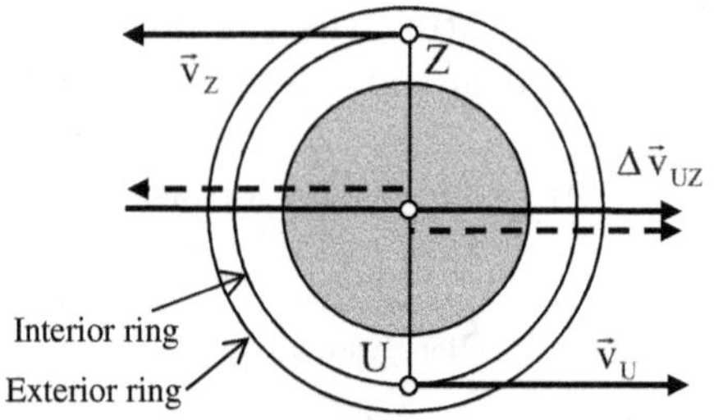

Fig. 36.12

Using the drawing in Figure 36.10, measuring the distance Δy between the spectral lines of the radiation reflected at points U and Z of the inner ring of Saturn, having the wavelengths λ_U and λ_Z, respectively, we find that:

$$\Delta y = 12 \text{ mm};$$

$$S_{\text{speed(U-Z)}} = \frac{v_{\text{interior ring,U-Z}}}{\Delta y};$$

$$v_{\text{interior ring,U-Z}} = S_{\text{speed(U-Z)}} \cdot \Delta y = 1.96 \ \frac{\text{km/s}}{\text{mm}} \cdot 12 \text{ mm}$$

$$= 23.52 \ \frac{\text{km}}{\text{s}};$$

$$2 \cdot v_{\text{interior ring}} = \Delta v_{\text{interior ring,U-Z}} = 47.04 \ \frac{\text{km}}{\text{s}};$$

$$v_{\text{interior ring}} = 23.52 \ \frac{\text{km}}{\text{s}},$$

representing the velocity of a point on Saturn's inner ring.

The length of the circumference of the inner ring of Saturn, with the diameter $D_{\text{interior ring}} = 194629.5$ km, is

$$L_{\text{interior ring}} = \pi D_{\text{interior ring}} \approx 611136.63 \text{ km},$$

so that the period of rotation of a point on the inner circle of Saturn's ring is:

$$T_{\text{interior ring}} = \frac{L_{\text{interior ring}}}{v_{\text{interior ring}}} = \frac{\pi D_{\text{interior ring}}}{\frac{1}{2} \cdot \Delta v_{\text{interior ring}}};$$

$$T_{\text{interior inel}} = \frac{3.14 \cdot 194629.5 \text{ km}}{23.52 \ \frac{\text{km}}{\text{s}}} = \frac{611136.63 \text{ km}}{23.52 \ \frac{\text{km}}{\text{s}}};$$

$$T_{\text{interior ring}} \approx 25983.70 \text{ s};$$
$$T_{\text{Saturn}} = 10.66 \text{ h} = 38376 \text{ s};$$
$$T_{\text{interior ring}} < T_{\text{Saturn}}.$$

From Kepler's third law, it follows that:

$$T_{\text{Earth}}^2 = 4\pi^2 \frac{a_{\text{Earth}}^3}{GM_{\text{Sun}}}; \quad T_{\text{interior ring}}^2 = 4\pi^2 \frac{a_{\text{interior ring}}^3}{GM_{\text{Saturn}}};$$

$$\left(\frac{T_{\text{interior ring}}}{T_{\text{Earth}}}\right)^2 = \frac{\left(\frac{a_{\text{interior ring}}}{a_{\text{Earth}}}\right)^3}{\frac{M_{\text{Saturn}}}{M_{\text{Sun}}}};$$

$$M_{\text{Saturn}} = \frac{\left(\frac{a_{\text{interior ring}}}{a_{\text{Earth}}}\right)^3}{\left(\frac{T_{\text{interior ring}}}{T_{\text{Earth}}}\right)^2} \cdot M_{\text{Sun}};$$

$$a_{\text{interior ring}} = 95817.6 \text{ km}; \quad a_{\text{Earth}} = 1 \text{ AU} = 1496 \cdot 10^5 \text{ km};$$
$$T_{\text{interior ring}} \approx 25983.70 \text{ s}; \quad T_{\text{Earth}} = 1 \text{ year} = 3.16 \cdot 10^7 \text{ s};$$
$$M_{\text{Sun}} = 1.99 \cdot 10^{30} \text{ kg};$$

$$M_{\text{Saturn}} = \frac{\left(\frac{a_{\text{interior ring}}}{a_{\text{Earth}}}\right)^3}{\left(\frac{T_{\text{interior ring}}}{T_{\text{Earth}}}\right)^2} \cdot M_{\text{Sun}};$$

$$M_{\text{Saturn}} = \frac{\left(\frac{95817.6}{1496 \cdot 10^5}\right)^3}{\left(\frac{25983.70}{3.16 \cdot 10^7}\right)^2} \cdot 1.99 \cdot 10^{30} \text{ kg};$$

$$M_{\text{Saturn}} = 7.73 \cdot 10^{26} \text{ kg}.$$

Problem 37

Binary Stellar Systems

For three visual binary star systems, in which the secondary components, S_1, evolve on relative elliptical orbits around the primary components, S_2, the data in Table 37.1 are known, where: α_{12} – the angular value of the large semiaxis, a_{12}, of the relative orbit, seen from the center of the Sun, as indicated by the drawing in Figure 37.1, expressed in seconds of arc, at a time when the secondary star is in the plane of the sky, on the line of nodes, and the distance between the two components is equal to the large semi-axis of the relative ellipse; τ – the period of rotation of the secondary star around the main star, expressed in years; M_1 – the mass of the secondary component of the binary star system, expressed in solar masses, M_S; M_2 – the mass of the main component of the binary star system, expressed in solar masses, M_S.

a) *Determine* the annual parallax of the main stars, p_{year}.

b) *Determine* the distances between the Sun and the main stars of the binary star systems, Δ_0.

c) *Determine* the large and small semi-axes of each relative orbit, a_{12} and b_{12}.

d) For each stellar system, we know the angle λ in the plane of the sky, between the direction of the node line, N_aN_d, and the line of the apsides AP, as well as e, the eccentricity of each relative orbit.

Table 37.1 Observational data.

No.	Binary stellar system	α_{12} (arcsec)	τ (years)	M_1 (M_S)	M_2 (M_S)	e	λ (degree)
1	Sirius	7.62	49.94	0.98	2.28	0.591	5
2	Procyon	4.55	40.65	0.65	1.76	0.407	6
3	η Cass.	11.09	480.00	0.47	0.73	0.497	8

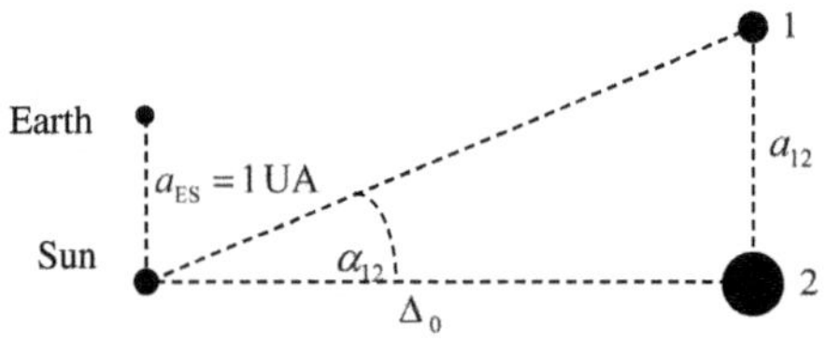

1 rad = 206265 arcsec
1 pc = 206265 AU
1 ly = 63240 AU

Fig. 37.1

For each binary stellar system, *determine*: the angle ω between the line of apsides, AP, and the line of nodes, N_aN_d; the length of the segment $A'P' = d_{12}$, representing the projection of the line of apsides, $AP = 2a_{12}$, in the plane of the sky, Π_s.

Solution

a) We write Kepler's third law for the binary system Earth–Sun and for a binary star system (S_1 – secondary star; S_2 – main star) as follows:

$$\frac{T^2(M_E + M_S)}{a_{ES}^3} = \frac{4\pi^2}{G},$$

where: T – period of rotation of the Earth around the Sun; M_S – the mass of the Sun; M_E – the mass of the Earth; a_{ES} – the large semi-axis of the Earth's relative orbit around the Sun; G – the gravitational attraction constant;

$$\frac{\tau^2(M_1 + M_2)}{a_{12}^3} = \frac{4\pi^2}{G},$$

where: τ – the period of rotation of the secondary star around the main star; M_1 – the mass of the secondary star; M_2 – the mass of the main star; a_{12} – the large semi-axis of the relative orbit of the secondary star around the main star.

From the two expressions, the result is:

$$M_1 + M_2 = \frac{T^2}{\tau^2} \cdot \frac{a_{12}^3}{a_{\mathrm{ES}}^3} \cdot (M_{\mathrm{E}} + M_{\mathrm{S}});$$

$$M_{\mathrm{E}} \ll M_{\mathrm{S}};$$

$$M_1 + M_2 = \frac{T^2}{\tau^2} \cdot \frac{a_{12}^3}{a_{\mathrm{ES}}^3} \cdot M_{\mathrm{S}},$$

representing the total mass of the binary star system, expressed in solar masses.

Under these conditions, according to the notation in Figures 37.2 and 37.3, where the annual stellar parallax is shown, p_{year}, the

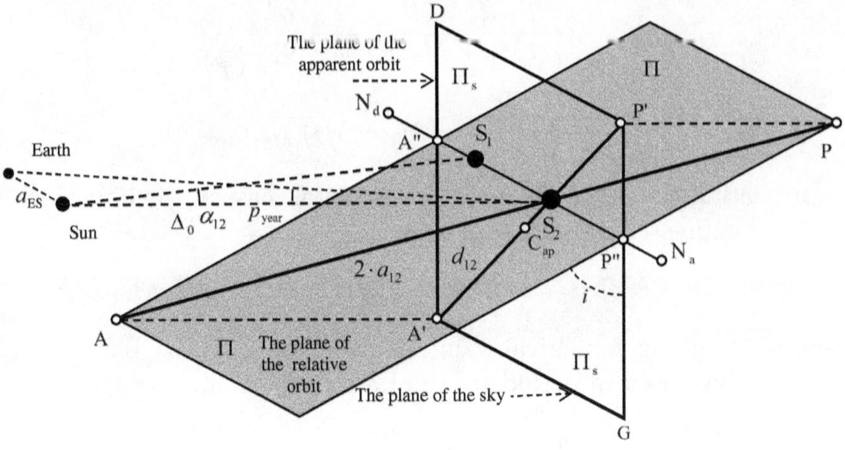

Fig. 37.2

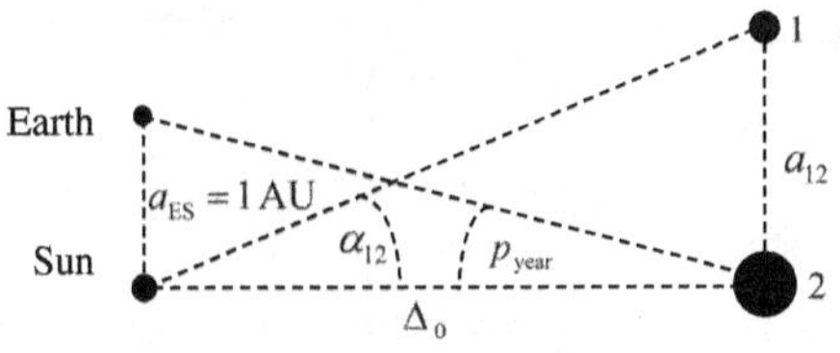

Fig. 37.3

following results:

$$\tan\alpha_{12} = \frac{a_{12}}{\Delta_0}; \quad a_{12} \ll \Delta_0; \quad \tan\alpha_{12} \approx \alpha_{12};$$

$$a_{12} \approx \alpha_{12} \cdot \Delta_0;$$

$$\tan p_{\text{year}} = \frac{a_{\text{ES}}}{\Delta_0}; \quad a_{\text{ES}} \ll \Delta_0; \quad \tan p_{\text{year}} \approx p_{\text{year}};$$

$$a_{\text{ES}} \approx p_{\text{year}} \cdot \Delta_0;$$

$$M_1 + M_2 = \frac{T^2}{\tau^2} \cdot \frac{a_{12}^3}{a_{\text{ES}}^3} \cdot M_{\text{S}};$$

$$M_1 + M_2 = \frac{T^2}{\tau^2} \cdot \frac{\alpha_{12}^3 \cdot \Delta_0^3}{p_{\text{year}}^3 \cdot \Delta_0^3} \cdot M_{\text{S}};$$

$$M_1 + M_2 = \frac{T^2}{\tau^2} \cdot \frac{\alpha_{12}^3}{p_{\text{year}}^3} \cdot M_{\text{S}};$$

$$p_{\text{year}}^3 = \frac{T^2\alpha_{12}^3 M_{\text{S}}}{\tau^2(M_1 + M_2)} = \frac{\alpha_{12}^3}{\frac{M_1+M_2}{M_{\text{S}}} \cdot \left(\frac{\tau}{T}\right)^2};$$

$$\frac{M_1 + M_2}{M_{\text{S}}} = |(M_1 + M_2)_{(M_{\text{S}})}|,$$

representing the numerical value of the total mass of the binary star system, expressed in solar masses;

$$\frac{\tau}{T} = |\tau_{(\text{years})}|,$$

representing the numerical value of the period of the binary star system, expressed in periods of the Earth's rotation, $T = 1$ year;

$$p_{\text{year}}^3 = \frac{\alpha_{12}^3}{|M_1 + M_2| \cdot (|\tau|)^2};$$

$$p_{\text{year(rad)}} = \frac{\alpha_{12(\text{rad})}}{\sqrt[3]{|(M_1 + M_2)_{(M_{\text{S}})}| \cdot |\tau_{(\text{years})}|^2}};$$

$$1\,\text{rad} = 206265\,\text{arcsec};$$

$$p_{\text{year(arcsec)}} = \frac{\alpha_{12(\text{arcsec})}}{\sqrt[3]{|(M_1 + M_2)_{(M_{\text{S}})}| \cdot |\tau_{(\text{years})}|^2}}.$$

Table 37.2 Observational Data.

No.	Binary stellar system	p_{year} (arcsec)	p_{year} (rad)
1	Sirius	0.379	0.379/206265
2	Procyon	0.286	0.286/206265
3	η Cass.	0.170	0.170/206265

The results of the calculations are presented in Table 37.2.

b) From Figure 37.3, it follows that:

$$\tan p_{year} = \frac{1\,\text{AU}}{\Delta_{0(\text{AU})}} \approx p_{year(rad)} = \frac{\frac{1\,\text{pc}}{206265}}{\Delta_{0(pc)}} = \frac{\frac{1\,\text{pc}}{206265}}{|\Delta_{0(pc)}| \cdot 1\,\text{pc}}$$

$$= \frac{1}{206265 \cdot |\Delta_{0(pc)}|} = |p_{year(rad)}| \cdot 1\,\text{rad} = |p_{year(rad)}|;$$

$$\frac{1}{206265 \cdot |\Delta_{0(pc)}|} = |p_{year(rad)}|;$$

$$\frac{1}{|\Delta_{0(pc)}|} = |p_{year(rad)}| \cdot 206265;$$

$$p_{year(rad)} = |p_{year(rad)}| \cdot 1\,\text{rad} = |p_{year(rad)}| \cdot 206265\,\text{arcsec}$$
$$= |p_{year(arcsec)}| \cdot \text{arcsec};$$

$$|p_{year(rad)}| \cdot 206265 = |p_{year(arcsec)}|;$$

$$|p_{year(rad)}| = \frac{|p_{year(arcsec)}|}{206265};$$

$$\frac{1}{206265 \cdot |\Delta_{0(pc)}|} = |p_{year(rad)}|;$$

$$\frac{1}{206265 \cdot |\Delta_{0(pc)}|} = \frac{|p_{year(arcsec)}|}{206265};$$

$$|p_{year(arcsec)}| = \frac{1}{|\Delta_{0(pc)}|}; \quad |\Delta_{0(pc)}| = \frac{1}{|p_{year(arcsec)}|};$$

$$\left|\Delta_{0(\mathrm{pc})}\right| \cdot \left|p_{\mathrm{year(arcsec)}}\right| = 1;$$

$$a_{\mathrm{ES}} \approx p_{\mathrm{year}} \Delta_0;$$

$$1\,\mathrm{rad} = 206265\,\mathrm{arcsec}; \quad 1\,\mathrm{arcsec} = \frac{1}{206265}\,\mathrm{rad};$$

$$\Delta_{0(\mathrm{AU})} = \frac{a_{\mathrm{ES(AU)}}}{p_{\mathrm{year(rad)}}} = \frac{1\,\mathrm{AU}}{p_{\mathrm{year(rad)}}},$$

the numerical results being listed in Table 37.3.

c)

$$a_{12} \approx \alpha_{12} \Delta_0; \quad a_{12(\mathrm{AU})} = \alpha_{12(\mathrm{rad})} \cdot \Delta_{0(\mathrm{AU})};$$

$$e = \sqrt{1 - \frac{b_{12}^2}{a_{12}^2}};$$

$$b_{12} = a_{12} \cdot \sqrt{1 - e^2},$$

the numerical results being listed in Table 37.4.

Table 37.3 Observational data.

No.	Binary stellar system	p_{year} (arcsec)	p_{year} (rad)	Δ_0 (pc)	Δ_0 (AU)	Δ_0 (ly)
1	Sirius	0.379	0.379/206265	2.638	544127.070	8.604
2	Procyon	0.286	0.286/206265	3.496	721102.440	11.402
3	η Cass.	0.170	0.170/206265	5.882	1213250.730	19.184

Table 37.4 Observational data.

No.	Binary stellar system	α_{12} (arcsec)	α_{12} (rad)	e	Δ_0 (AU)	a_{12} (AU)	b_{12} (AU)
1	Sirius	7.62	7.62/206265	0.591	544127.070	20.101	16.214
2	Procyon	4.55	4.55/206265	0.407	721102.440	15.906	14.528
3	η Cass.	11.09	11.09/206265	0.497	1213250.732	65.231	56.604

d) In the drawing in Figure 37.4, we note: S_1 – the secondary star; S_2 – the main star; A – the apastron (of the star S_1 on the ellipse E in relation to the star S_2 in the focus of the ellipse); P – the peristron (of the star S_1 on the ellipse E with respect to the star S_2); AP – the apsides line (large axis of the ellipse, representing the relative orbit); $A'P' = d_{12}$ – the projection of the apsides line, $AP = 2a_{12}$, in the sky plane, Π_s; N_aN_d – the line of nodes; i – the angle between the plane of the relative orbit and the plane of the sky; Π_s; ω – the angle between the line of apsides, AP, and the line of the nodes, N_aN_d; $\Pi(ABPC)$ – the plane of the relative orbit; $\Pi_s(DP'GA')$ – the plane of the sky (the plane of the apparent orbit); $2a_{12}$ – the distance between the apastron and the periastron (the large axis of the relative ellipse); $A''P$ – the projection of the large axis of the orbit (the ellipse) relative to the line of the nodes.

The relative orbit of the star S_1 in relation to the star S_2 is an ellipse (E) located in the plane $\Pi(ABPC)$.

The apparent orbit of the star S_1 is an ellipse, E_{ap}, located in the plane of the sky, $\Pi_s(DP'GA')$ The ellipse E_{ap} is the projection of the ellipse E.

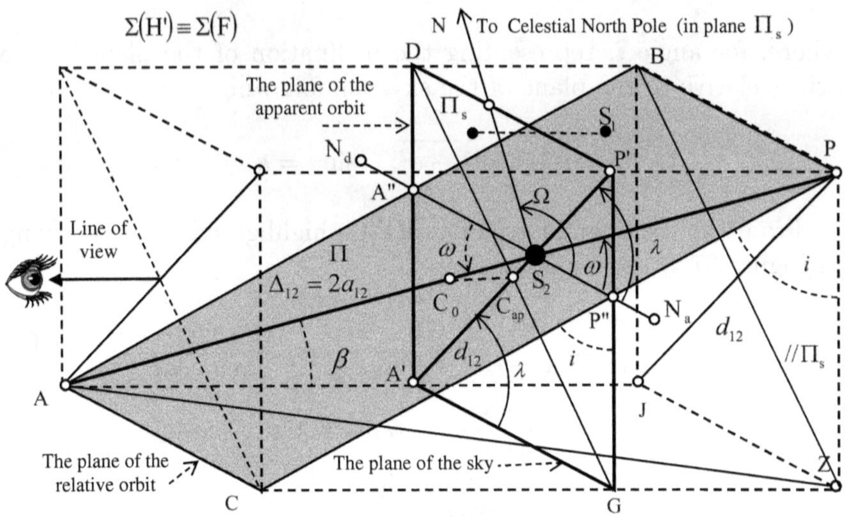

Fig. 37.4

From Figure 37.4, it follows that:

$$A\Sigma = r_{\max} = a(1+e); \quad \mathrm{AA}' = r_{\max}\cos\alpha;$$
$$\mathrm{P}\Sigma = r_{\min} = a(1-e); \quad \mathrm{JA}' = r_{\min}\cos\alpha;$$
$$\mathrm{AA}' + \mathrm{JA}' = \mathrm{AJ} = (r_{\max} + r_{\min})\cos\alpha = 2a_{12}\cdot\cos\alpha;$$
$$\mathrm{AJ} = 2a_{12}\cdot\cos\alpha = \mathrm{CZ};$$
$$\mathrm{CZ} = \mathrm{CP}\cdot\sin i = \mathrm{AB}\cdot\sin i = \mathrm{AP}\cdot\sin\omega\cdot\sin i;$$
$$\mathrm{AP} = 2a_{12};$$
$$\mathrm{CZ} = 2a_{12}\cdot\sin\omega\cdot\sin i;$$
$$\mathrm{CZ} = 2a_{12}\cdot\cos\beta;$$
$$2a_{12}\cdot\cos\beta = 2a_{12}\cdot\sin\omega\cdot\sin i;$$
$$\cos\beta = \sin\omega\cdot\sin i;$$
$$\mathrm{A'P'} = \mathrm{JP} = \mathrm{AP}\cdot\sin\beta = 2a_{12}\cdot\sin\beta = d_{12};$$
$$d_{12} = 2a_{12}\cdot\sin\beta;$$
$$d_{12} = 2a_{12}\cdot\sqrt{1-\cos^2\beta};$$
$$d_{12} = 2a_{12}\cdot\sqrt{1-\sin^2\omega\cdot\sin^2 i},$$

where, for angle i, representing the inclination of the plane of the orbit relative to the plane of the sky, the following relations hold:

$$\cos i = \sqrt{1-e^2}; \quad \sin i = e.$$

From the rectangular triangle $\mathrm{A'P'}G$, highlighted in the drawing in Figure 37.4, the result is:

$$\tan\lambda = \frac{\mathrm{P'G}}{\mathrm{A'G}} = \frac{\mathrm{PZ}}{\mathrm{A''P''}} = \frac{\mathrm{CP}\cdot\cos i}{\mathrm{AP}\cdot\cos\omega} = \frac{\mathrm{AP}\cdot\sin\omega\cdot\cos i}{\mathrm{AP}\cdot\cos\omega};$$
$$\tan\lambda = \frac{\sin\omega\cdot\cos i}{\cos\omega}; \quad \tan\lambda = \tan\omega\cdot\cos i;$$
$$\tan\omega = \frac{\tan\lambda}{\cos i}.$$

Using the drawings in Figures 37.5 and 37.6, writing one of the formulas of the cotangents for the sides and angles of the spherical triangle ABC, the same expression is obtained:

$$\sin a \cdot \cot c = \cos a \cdot \cos B + \sin B \cdot \cot C;$$

$$\sin \lambda \cdot \cot \omega = \cos \lambda \cdot \cos i + \sin i \cdot \cot 90^\circ;$$

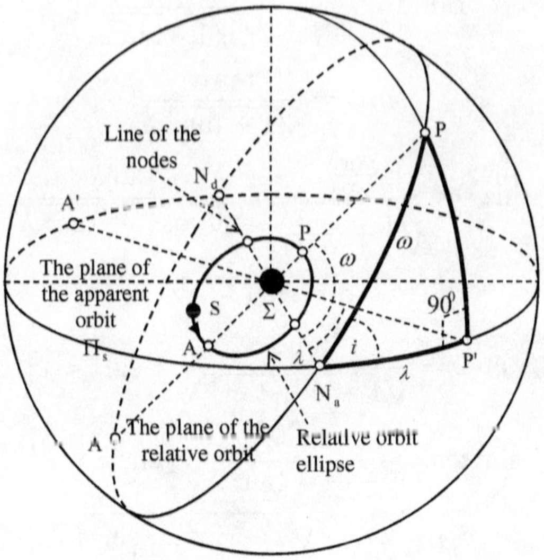

Fig. 37.5

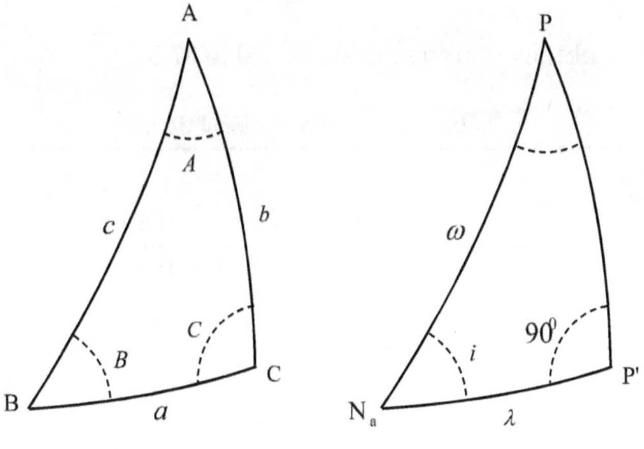

Fig. 37.6

$$\sin\lambda \cdot \cot\omega = \cos\lambda \cdot \cos i;$$

$$\frac{\sin\lambda}{\cos\lambda} = \frac{\cos i}{\cot\omega} = \cos i \cdot \tan\omega = \tan\lambda;$$

$$\tan\omega = \frac{\tan\lambda}{\cos i};$$

$$\tan\omega = \frac{\sin\omega}{\cos\omega} = \frac{\sin\omega}{\sqrt{1-\sin^2\omega}};$$

$$\sin\omega = \frac{\tan\omega}{\sqrt{1+\tan^2\omega}};$$

$$\sin\omega = \frac{\frac{\tan\lambda}{\cos i}}{\sqrt{1+\frac{\tan^2\lambda}{\cos^2 i}}} = \frac{\tan\lambda}{\sqrt{\cos^2 i + \tan^2\lambda}};$$

$$\cos^2 i = 1 - e^2;$$

$$\sin\omega = \frac{\tan\lambda}{\sqrt{1-e^2+\tan^2\lambda}}; \quad \sin i = e;$$

$$\sin^2\omega = \frac{\tan^2\lambda}{1-e^2+\tan^2\lambda}; \quad \sin^2 i = e^2;$$

$$d_{12} = 2a_{12} \cdot \sqrt{1-\sin^2\omega \cdot \sin^2 i};$$

$$d_{12} = 2a_{12} \cdot \sqrt{1 - \frac{e^2 \cdot \tan^2\lambda}{1-e^2+\tan^2\lambda}}.$$

The results obtained are recorded in Table 37.5.

Table 37.5 Observational data.

No.	Binary stellar system	a_{12} (AU)	e	λ (degree)	$\tan\lambda$	$\sin\omega$	ω (degree)	d_{12} (AU)
1	Sirius	20.101	0.591	5	0.087488663	0.107824062	$6^\circ.1$	40.120
2	Procyon	15.906	0.407	6	0.105104235	0.114311423	$6^\circ.2$	31.777
5	η Cass.	65.231	0.497	8	0.140540834	0.159876659	$9^\circ.2$	130.049

Problem 38

Big Dipper Asterism

The data in Table 38.1 are known for the stars of the Big Dipper asterism, shown in the drawing in Figure 38.1.

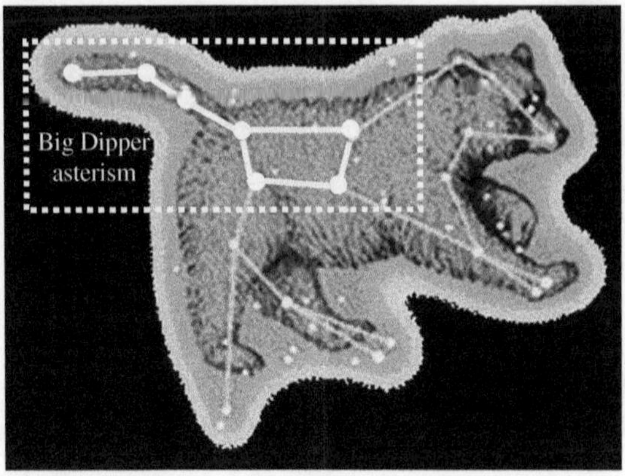

Fig. 38.1

Determine:

a) the brightness of the Big Dipper asterism, L_{BD}, expressing it in relation to the brightness of the Sun, L_{S};
b) the absolute magnitude of the Big Dipper asterism, M_{BD};
c) the apparent visual magnitude of the Big Dipper asterism, m_{BD}.

Table 38.1

Star number	Star name	Apparent magnitude	Distance $\Delta_{\text{Star, Earth}}$ (in light years)
1	Dubhe	1.8	124
2	Merak	2.4	79
3	Phecda	2.4	84
4	Megrez	3.3	58
5	Alioth	1.8	81
6	Mizar	2.1	78
7	Alkaid	1.9	101

The following are known: the apparent visual magnitude of the Sun, $m_S = -26.78^m$; the absolute magnitude of the Sun, $M_S = 4.76$; the distance between the Earth and the Sun, $\Delta_S = 1\text{AU}$; $1\,\text{ly} = 63240\,\text{AU}$.

Solution

a) The luminosity of the Big Dipper asterism is calculated as follows:

$$L_{\text{Big Dipper}} = L_{\text{BD}} = L_1 + L_2 + L_3 + L_4 + L_5 + L_6 + L_7,$$

where $L_1, L_2, \ldots, L_7$ are the luminosities of the seven stars in the asterism.

According to Pogson's formula:

$$\log \frac{E_S}{E_1} = -0.4(m_S - m_1),$$

where: E_S – the Sun's apparent brightness at a distance corresponding to the observer on the Earth; E_1 – the apparent brightness of star 1 (Dubhe) in the Big Dipper asterism at the distance corresponding to the observer on the Earth; m_S – the apparent magnitude of the Sun for the observer on the Earth; m_1 – the apparent magnitude of star 1 (Dubhe) in the Big Dipper asterism for the observer on the Earth;

$$\log \frac{\frac{L_S}{4\pi\Delta_S^2}}{\frac{L_1}{4\pi\Delta_1^2}} = -0.4(m_S - m_1);$$

$$\log\left(\frac{L_S}{L_1}\cdot\left(\frac{\Delta_1}{\Delta_S}\right)^2\right) = -0.4\,(m_S - m_1);$$

$$\Delta_S = 1\,\mathrm{AU};\quad \Delta_1 = 124\,\mathrm{ly} = 124\cdot 63240\,\mathrm{AU};$$

$$m_S = -26.78^{\mathrm{m}};\quad m_1 = 1.8^{\mathrm{m}};$$

$$m_S - m_1 = -28.58;$$

$$-0.4\,(m_S - m_1) = 11.432;$$

$$\log\left(\frac{L_S}{L_1}\cdot\left(\frac{\Delta_1}{\Delta_S}\right)^2\right) = 11.432;$$

$$\log\left(\left(\sqrt{\frac{L_S}{L_1}}\frac{\Delta_1}{\Delta_S}\right)^2\right) = 11.432;$$

$$2\cdot\log\left(\sqrt{\frac{L_S}{L_1}}\cdot\frac{\Delta_1}{\Delta_S}\right) = 11.432;$$

$$\log\left(\sqrt{\frac{L_S}{L_1}}\cdot\frac{\Delta_1}{\Delta_S}\right) = 5.716;$$

$$\log\left(\sqrt{\frac{L_S}{L_1}}\cdot\frac{\Delta_1}{\Delta_S}\right) = \log 10^{5.716};$$

$$\left(\sqrt{\frac{L_S}{L_1}}\cdot\frac{\Delta_1}{\Delta_S}\right) = 10^{5.716};$$

$$\sqrt{\frac{L_S}{L_1}}\cdot\frac{\Delta_1}{\Delta_S} \approx 52\cdot 10^4;$$

$$\frac{L_S}{L_1}\cdot\left(\frac{\Delta_1}{\Delta_S}\right)^2 = 2704\cdot 10^8;$$

$$L_1 = \frac{L_S}{2704\cdot 10^8}\cdot\left(\frac{\Delta_1}{\Delta_S}\right)^2;$$

$$\Delta_1 = 124\,\mathrm{ly} = 124\cdot 63240\,\mathrm{UA};$$

$$L_1 \approx 227.41\cdot L_S,$$

representing the luminosity of star 1 (Dubhe) in the Big Dipper asterism.

Similarly, for the luminosities of the other stars in the Big Dipper asterism:

$$\log \frac{\frac{L_S}{4\pi\Delta_S^2}}{\frac{L_2}{4\pi\Delta_2^2}} = -0.4\,(m_S - m_2);$$

$$\log\left(\frac{L_S}{L_2}\cdot\left(\frac{\Delta_2}{\Delta_S}\right)^2\right) = -0.4\,(m_S - m_2);$$

$$m_S = -26.78^{\mathrm{m}}; \quad m_2 = 2.4^{\mathrm{m}};$$

$$m_S - m_2 = -29.18;$$

$$-0.4\,(m_S - m_2) = 11.672;$$

$$\log\left(\frac{L_S}{L_2}\cdot\left(\frac{\Delta_2}{\Delta_S}\right)^2\right) = 11.672;$$

$$\log\left(\left(\sqrt{\frac{L_S}{L_2}}\cdot\frac{\Delta_2}{\Delta_S}\right)^2\right) = 11.672;$$

$$2\cdot\log\left(\sqrt{\frac{L_S}{L_2}}\cdot\frac{\Delta_2}{\Delta_S}\right) = 11.672;$$

$$\log\left(\sqrt{\frac{L_S}{L_2}}\cdot\frac{\Delta_2}{\Delta_S}\right) = 5.836;$$

$$\log\left(\sqrt{\frac{L_S}{L_2}}\cdot\frac{\Delta_2}{\Delta_S}\right) = \log 10^{5.836};$$

$$\left(\sqrt{\frac{L_S}{L_2}}\cdot\frac{\Delta_2}{\Delta_S}\right) = 10^{5.836};$$

$$\sqrt{\frac{L_S}{L_2}}\cdot\frac{\Delta_2}{\Delta_S} \approx 68\cdot 10^4;$$

$$\frac{L_S}{L_2}\cdot\left(\frac{\Delta_2}{\Delta_S}\right)^2 = 4624\cdot 10^8;$$

$$L_2 = \frac{L_S}{4624 \cdot 10^8} \cdot \left(\frac{\Delta_2}{\Delta_S}\right)^2;$$

$$\Delta_2 = 79\,\text{ly} = 79 \cdot 63240\,\text{AU};$$

$$L_2 \approx 53.97 \cdot L_S;$$

$$\log \frac{\frac{L_S}{4\pi\Delta_S^2}}{\frac{L_3}{4\pi\Delta_3^2}} = -0.4\,(m_S - m_3);$$

$$\log\left(\frac{L_S}{L_3} \cdot \left(\frac{\Delta_3}{\Delta_S}\right)^2\right) = -0.4\,(m_S - m_3);$$

$$m_S = -26.78^{\text{m}}; \quad m_3 = 2.4^{\text{m}};$$

$$m_S - m_3 = -29.18;$$

$$-0.4\,(m_S - m_3) = 11.672;$$

$$\log\left(\frac{L_S}{L_3} \cdot \left(\frac{\Delta_3}{\Delta_S}\right)^2\right) = 11.672;$$

$$\log\left(\left(\sqrt{\frac{L_S}{L_3}}\frac{\Delta_3}{\Delta_S}\right)^2\right) = 11.672;$$

$$2 \cdot \log\left(\sqrt{\frac{L_S}{L_3}}\frac{\Delta_3}{\Delta_S}\right) = 11.672;$$

$$\log\left(\sqrt{\frac{L_S}{L_3}} \cdot \frac{\Delta_3}{\Delta_S}\right) = 5.836;$$

$$\log\left(\sqrt{\frac{L_S}{L_3}} \cdot \frac{\Delta_3}{\Delta_S}\right) = \log 10^{5.836};$$

$$\left(\sqrt{\frac{L_S}{L_3}} \cdot \frac{\Delta_3}{\Delta_S}\right) = 10^{5.836};$$

$$\sqrt{\frac{L_S}{L_3}} \cdot \frac{\Delta_3}{\Delta_S} \approx 68 \cdot 10^4;$$

$$\frac{L_S}{L_3} \cdot \left(\frac{\Delta_3}{\Delta_S}\right)^2 = 4624 \cdot 10^8;$$

$$L_3 = \frac{L_S}{4624 \cdot 10^8} \cdot \left(\frac{\Delta_3}{\Delta_S}\right)^2;$$

$$\Delta_3 = 84\,\text{ly} = 84 \cdot 63240\,\text{AU};$$

$$L_3 \approx 61.02 \cdot L_S;$$

$$\log \frac{\frac{L_S}{4\pi\Delta_S^2}}{\frac{L_4}{4\pi\Delta_4^2}} = -0.4\,(m_S - m_4);$$

$$\log\left(\frac{L_S}{L_4} \cdot \left(\frac{\Delta_4}{\Delta_S}\right)^2\right) = -0.4\,(m_S - m_4);$$

$$m_S = -26.78^{\text{m}}; m_4 = 3.3^{\text{m}};$$

$$m_S - m_4 = -30.08;$$

$$-0.4\,(m_S - m_4) = 12.032;$$

$$\log\left(\frac{L_S}{L_4} \cdot \left(\frac{\Delta_4}{\Delta_S}\right)^2\right) = 12.032;$$

$$\log\left(\left(\sqrt{\frac{L_S}{L_4}}\frac{\Delta_4}{\Delta_S}\right)^2\right) = 12.032;$$

$$2 \cdot \log\left(\sqrt{\frac{L_S}{L_4}}\frac{\Delta_4}{\Delta_S}\right) = 12.032;$$

$$\log\left(\sqrt{\frac{L_S}{L_4}} \cdot \frac{\Delta_4}{\Delta_S}\right) = 6.016;$$

$$\log\left(\sqrt{\frac{L_S}{L_4}} \cdot \frac{\Delta_4}{\Delta_S}\right) = \log 10^{6.016};$$

$$\left(\sqrt{\frac{L_S}{L_4}} \cdot \frac{\Delta_4}{\Delta_S}\right) = 10^{6.016};$$

$$\sqrt{\frac{L_S}{L_4}} \cdot \frac{\Delta_4}{\Delta_S} \approx 100 \cdot 10^4;$$

$$\frac{L_S}{L_4} \cdot \left(\frac{\Delta_4}{\Delta_S}\right)^2 = 10000 \cdot 10^8;$$

$$L_4 = \frac{L_S}{10000 \cdot 10^8} \cdot \left(\frac{\Delta_4}{\Delta_S}\right)^2;$$

$$\Delta_4 = 58\,\text{ly} = 58 \cdot 63240\,\text{AU};$$

$$L_4 \approx 13.45 \cdot L_S;$$

$$\log \frac{\frac{L_S}{4\pi\Delta_S^2}}{\frac{L_5}{4\pi\Delta_5^2}} = -0.4\,(m_S - m_5);$$

$$\log \left(\frac{L_S}{L_5} \cdot \left(\frac{\Delta_5}{\Delta_S}\right)^2\right) = -0.4\,(m_S - m_5);$$

$$m_S = -26.78^{\text{m}}; \quad m_5 = 1.8^{\text{m}};$$

$$m_S - m_5 = -28.58;$$

$$-0.4\,(m_S - m_5) = 11.432;$$

$$\log \left(\frac{L_S}{L_5} \cdot \left(\frac{\Delta_5}{\Delta_S}\right)^2\right) = 11.432;$$

$$\log \left(\left(\sqrt{\frac{L_S}{L_5}} \frac{\Delta_5}{\Delta_S}\right)^2\right) = 11.432;$$

$$2 \cdot \log \left(\sqrt{\frac{L_S}{L_5}} \frac{\Delta_5}{\Delta_S}\right) = 11.432;$$

$$\log \left(\sqrt{\frac{L_S}{L_5}} \cdot \frac{\Delta_5}{\Delta_S}\right) = 5.716;$$

$$\log \left(\sqrt{\frac{L_S}{L_5}} \cdot \frac{\Delta_5}{\Delta_S}\right) = \log 10^{5.716};$$

$$\left(\sqrt{\frac{L_S}{L_5}} \cdot \frac{\Delta_5}{\Delta_S}\right) = 10^{5.716};$$

$$\sqrt{\frac{L_S}{L_5}} \cdot \frac{\Delta_5}{\Delta_S} \approx 52 \cdot 10^4;$$

$$\frac{L_S}{L_5} \cdot \left(\frac{\Delta_5}{\Delta_S}\right)^2 = 2704 \cdot 10^8;$$

$$L_5 = \frac{L_S}{2704 \cdot 10^8} \cdot \left(\frac{\Delta_5}{\Delta_S}\right)^2;$$

$$\Delta_5 = 81\,\mathrm{ly} = 81 \cdot 63240\,\mathrm{AU};$$

$$L_5 \approx 97.03 \cdot L_S;$$

$$\log \frac{\frac{L_S}{4\pi\Delta_S^2}}{\frac{L_6}{4\pi\Delta_6^2}} = -0.4\,(m_S - m_6);$$

$$\log \left(\frac{L_S}{L_6} \cdot \left(\frac{\Delta_6}{\Delta_S}\right)^2\right) = -0.4\,(m_S - m_6);$$

$$m_S = -26.78^{\mathrm{m}}; \quad m_6 = 2.1^{\mathrm{m}};$$

$$m_S - m_6 = -28.88;$$

$$-0.4\,(m_S - m_6) = 11.552;$$

$$\log \left(\frac{L_S}{L_6} \cdot \left(\frac{\Delta_6}{\Delta_S}\right)^2\right) = 11.552;$$

$$\log \left(\left(\sqrt{\frac{L_S}{L_6}} \frac{\Delta_6}{\Delta_S}\right)^2\right) = 11.552;$$

$$2 \cdot \log \left(\sqrt{\frac{L_S}{L_6}} \frac{\Delta_6}{\Delta_S}\right) = 11.552;$$

$$\log \left(\sqrt{\frac{L_S}{L_6}} \cdot \frac{\Delta_6}{\Delta_S}\right) = 5.776;$$

$$\log\left(\sqrt{\frac{L_{\text{S}}}{L_6}}\cdot\frac{\Delta_6}{\Delta_{\text{S}}}\right)=\log 10^{5.776};$$

$$\left(\sqrt{\frac{L_{\text{S}}}{L_6}}\cdot\frac{\Delta_6}{\Delta_{\text{S}}}\right)=10^{5.776};$$

$$\sqrt{\frac{L_{\text{S}}}{L_6}}\cdot\frac{\Delta_6}{\Delta_{\text{S}}}\approx 60\cdot 10^4;$$

$$\frac{L_{\text{S}}}{L_6}\cdot\left(\frac{\Delta_6}{\Delta_{\text{S}}}\right)^2=3600\cdot 10^8;$$

$$L_6=\frac{L_{\text{S}}}{3600\cdot 10^8}\cdot\left(\frac{\Delta_6}{\Delta_{\text{S}}}\right)^2;$$

$$\Delta_6=78\,\text{ly}=78\cdot 63240\,\text{AU};$$

$$L_6\approx 67.58\cdot L_{\text{S}};$$

$$\log\frac{\frac{L_{\text{S}}}{4\pi\Delta_{\text{S}}^2}}{\frac{L_7}{4\pi\Delta_7^2}}=-0.4\,(m_{\text{S}}-m_7);$$

$$\log\left(\frac{L_{\text{S}}}{L_7}\cdot\left(\frac{\Delta_7}{\Delta_{\text{S}}}\right)^2\right)=-0.4\,(m_{\text{S}}-m_7);$$

$$m_{\text{S}}=-26.78^{\text{m}};\quad m_7=1.9^{\text{m}};$$

$$m_{\text{S}}-m_7=-28.68,$$

$$-0.4\,(m_{\text{S}}-m_7)=11.472;$$

$$\log\left(\frac{L_{\text{S}}}{L_7}\cdot\left(\frac{\Delta_7}{\Delta_{\text{S}}}\right)^2\right)=11.472;$$

$$\log\left(\left(\sqrt{\frac{L_{\text{S}}}{L_7}}\cdot\frac{\Delta_7}{\Delta_{\text{S}}}\right)^2\right)=11.472;$$

$$2\cdot\log\left(\sqrt{\frac{L_{\text{S}}}{L_7}}\cdot\frac{\Delta_7}{\Delta_{\text{S}}}\right)=11.472;$$

$$\log\left(\sqrt{\frac{L_S}{L_7}}\cdot\frac{\Delta_7}{\Delta_S}\right) = 5.736;$$

$$\log\left(\sqrt{\frac{L_S}{L_7}}\cdot\frac{\Delta_7}{\Delta_S}\right) = \log 10^{5.736};$$

$$\left(\sqrt{\frac{L_S}{L_7}}\cdot\frac{\Delta_7}{\Delta_S}\right) = 10^{5.736};$$

$$\sqrt{\frac{L_S}{L_7}}\cdot\frac{\Delta_7}{\Delta_S} \approx 54\cdot 10^4;$$

$$\frac{L_S}{L_7}\cdot\left(\frac{\Delta_7}{\Delta_S}\right)^2 = 2916\cdot 10^8;$$

$$L_7 = \frac{L_S}{2916\cdot 10^8}\cdot\left(\frac{\Delta_7}{\Delta_S}\right)^2;$$

$$\Delta_7 = 101\,\text{ly} = 101\cdot 63240\,\text{AU};$$

$$L_7 \approx 139.90\cdot L_S.$$

Under these conditions, the luminosity of the Big Dipper asterism is:

$$L_{BD} = L_1 + L_2 + L_3 + L_4 + L_5 + L_6 + L_7;$$

$$L_{BD} = (227.41 + 53.97 + 61.02 + 13.45 + 97.03 + 67.58 + 139.90)L_{Sun};$$

$$L_{\text{Big Dipper}} = 660.36\cdot L_S.$$

b) If the Big Dipper asterism is equivalent to a single star, whose luminosity is that calculated previously, located at a distance Δ_{BD} from the Earth, then, according to Pogson's formula:

$$\log\frac{\frac{L_{BD}}{4\pi\Delta_{BD}^2}}{\frac{L_S}{4\pi\Delta_S^2}} = -0.4\,(m_{BD} - m_S),$$

where m_{BD} is the Big Dipper's apparent visual magnitude;

$$\log\frac{L_{BD}}{L_S}\cdot\left(\frac{\Delta_S}{\Delta_{BD}}\right)^2 = -0.4\,(m_{BD} - m_S).$$

We define the absolute magnitude of the Big Dipper and the absolute magnitude of the Sun, respectively, as follows:

$$M_{\rm BD} = m_{\rm BD} + 5 - 5 \cdot \log\left|\Delta_{\rm BD\,(pc)}\right|;$$
$$m_{\rm BD} = M_{\rm BD} - 5 + 5 \cdot \log\left|\Delta_{\rm BD\,(pc)}\right|;$$
$$M_{\rm S} = m_{\rm S} + 5 - 5 \cdot \log\left|\Delta_{\rm S\,(pc)}\right|;$$
$$m_{\rm S} = M_{\rm S} - 5 + 5 \cdot \log\left|\Delta_{\rm S\,(pc)}\right|.$$

This results in:

$$m_{\rm BD} - m_{\rm S} = M_{\rm BD} - M_{\rm S} + 5 \cdot \left(\log\left|\Delta_{\rm BD\,(pc)}\right| - \log\left|\Delta_{\rm S\,(pc)}\right|\right);$$

$$m_{\rm BD} - m_{\rm S} = M_{\rm BD} - M_{\rm S} + 5 \cdot \log\frac{\left|\Delta_{\rm BD\,(pc)}\right|}{\left|\Delta_{\rm S\,(pc)}\right|};$$

$$\log\frac{L_{\rm BD}}{L_{\rm S}} \cdot \left(\frac{\Delta_{\rm S}}{\Delta_{\rm BD}}\right)^2 = -0.4\,(m_{\rm BD} - m_{\rm S});$$

$$\log\frac{L_{\rm BD}}{L_{\rm S}} + \log\left(\frac{\Delta_{\rm S}}{\Delta_{\rm BD}}\right)^2 = -0.4\,(M_{\rm BD} - M_{\rm S}) - 0.4 \cdot 5 \cdot \log\frac{\left|\Delta_{\rm BD\,(pc)}\right|}{\left|\Delta_{S\,(pc)}\right|};$$

$$\log\frac{L_{\rm BD}}{L_{\rm S}} + 2 \cdot \log\left(\frac{\Delta_{\rm S}}{\Delta_{\rm BD}}\right) = -0.4\,(M_{\rm BD} - M_{\rm S}) - 2 \cdot \log\frac{\left|\Delta_{\rm BD\,(pc)}\right|}{\left|\Delta_{S\,(pc)}\right|};$$

$$\log\frac{L_{\rm BD}}{L_{\rm S}} - 2 \cdot \log\left(\frac{\Delta_{\rm BD}}{\Delta_{\rm S}}\right) = -0.4\,(M_{\rm BD} - M_{\rm S}) - 2 \cdot \log\frac{\left|\Delta_{\rm BD\,(pc)}\right|}{\left|\Delta_{S\,(pc)}\right|};$$

$$-2 \cdot \log\left(\frac{\Delta_{\rm BD}}{\Delta_{\rm S}}\right) = -2 \cdot \log\frac{\left|\Delta_{\rm BD\,(pc)}\right|}{\left|\Delta_{S\,(pc)}\right|};$$

$$\log\frac{L_{\rm BD}}{L_{\rm S}} = -0.4\,(M_{\rm BD} - M_{\rm S});$$

$$M_{\rm BD} = M_{\rm S} - 2.5 \cdot \log\frac{L_{\rm BD}}{L_{\rm S}};$$

$$L_{\text{BD}} = 660.36 \cdot L_{\text{S}};\, M_{\text{S}} = 4.76;$$

$$M_{\text{BD}} = 4.76 - 2.5 \cdot \log(660.36) = 4.76 - 7.04;$$

$$M_{\text{BD}} = -2.28.$$

c) The seven stars of the Big Dipper asterism ($\sigma_1, \sigma_2, \ldots, \ldots, \ldots, \sigma_7$) are equivalent to the star shown in the diagram in Figure 38.2.

If E_1 and E_2, respectively, are the apparent luminosities (illuminations) of the stars σ_1 and σ_2 of the Big Dipper asterism, then, according to Pogson's formula, written first for the two components:

$$\log\frac{E_1}{E_2} = -0.4\,(m_1 - m_2) = 0.4\,(m_2 - m_1) = \log 10^{0.4(m_2 - m_1)};$$

$$\frac{E_1}{E_2} = 10^{0.4(m_2 - m_1)};\quad E_1 = E_2 \cdot 10^{0.4(m_2 - m_1)}.$$

Writing the same formula for the system ($\sigma_1; \sigma_2$) and for the star σ_2:

$$\log\frac{E_{12}}{E_2} = -0.4\,(m_{12} - m_2) = 0.4\,(m_2 - m_{12}),$$

where E_{12} is the equivalent illumination of the two stars ($\sigma_{1,2} \equiv \sigma_a$), and m_{12} is the apparent equivalent magnitude of the system formed

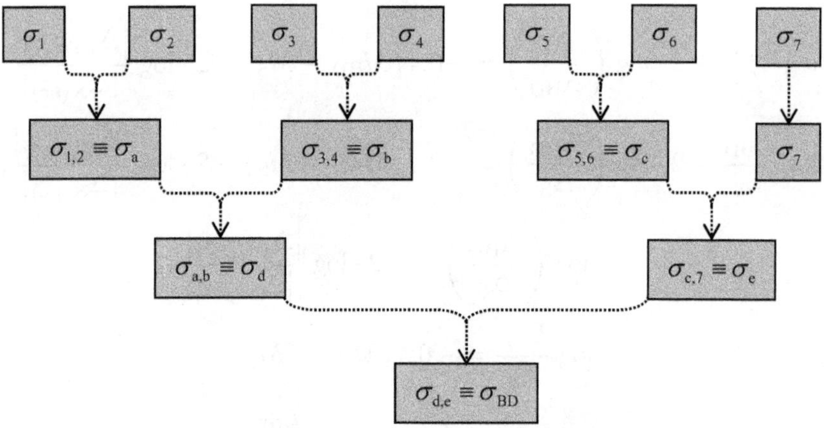

Fig. 38.2

by the stars σ_1 and σ_2;

$$E_{12} = E_1 + E_2;$$

$$\log \frac{E_1 + E_2}{E_2} = 0.4\,(m_2 - m_{12}) = \log 10^{0.4(m_2 - m_{12})};$$

$$\log\left(1 + \frac{E_1}{E_2}\right) = \log 10^{0.4(m_2 - m_{12})};$$

$$1 + \frac{E_1}{E_2} = 10^{0.4(m_2 - m_{12})};$$

$$E_1 = E_2 \cdot 10^{0.4(m_2 - m_1)};$$

$$\frac{E_1}{E_2} = 10^{0.4(m_2 - m_1)};$$

$$1 + 10^{0.4(m_2 - m_1)} = 10^{0.4(m_2 - m_{12})};$$

$$10^{0.4(m_2 - m_{12})} = 10^{0.4(m_2 - m_1)} + 1;$$

$$x = 10^{0.4(m_2 - m_1)}; \quad m_1 = 1.8; \quad m_2 = 2.4;$$

$$x = 10^{0.4(2.4 - 1.8)} = 10^{0.24};$$

$$\log x = 0.24; \quad x = 1.74;$$

$$10^{0.4(m_2 - m_1)} = 1.74;$$

$$10^{0.4(m_2 - m_{12})} = 10^{0.4(m_2 - m_1)} + 1;$$

$$10^{0.4(m_2 - m_{12})} = 1.74 + 1;$$

$$10^{0.4(m_2 - m_{12})} = 2.74;$$

$$\log 10^{0.4(m_2 - m_{12})} = \log 2.74;$$

$$0.4\,(m_2 - m_{12}) = 0.437;$$

$$m_2 - m_{12} = 1.0925; m_2 = 2.4;$$

$$m_{12} = 1.3075.$$

Similarly, for the stars σ_3 and σ_4, we get:

$$10^{0.4(m_4 - m_{34})} = 10^{0.4(m_4 - m_3)} + 1;$$

$$x = 10^{0.4(m_4 - m_3)}; m_3 = 2.4; m_4 = 3.3;$$

$$x = 10^{0.4(3.3-2.4)} = 10^{0.36};$$

$$\log x = 0.36; x = 2.3;$$

$$10^{0.4(m_4-m_3)} = 2.3;$$

$$10^{0.4(m_4-m_{34})} = 10^{0.4(m_4-m_3)} + 1;$$

$$10^{0.4(m_4-m_{34})} = 2.3 + 1;$$

$$10^{0.4(m_4-m_{34})} = 3.3;$$

$$\log 10^{0.4(m_4-m_{34})} = \log 3.3;$$

$$m_4 - m_{34} = 1.295; m_4 = 3.3;$$

$$m_{34} = 2.005.$$

For the stars σ_5 and σ_6, we get:

$$10^{0.4(m_6-m_{56})} = 10^{0.4(m_6-m_5)} + 1;$$

$$x = 10^{0.4(m_6-m_5)}; m_5 = 1.8; m_6 = 2.1;$$

$$x = 10^{0.4(2.1-1.8)} = 10^{0.12};$$

$$\log x = 0.12; x = 1.32;$$

$$10^{0.4(m_6-m_5)} = 1.32;$$

$$10^{0.4(m_6-m_{56})} = 10^{0.4(m_6-m_5)} + 1;$$

$$10^{0.4(m_6-m_{56})} = 1.32 + 1;$$

$$10^{0.4(m_6-m_{56})} = 2.32;$$

$$\log 10^{0.4(m_6-m_{56})} = \log 2.32;$$

$$0.4\,(m_6 - m_{56}) = 0.365;$$

$$m_6 - m_{56} = 0.9125; m_6 = 2.1;$$

$$m_{56} = 1.1875.$$

The stars σ_1 and σ_2 are equivalent to a star, σ_a, whose apparent magnitude is

$$m_a = m_{12} = 1.3075.$$

The stars σ_3 and σ_4 are equivalent to a star, σ_b, whose apparent magnitude is

$$m_b = m_{34} = 2.548.$$

The stars σ_5 and σ_6 are equivalent to a star, σ_{c}, whose apparent magnitude is

$$m_{\mathrm{c}} = m_{56} = 1.1875.$$

For the stars σ_a and σ_b, we can write that:

$$1 + 10^{0.4(m_{\mathrm{b}}-m_{\mathrm{a}})} = 10^{0.4(m_{\mathrm{b}}-m_{\mathrm{ab}})};$$

$$10^{0.4(m_{\mathrm{b}}-m_{\mathrm{ab}})} = 10^{0.4(m_{\mathrm{b}}-m_{\mathrm{a}})} + 1;$$

$$x = 10^{0.4(m_{\mathrm{b}}-m_{\mathrm{a}})}; \quad m_{\mathrm{a}} = 1.3075; \quad m_{\mathrm{b}} = 2.548;$$

$$x = 10^{0.4(2.548-1.3075)} = 10^{0.4962};$$

$$\log x = 0.4962; \quad x = 3.11,$$

$$10^{0.4(m_{\mathrm{b}}-m_{\mathrm{a}})} = 3.11;$$

$$10^{0.4(m_{\mathrm{b}}-m_{\mathrm{ab}})} = 10^{0.4(m_{\mathrm{b}}-m_{\mathrm{a}})} + 1;$$

$$10^{0.4(m_{\mathrm{b}}-m_{\mathrm{ab}})} = 3.11 + 1;$$

$$10^{0.4(m_{\mathrm{b}}-m_{\mathrm{ab}})} = 4.11;$$

$$\log 10^{0.4(m_{\mathrm{b}}-m_{\mathrm{ab}})} = \log 4.11;$$

$$0.4(m_{\mathrm{b}} - m_{\mathrm{ab}}) = 0.613;$$

$$m_{\mathrm{b}} - m_{\mathrm{ab}} = 1.5325; m_{\mathrm{b}} = 2.548;$$

$$m_{\mathrm{ab}} = 1.0155.$$

For the stars σ_{c} and σ_7, we can write that:

$$1 + 10^{0.4(m_7-m_{\mathrm{c}})} = 10^{0.4(m_7-m_{\mathrm{c7}})};$$

$$10^{0.4(m_7-m_{\mathrm{c7}})} = 10^{0.4(m_7-m_{\mathrm{c}})} + 1;$$

$$x = 10^{0.4(m_7-m_{\mathrm{c}})}; \quad m_{\mathrm{c}} = 1.1875; \quad m_7 = 1.9;$$

$$x = 10^{0.4(1.9-1.1875)} = 10^{0.285};$$

$$\log x = 0.285; \quad x = 1.94;$$

$$10^{0.4(m_7-m_c)} = 1.94;$$

$$10^{0.4(m_7-m_{c7})} = 10^{0.4(m_7-m_c)} + 1;$$

$$10^{0.4(m_7-m_{c7})} = 1.94 + 1;$$

$$10^{0.4(m_7-m_{c7})} = 2.94;$$

$$\log 10^{0.4(m_7-m_{c7})} = \log 2.94;$$

$$0.4(m_7 - m_{c7}) = 0.468;$$

$$m_7 - m_{c7} = 1.17; \quad m_7 = 1.9;$$

$$m_{c7} = 0.73.$$

The equivalent to star $\sigma_{\mathrm{a,b}}$ we will call star σ_{d}, with the apparent magnitude

$$m_{\mathrm{d}} = m_{\mathrm{ab}} = 4.0805.$$

The equivalent to star $\sigma_{\mathrm{c,7}}$ we call star σ_{e}, with the apparent magnitude

$$m_{\mathrm{e}} = m_{\mathrm{c7}} = 0.73.$$

For the stars σ_{d} and σ_{e}, we can write that:

$$1 + 10^{0.4(m_{\mathrm{e}}-m_{\mathrm{d}})} = 10^{0.4(m_{\mathrm{e}}-m_{\mathrm{de}})},$$

where

$$m_{\mathrm{de}} = m_{\mathrm{BD}},$$

representing the apparent magnitude of the Big Dipper asterism;

$$10^{0.4(m_{\mathrm{e}}-m_{\mathrm{de}})} = 10^{0.4(m_{\mathrm{e}}-m_{\mathrm{d}})} + 1;$$

$$10^{0.4(m_{\mathrm{e}}-m_{\mathrm{BD}})} = 10^{0.4(m_{\mathrm{e}}-m_{\mathrm{d}})} + 1;$$

$$x = 10^{0.4(m_{\mathrm{e}}-m_{\mathrm{d}})}; \quad m_{\mathrm{e}} = 0.73; \quad m_{\mathrm{d}} = 4.0805;$$

$$x = 10^{0.4(0.73-4.0805)} = 10^{-1.3402};$$

$$\log x = -1.3402; x = 0.045;$$

$$10^{0.4(m_e - m_d)} = 0.045;$$

$$10^{0.4(m_e - m_{BD})} = 10^{0.4(m_e - m_d)} + 1;$$

$$10^{0.4(m_e - m_{BD})} = 0.045 + 1;$$

$$10^{0.4(m_e - m_{BD})} = 1.045;$$

$$\log 10^{0.4(m_e - m_{BD})} = \log 1.045;$$

$$0.4(m_e - m_{BD}) = 0.0191;$$

$$m_e - m_{BD} = 0.04775; \quad m_e = 0.73;$$

$$m_{BD} = 0.68,$$

representing the apparent magnitude of the Big Dipper asterism.

Problem 39

From the North Pole to the Earth's Equator

Determine the elements of the minimum velocity vector with which a projectile must be launched from the North Pole of the Earth for it to land at a point on the Equator.

It is known that the second cosmic velocity required for a projectile to leave the Earth is $v_{II} = v_0$. It will be considered that the rotation of the Earth is very slow.

Solution

A parabola is the geometric location of the points in a plane at equal distances to a fixed point, F, *called the focal point, and to a fixed line,* Δ, *called the directrix.*

The projectile must evolve on a parabolic trajectory, with the Earth's center in its focus, in order to escape from the Earth's gravitational field and reach a very distant point, where its velocity relative to Earth shall be zero.

Suppose that the projectile was "prepared", by calculation, to escape from the gravitational field of the Earth on the parabola represented in the drawing in Figure 39.1, in which the focal point is the center of the Earth. The equation of the parabola in Cartesian coordinates $(x; y)$ is $y^2 = 2px$, for which the parameter of the parabola is known, $p = 2r_{\text{min}}$.

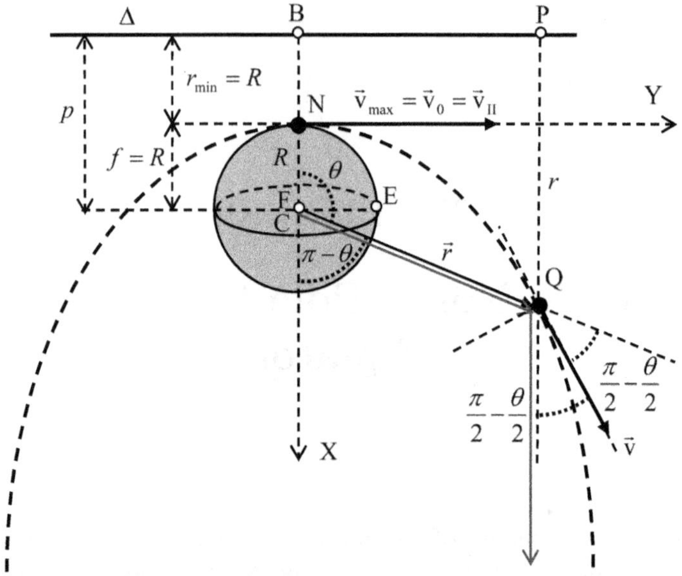

Fig. 39.1

To achieve such an escape, when the projectile has reached the altitude of point Q, its velocity, $\vec{v}$, on the tangent to the parabola, must be such that the total mechanical energy of the projectile–Earth system is

$$E = \frac{mv^2}{2} - K\frac{mM}{r} = 0.$$

When the "escape" has succeeded and the projectile has reached a location very far from Earth ($r \to \infty$), it is at rest in relation to Earth ($v_\infty = 0$).

The optical properties of the parabola are demonstrated as follows: All light rays emitted from the focal point of a concave parabolic mirror, after reflection, become parallel to the main optical axis and to each other. The incident rays parallel to the main optical axis are reflected through the focal point.

As a result, the tangent to the parabola at point Q is the bisector of the angle CQP.

Given the definition of the parabola, it follows that:

$$\mathrm{CQ} = \mathrm{QP};$$

$$r = \mathrm{CQ} = \mathrm{QP} = \mathrm{BC} + \mathrm{CQ}\cos(\pi - \theta);$$

$$r = 2r_{\min} + r \cdot \cos(\pi - \theta);$$

$$2r_{\min} = r(1 + \cos\theta);$$

$$r_{\min} = r \cdot \cos\frac{\theta}{2} = R.$$

Because the projectile is launched from the surface of the Earth, from point N, representing the North Pole of the Earth (the tip of the parabola), where $r = r_{\min}$, the launch velocity must be $v = v_{\max}$, so that we have:

$$E = \frac{mv_{\max}^2}{2} - K\frac{mM}{r_{\min}} = 0;$$

$$v_{\max} = \sqrt{2\frac{KM}{r_{\min}}}; \quad r_{\min} = R;$$

$$v_{\max} = \sqrt{2\frac{KM}{R}} = v_0,$$

representing the second cosmic velocity (v_{II}), that is, the velocity necessary for the projectile, at the moment of launch from the surface of the Earth, to leave the Earth, reach a very distant location, and be at rest in relation to the Earth.

From the law of conservation of angular momentum, it follows that:

$$L_N = L_Q;$$

$$r_{\min} v_{\max} = rv\sin\left(\frac{\pi}{2} - \frac{\theta}{2}\right) = rv\cos\frac{\theta}{2};$$

$$r\cos^2\frac{\theta}{2} v_{\max} = rv\cos\frac{\theta}{2}; \quad v_{\max} = \frac{v}{\cos\frac{\theta}{2}}.$$

Under these conditions, the only possibilities for launching the projectile from one of the Earth's poles, N, so that it reaches a point E on the Earth's Equator, remain those shown in drawings a and b in Figure 39.2: a – evolution on a very low circular trajectory, its radius being equal to the radius of the Earth; b – evolution along an elliptical trajectory, having the center of the Earth in one of its foci.

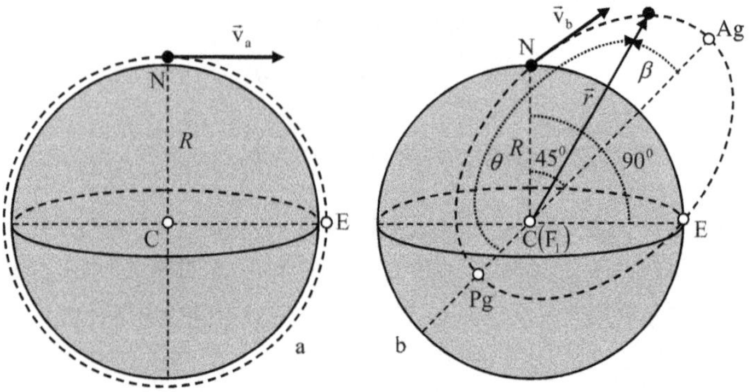

Fig. 39.2

a) In this case, the speed required when launching the projectile is

$$\mathrm{v_a} = \sqrt{\frac{KM}{R}} = \mathrm{v_I} = \frac{\mathrm{v_{II}}}{\sqrt{2}} = \frac{\mathrm{v_0}}{\sqrt{2}} < \mathrm{v_0},$$

representing the first cosmic velocity.

b) If the orbit of the projectile is the ellipse represented in drawing b, it results in:

$$\mathrm{v_b} = \sqrt{KM\left(\frac{2}{R} - \frac{1}{a}\right)};$$

$$\mathrm{v_b^2} = 2\frac{KM}{R} - \frac{KM}{a};$$

$$\mathrm{v_0} = \sqrt{2\frac{KM}{R}}; \quad \mathrm{v_0^2} = 2\frac{KM}{R};$$

$$\mathrm{v_b^2} = \mathrm{v_0^2} - \frac{KM}{a};$$

$$r = \frac{p}{1 + e\cos\theta} = \frac{a \cdot \left(1 - e^2\right)}{1 + e\cos\theta};$$

$$\theta = 180° - \beta;$$

$$r = \frac{a \cdot (1 - e^2)}{1 - e \cos \beta};$$

$$\beta = 45^\circ; \quad r = R; \quad R = \frac{a \cdot (1 - e^2)}{1 - e \cos \beta};$$

$$a = R \cdot \frac{1 - e \cdot \cos \beta}{1 - e^2};$$

$$\beta = 45^\circ;$$

$$a = R \cdot \frac{1 - \frac{\sqrt{2}}{2} \cdot e}{1 - e^2} = f(e);$$

$$\mathrm{v_b} = \sqrt{KM \left(\frac{2}{R} - \frac{1}{a} \right)};$$

$$\mathrm{v_b^2} = 2 \frac{KM}{R} - \frac{KM}{a};$$

$$\mathrm{v_0} = \sqrt{2 \frac{KM}{R}}; \quad \mathrm{v_0^2} = 2 \frac{KM}{R};$$

$$\mathrm{v_b^2} = \mathrm{v_0^2} - \frac{KM}{a};$$

$$\mathrm{v_{b,min}^2} = \mathrm{v_0^2} - \frac{KM}{a_{\min}};$$

$$\frac{\mathrm{d}a}{\mathrm{d}e} = R \cdot \frac{\frac{\mathrm{d}}{\mathrm{d}e} \left(1 - \frac{\sqrt{2}}{2} \cdot e\right) \cdot (1 - e^2) - \left(1 - \frac{\sqrt{2}}{2} \cdot e\right) \cdot \frac{\mathrm{d}}{\mathrm{d}e} (1 - e^2)}{(1 - e^2)^2} = 0;$$

$$\frac{-\frac{\sqrt{2}}{2} \cdot (1 - e^2) + 2e \cdot \left(1 - \frac{\sqrt{2}}{2} \cdot e\right)}{(1 - e^2)^2} = 0;$$

$$-\frac{\sqrt{2}}{2} + \frac{\sqrt{2}}{2} \cdot e^2 + 2e - \sqrt{2} \cdot e^2 = 0;$$

$$-\frac{\sqrt{2}}{2} \cdot e^2 + 2e - \frac{\sqrt{2}}{2} = 0;$$

$$\sqrt{2} \cdot e^2 - 4e + \sqrt{2} = 0;$$

$$e = \sqrt{2} \pm 1;$$

$$e_{\min} = \sqrt{2} - 1;$$

$$a_{\min} = R \cdot \frac{1 - \frac{\sqrt{2}}{2} \cdot e_{\min}}{1 - e_{\min}^2} = 0.85 \cdot R;$$

$$v_{b,\min}^2 = v_0^2 - \frac{KM}{a_{\min}};$$

$$v_{b,\min}^2 = v_0^2 - \frac{KM}{0.85 \cdot R};$$

$$v_{b,\min}^2 = v_0^2 - \frac{1}{0.85} \cdot \frac{KM}{R};$$

$$v_0^2 = 2\frac{KM}{R}; \quad v_0^2 = \frac{KM}{R} = \frac{v_0^2}{2};$$

$$v_{b,\min}^2 = v_0^2 - \frac{1}{0.85} \cdot \frac{v_0^2}{2};$$

$$v_{b,\min}^2 = v_0^2 \left(1 - \frac{1}{1.7}\right) = \frac{0.7}{1.7} \cdot v_0^2;$$

$$v_{b,\min} = 0.64 \cdot v_0;$$

$$v_a = \frac{v_0}{\sqrt{2}} = 0.71 \cdot v_0;$$

$$v_{b,\min} < v_a.$$

If α is the angle between the geometric tangent to the ellipse at point N and the axis OX, as indicated by the drawing in Figure 39.3, from the geometric interpretation of the derivative of a function, the result is:

$$\text{tg}\ \alpha = \frac{dy}{dx};$$

$$\frac{x^2}{a_{\min}^2} + \frac{y^2}{b^2} = 1;$$

$$y^2 = b^2 \left(1 - \frac{x^2}{a_{\min}^2}\right);$$

$$y \cdot dy = -\frac{b^2}{a_{\min}^2} \cdot x \cdot dx;$$

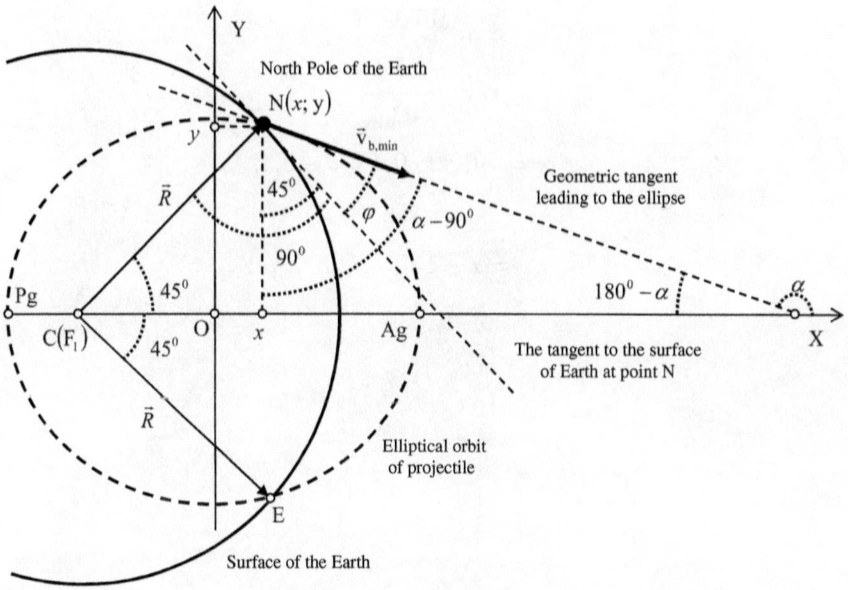

Fig. 30.3

$$\frac{dy}{dx} = -\frac{b^2}{a^2} \cdot \frac{x}{y};$$

$$\operatorname{tg} \alpha = -\frac{b^2}{a_{\min}^2} \cdot \frac{x}{y};$$

$$x = R \cdot \cos 45^\circ - c = R \cdot \frac{\sqrt{2}}{2} - c;$$

$$a_{\min} = 0.85 \cdot R; \quad e_{\min} = 0.41;$$

$$e_{\min} = \sqrt{1 - \frac{b^2}{a_{\min}^2}}; \quad b = a_{\min} \cdot \sqrt{1 - e_{\min}^2} = 0.77 \cdot R;$$

$$c = \sqrt{a_{\min}^2 - b^2} = 0.36 \cdot R;$$

$$x = R \cdot \frac{\sqrt{2}}{2} - c = 0.70 \cdot R - 0.36 \cdot R = 0.34 \cdot R;$$

$$y = R \cdot \sin 45^\circ = R \cdot \frac{\sqrt{2}}{2};$$

$$y = 0.70 \cdot R;$$

$$\text{tg } \alpha = -\frac{b^2}{a_{\min}^2} \cdot \frac{x}{y};$$

$$\text{tg } \alpha = -0.82 \cdot 0.48 = -0.4;$$

$$\alpha = 158^\circ;$$

$$\varphi = \alpha - 135^\circ = 23^\circ.$$

Problem 40

Solar Explosions

A. Solar Protuberances

a) *Determine* the heights H_0, H_1 and H_2 of the solar protuberances, with solar radial-axial symmetry, observed in the plane of the Sun's disk, perpendicular to the plane of the Earth's orbit around the Sun, represented in Figure 40.1, knowing that the Sun's radius is $R = 696\,000\,\text{km}$.

b) On the day of the spring equinox, a terrestrial observer located at a point on the Earth's Equator follows the evolution of the image of the Sun's apparent disk with two diametrically opposed, symmetrical solar protuberances whose heights are identical, H_0 (determined previously), located in the plane of the celestial equator. In the drawing in Figure 40.2, the Sun's disk is below the horizon of the Earth observer, at its limit.

Determine the duration of the sunrise, τ, so that both protuberances are visible to the observer, O, if the effect of atmospheric refraction on the sunrise is neglected.

During the sunrise, the sun's rotation around its own axis is neglected. It will be considered that the terrestrial Equator is in the plane of the celestial equator.

Given: the distance between the center of the Earth (observer O) and the center C of the Sun, $d = 150\,000\,000\,\mathrm{km}$; the radius of the Sun, $R = 696\,000\,\mathrm{km}$; the length of a day, $T = 24\,\mathrm{h}$.

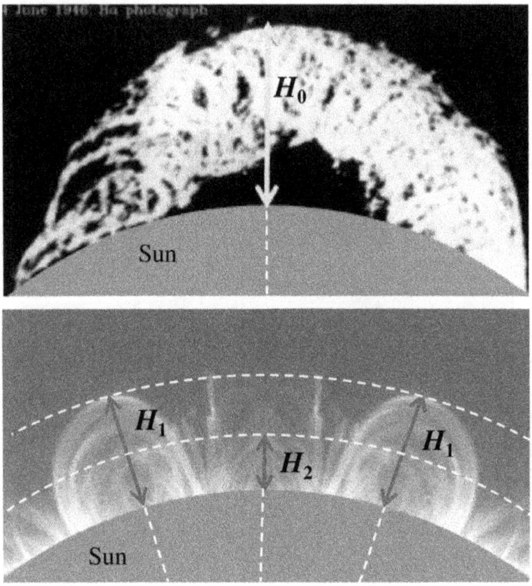

40.1.1

40.1.2

Solar protuberances

Fig. 40.1

B. Coronal Mass Explosions (CME)

The images in Figure 40.3 show three moments of a coronal mass explosion recorded by the LASCO/C2 satellite. Such coronal explosions were first observed in 1970 by the Orbiting Solar Observatory (OSO 7). In the three images, three moments in the propagation of the explosion front are shown, having the shape of an expanding arc. The images were taken with a coronagraph in which the image of the Sun is not visible (being obscured), but the central circle indicates the edge of the Sun. The Sun's radius is $R_S = 6.96 \cdot 10^8\,\mathrm{m}$. The time τ of the photograph is noted on each image.

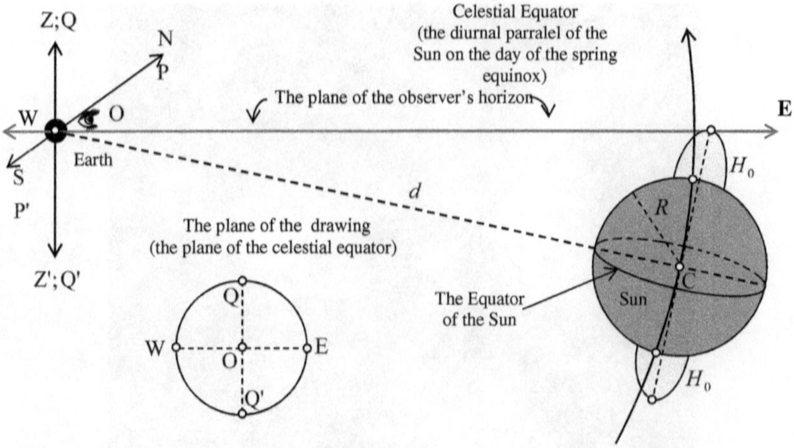

Fig. 40.2

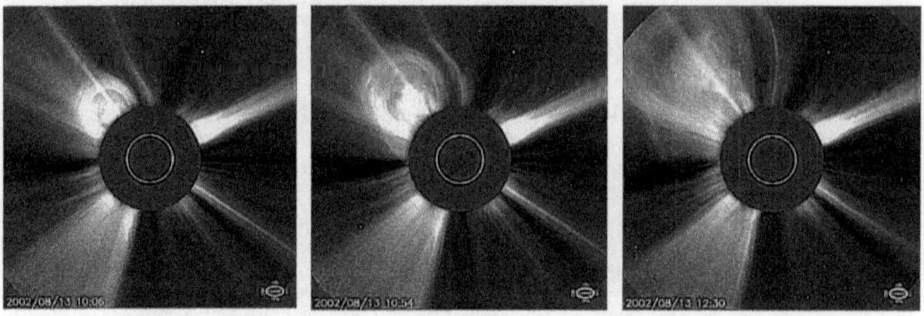

Fig. 40.3 (Credit: NASA and ESA)

Determine:

c) the speed of movement of the front of the jet resulting from the explosion;
d) the moment of the explosion;
e) the coordinates of the position of the jet front at the time $\tau_4 = 14^h 00^{\text{min}}$, using the magnified images corresponding to the three moments represented in Figures 40.4, 40.5, and 40.6.

The distances on the images are measured with a ruler and expressed in divisions (or mm).

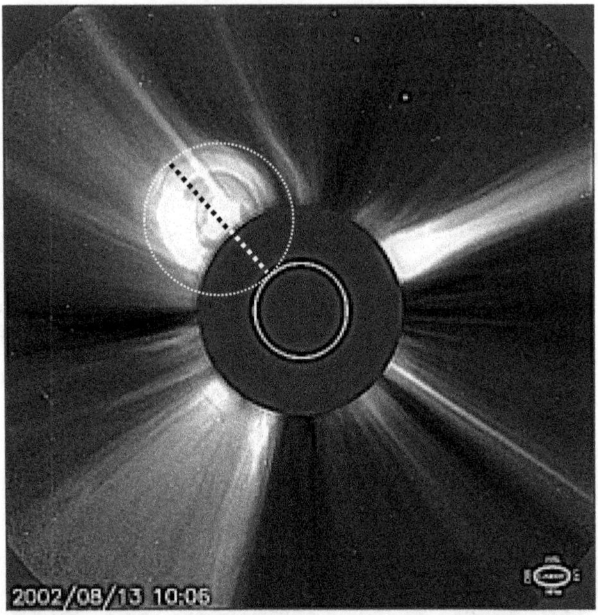

Fig. 40.4 (Credit: NASA and ESA)

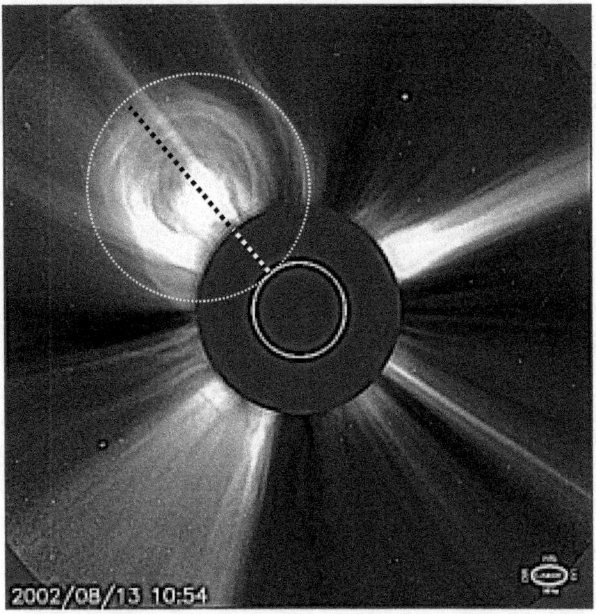

Fig. 40.5 (Credit: NASA and ESA)

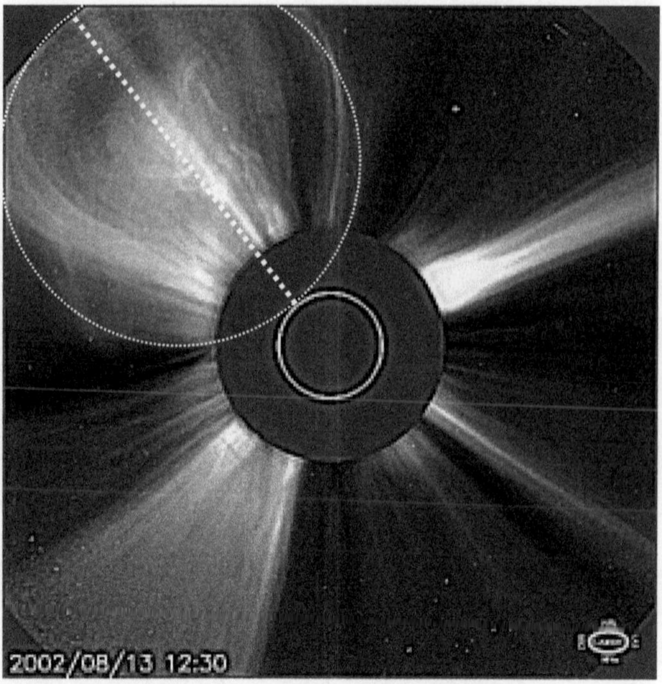

Fig. 40.6 (Credit: NASA and ESA)

Solution

a)

1) According to the notation in the drawing in Figure 40.7, measurements made on the same drawing show that:

$$R = kr; \quad H = kh; \quad D = kd; \quad R - H = k(r - h);$$

$$R^2 = (R - H)^2 + D^2;$$

$$k^2 r^2 = (kr - kh)^2 + k^2 d^2;$$

$$r^2 = (r - h)^2 + d^2;$$

$$0 = -2rh + h^2 + d^2;$$

$$r = \frac{d^2 + h^2}{2h}; \quad d = 45 \text{ mm}; \quad h = 15 \text{ mm};$$

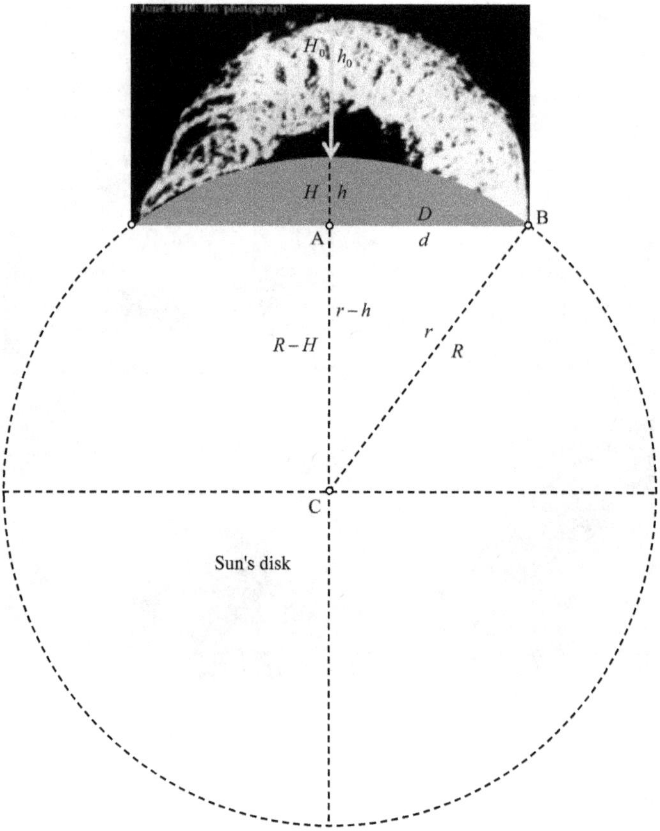

Fig. 40.7

$$r = \frac{2025 + 225}{30}\ \text{mm}; \quad r = 75\text{mm};$$

$$R = 696\,000\,\text{km}; \quad r = 75\,\text{mm};$$

$$R = kr;$$

$$696\,000\,\text{km} = k \cdot 75\,\text{mm};$$

$$k = \frac{696\,000}{75}\frac{\text{km}}{\text{mm}};$$

$$H_0 = kh_0;$$

$$h_0 = 30\,\text{mm};$$

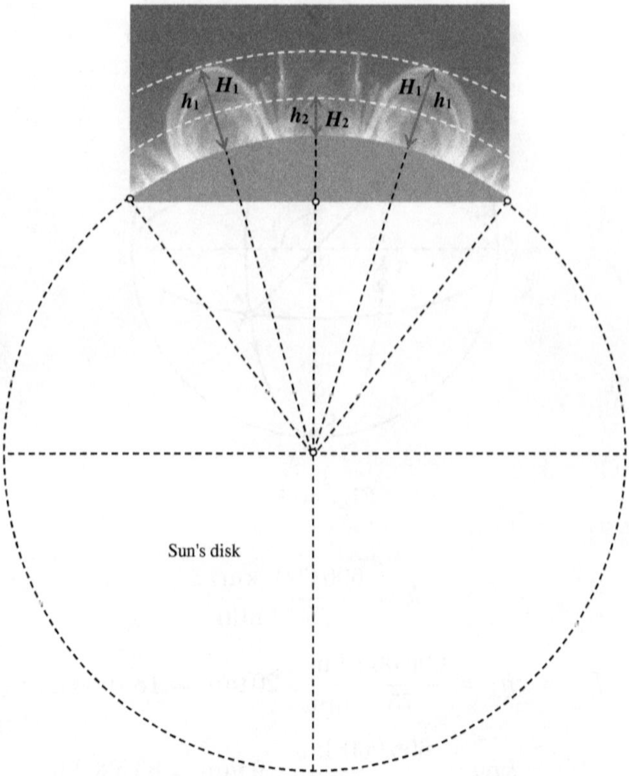

Fig. 40.8

$$H_0 = \frac{696\,000}{75}\,\frac{\text{km}}{\text{mm}} \cdot 30\,\text{mm};$$

$$H_0 = 278\,400\,\text{km},$$

representing the height of the solar protuberance shown in image 40.1.1 of Figure 40.1 and Figure 40.7.

2) According to the existing notations in the drawing in Figure 40.8, after measurements on the same drawing, given that both images in Figure 40.1 have the same scale, we find that:

$$h_1 = 20\,\text{mm};$$

$$h_2 = 9\,\text{mm};$$

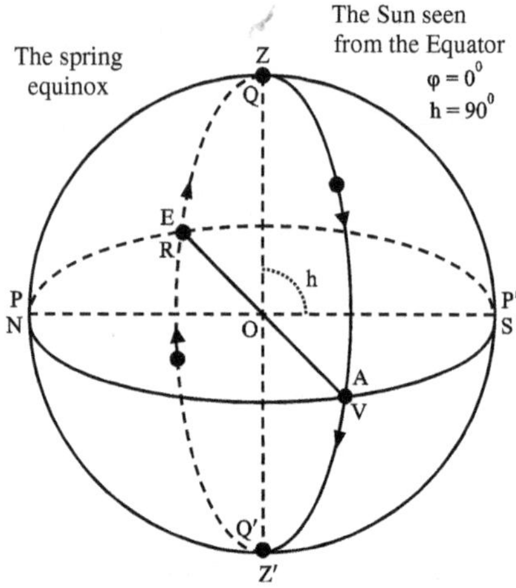

Fig. 40.9

$$k = \frac{696\,000}{75}\frac{\text{km}}{\text{mm}};$$

$$H_1 = kh_1 = \frac{696\,000}{75}\frac{\text{km}}{\text{mm}} \cdot 20\,\text{mm} = 185\,600\,\text{km};$$

$$H_2 = kh_2 = \frac{696\,000}{75}\frac{\text{km}}{\text{mm}} \cdot 9\,\text{mm} = 83\,520\,\text{km}.$$

b) The evolution of the Sun on the day of the spring equinox for the observer on the Equator is shown in the drawing in Figure 40.9, where the diurnal parallel of the Sun coincides with the celestial equator.

From the initial (lower) position of the center of the Sun, at the point C_0, below the horizon of the observer, as shown in the drawing in Figure 41.10, the center of the Sun must reach the upper position, C, above the horizon of the observer under the specified conditions.

Under these conditions, the angle at the center described by the center of the solar disk is:

$$2\theta = \omega\tau; \quad \omega = \frac{2\pi}{T};$$

$$\tau = \frac{2\theta}{\omega} = \frac{2\theta}{\frac{2\pi}{T}}; \quad \tau = \frac{\theta}{\pi}T.$$

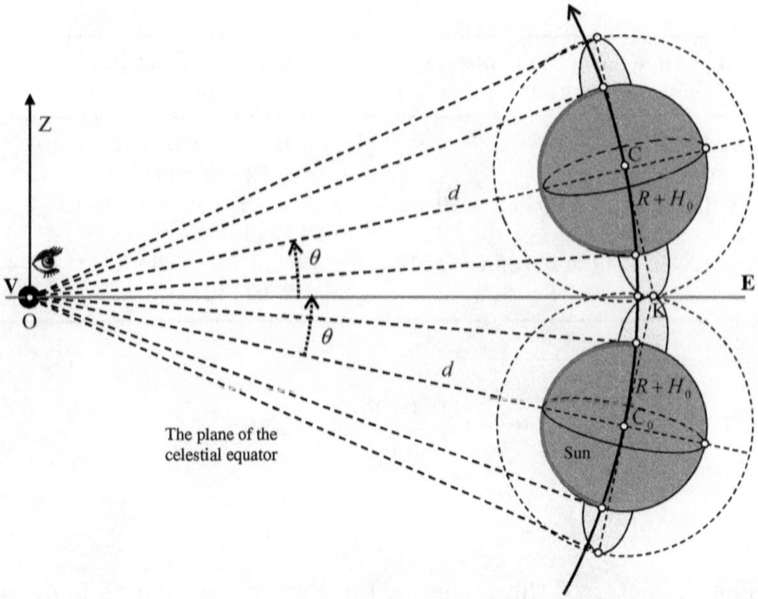

Fig. 40.10

For the calculation of the angle θ in Figure 40.10, from the identical right triangles, OKC_0 and OKC, we find:

$$\tan\theta = \frac{R + H_0}{d};$$

$$d = 150\,000\,000\,\text{km};\ \ R = 696\,000\,\text{km};\ \ T = 24\text{h};\ \ H_0 = 278\,400\,\text{km};$$

$$\tan\theta = \frac{696\,000\,\text{km} + 278\,400\,\text{km}}{150\,000\,000\,\text{km}} = \frac{974\,400}{150\,000\,000};$$

$$\tan\theta = 0.006496;$$

$$\tan\theta \approx \theta;$$

$$\theta = 0.006496\,\text{rad}.$$

Thus, we get:

$$2\theta = \omega\tau = \frac{2\pi}{T}\tau;$$

$$\tau = \frac{\theta}{\pi}T;$$

Table 40.1

The time when the photo was taken t(min)	Explosion front position coordinates $x(m)$
$\tau_1 = 10^{\mathrm{h}}05^{\mathrm{min}}$ $\mathrm{t}_1 = 0$	$x_1 = n_1 \cdot S = 26\,\mathrm{div} \cdot 0.82 \cdot 10^8 \frac{\mathrm{m}}{\mathrm{div}} \approx 21.32 \cdot 10^8\,\mathrm{m}$
$\tau_2 = 10^{\mathrm{h}}54^{\mathrm{min}}$ $\mathrm{t}_2 = 49\,\mathrm{min} = 2940\,\mathrm{s}$	$x_2 = n_2 \cdot S = 36\,\mathrm{div} \cdot 0.82 \cdot 10^8 \frac{\mathrm{m}}{\mathrm{div}} \approx 29.52 \cdot 10^8\,\mathrm{m}$
$\tau_3 = 12^{\mathrm{h}}30^{\mathrm{min}}$ $\mathrm{t}_3 = 145\,\mathrm{min} = 8700\,\mathrm{s}$	$x_3 = n_3 \cdot S = 56\,\mathrm{div} \cdot 0.82 \cdot 10^8 \frac{\mathrm{m}}{\mathrm{div}} \approx 45.92 \cdot 10^8\,\mathrm{m}$

$$\tau = \frac{0.006496}{3.14} 24 \cdot 3600\mathrm{s};$$

$$\tau \approx 178.74\mathrm{s}; \quad \tau \approx 3\,\mathrm{min}.$$

c) The diameter of the image of the Sun in the photos is $d_{\mathrm{S}} = 17$ divisions (mm), and the diameter of the Sun is $D_{\mathrm{S}} = 2R_{\mathrm{S}} = 2 \cdot 6.96 \cdot 10^8\mathrm{m} = 13.92 \cdot 10^8\mathrm{m}$. As a result, the scale of the images in the photos is

$$S = \frac{13.92 \cdot 10^8\,\mathrm{m}}{17\,\mathrm{divisions}} \approx 0.82 \cdot 10^8 \frac{\mathrm{m}}{\mathrm{division}}.$$

The distance from the front of the explosion to the edge of the Sun's disk (number of divisions) is measured with a ruler on each of the given images. Then, using the image scale, S, the position coordinates of the explosion front are determined, corresponding to each specified moment. The results are listed in Table 40.1.

In the drawing in Figure 40.11, the linear form of the graph of the dependence $x = f(t)$ is drawn, allowing us to write:

$$x\,(t) = x_0 + \mathrm{v} \cdot t;$$

$$\mathrm{v} = \frac{\Delta x}{\Delta t} = \frac{45.92 \cdot 10^8\,\mathrm{m} - 21.32 \cdot 10^8\,\mathrm{m}}{8700\mathrm{s}} = \frac{24.6 \cdot 10^8}{8700} \frac{\mathrm{m}}{\mathrm{s}} \approx 282.75 \frac{\mathrm{km}}{\mathrm{s}};$$

$$\mathrm{v} = \frac{\Delta x}{\Delta t} = \frac{29.52 \cdot 10^8\,\mathrm{m} - 21.32 \cdot 10^8\,\mathrm{m}}{2940\mathrm{s}} = \frac{8.2 \cdot 10^8}{2940} \frac{\mathrm{m}}{\mathrm{s}} \approx 278.91 \frac{\mathrm{km}}{\mathrm{s}};$$

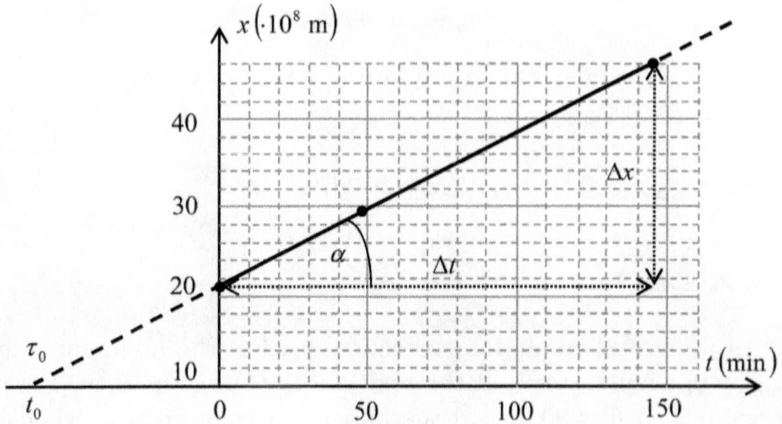

Fig. 40.11

$$\mathrm{v} = \frac{\Delta x}{\Delta t} = \frac{45.92 \cdot 10^8\,\mathrm{m} - 29/52 \cdot 10^8\,\mathrm{m}}{5760\mathrm{s}} = \frac{16.4 \cdot 10^8}{5760}\,\frac{\mathrm{m}}{\mathrm{s}} \approx 284.72\,\frac{\mathrm{km}}{\mathrm{s}};$$

$$\mathrm{v_m} = 282.12\,\frac{\mathrm{km}}{\mathrm{s}}.$$

d) Calculation of the moment of the solar explosion:

$$x = 0; \quad 0 = x_0 + \mathrm{v} \cdot t_0;$$

$$t_0 = -\frac{x_0}{\mathrm{v}} = -\frac{21.32 \cdot 10^8\,\mathrm{m}}{282.12\,\frac{\mathrm{km}}{\mathrm{s}}} = -\frac{21.32 \cdot 10^8}{282.12 \cdot 10^3}\,\mathrm{s} \approx -7202.6\,\mathrm{s}$$

$$\approx -120\ \mathrm{min} = -2\,\mathrm{h};$$

$$\tau_0 = 8^{\mathrm{h}}05^{\mathrm{min}}.$$

e)

$$x(t) = x_0 + \mathrm{v} \cdot t;$$

$$x_0 = x_1 = 21.32 \cdot 10^8\,\mathrm{m}; \quad \mathrm{v} = \mathrm{v_m} = 282.12\,\frac{\mathrm{km}}{\mathrm{s}};$$

$$\tau_4 = 14^{\text{h}}00^{\text{min}};$$

$$t = t_4 = t_3 + 90\ \text{min} = 145\ \text{min} + 90\ \text{min} = 235\ \text{min};$$

$$x_4 = x_0 + \text{v} \cdot t_4;$$

$$x_4 = 61.09 \cdot 10^8\ \text{m}.$$

Annex. The Sun on the Horizon

The astronomical refraction correction, ρ_{r}, has the minimum value, $\rho_{\text{r,min}} = 0$, when a star is at the zenith of the observer on the Earth's surface ($z = 0; h = 90°$), and the maximum value, $\rho_{\text{r,max}}$, when the star, whose light passes through the Earth's atmosphere to reach the eye of the observer on the Earth's surface. ($z = 90°; h = 0$), is on the horizon. That is, when the star rises or sets.

Determine the value $\rho_{\text{r,max}}$, knowing the radius of the Sun, $R_{\text{s}} = 6.96 \cdot 10^5$ km, and the distance between the center of the Earth and the center of the Sun, $d = 15 \cdot 10^7$ km.

It is known that, for an observer, atmospheric refraction raises the apparent disk of the Sun at the moments of sunrise and sunset, when it is below the horizon, approximately at its limit.

Solution

Atmospheric refraction raises the apparent disk of the Sun above the horizon at sunrise and sunset, when it is approximately at the horizon's limit, as shown in the drawing in Figure 40.12.

Therefore, in reality, the rising of the solar disk's upper arc begins after we have already seen the entire solar disk above the horizon. We see the sunrise and sunset later than they occur, which makes the day a little longer.

We can see the whole disk of the Sun above the horizon, at its limit, but the Sun has not yet risen; it is still below the horizon!

Equivalently, after the Sun has gone below the horizon, at its limit, we still see the whole solar disk above the horizon!

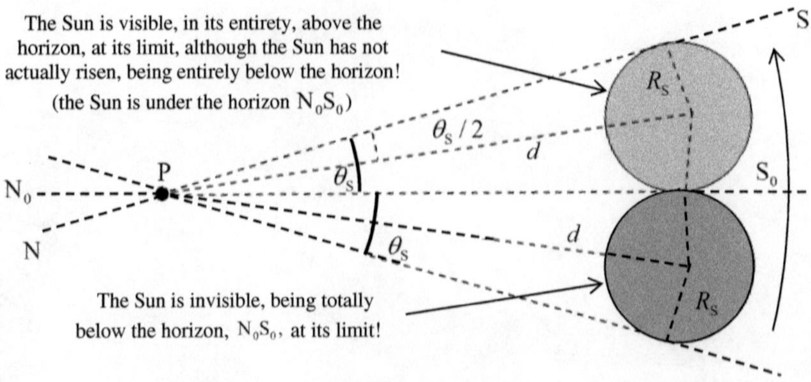

Fig. 40.12

Under these conditions, the astronomical refraction has the maximum value:

$$\rho_{\mathrm{r,max}} = \theta_{\mathrm{s}};$$

$$\sin\frac{\theta_{\mathrm{s}}}{2} = \frac{R_{\mathrm{s}}}{d} \approx \frac{\theta_{\mathrm{s}}}{2};$$

$$\theta_{\mathrm{s}} = \frac{2R_{\mathrm{s}}}{d} = \frac{2\cdot 6.96\cdot 10^5\,\mathrm{km}}{15\cdot 10^7\,\mathrm{km}} = 0.0098\,\mathrm{radians};$$

$$1\,\mathrm{rad} = \frac{180\cdot 60'}{3.14};$$

$$\theta_{\mathrm{s}} = 0.0098\cdot\frac{180\cdot 60'}{3.14} \approx 33.7' = \rho_{\mathrm{r,max}},$$

which proves that the value of the maximum astronomical refraction correction, $\rho_{\mathrm{r,max}}$, is approximately equal to the apparent angular diameter of the Sun, θ_{s}.

Problem 41

Gravitational Retardation of Light

Based on the existence of the gravitational refractive index, the physicist Irwin Shapiro proposed a test of the relativistic theory of gravity, observing that the duration of light propagation between two planets, considered material points, increases if on its way the light passes through the immediate vicinity of the Sun, a phenomenon known as the gravitational retardation of light.

Figure 41.1 represents the Sun (a sphere with radius R) as well as the planets Earth and Mercury in the positions (P1, M1) – lower conjunction; (P2, M2) – superior conjunction; and (P, M) – elongation.

If t_0 is the duration of light propagation between the two planets at superior conjunction, in the absence of the Sun, and if t is the duration of light propagation between the same two planets at superior conjunction, in the presence of the Sun, *determine* the gravitational retardation of light, $\Delta t = t - t_0$, due to the gravitational action of the Sun.

Given: the constant of universal attraction, $K = 6.67 \cdot 10^{-11}\,\mathrm{Nm^2\,kg^{-2}}$; the mass of the Sun, $M \approx 2 \cdot 10^{30}$ kg; the radius of the Sun, $R \approx 6.96 \cdot 10^5$ km; the radius of Mercury's orbit, $r_\mathrm{M} \approx 58 \cdot 10^6$ km; the radius of Earth's orbit, $r_\mathrm{P} \approx 149.6 \cdot 10^6$ km; the speed of light in the absence of the gravitational action of the Sun, $c = 3 \cdot 10^8\ \mathrm{ms^{-1}}$. The orbits of the two planets are coplanar circles.

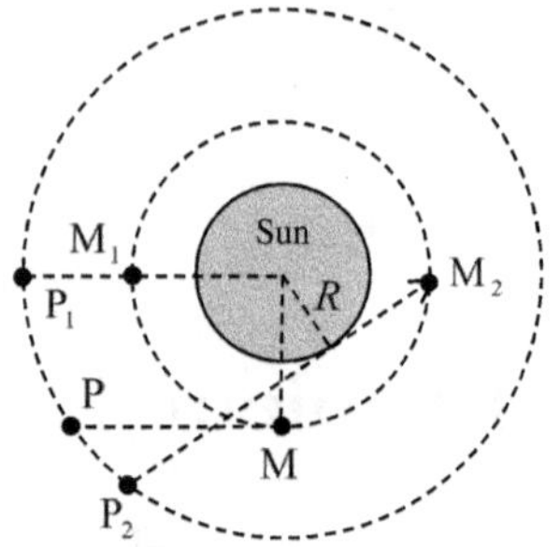

Fig. 41.1

Determine the numerical value of the gravitational retardation of light. You may simplify the expression obtained for Δt by using the fact that $R << r_P$ and $R << r_M$. You may find useful the following integral:

$$\int \frac{\mathrm{d}x}{\sqrt{a^2+x^2}} = \ln(x+\sqrt{a^2+x^2}).$$

Solution

Considering the OX axis, as indicated in the drawing in Figure 41.2, in the absence of the Sun, we find:

$$t_0 = \int_{-l_1}^{+l_2} \frac{\mathrm{d}x}{c} = \frac{1}{c}\int_{-l_1}^{+l_2} \mathrm{d}x;$$

$$l_1 = \mathrm{OP}_2; \quad l_2 = \mathrm{OM}_2;$$

$$l_1 = \sqrt{r_{\mathrm{P}}^2 - R^2}; \quad l_2 = \sqrt{r_{\mathrm{M}}^2 - R^2}.$$

In the presence of the Sun:

$$t = \int_{-l_1}^{+l_2} \frac{\mathrm{d}x}{\mathrm{v}} = \frac{1}{c}\int_{-l_1}^{+l_2} n\mathrm{d}x > t_0.$$

This results in:

$$\Delta t = \frac{1}{c}\int_{-l_1}^{l_2} (n-1)\mathrm{d}x;$$

$$n = e^2\frac{KM}{rc^2}; \quad KM << rc^2;$$

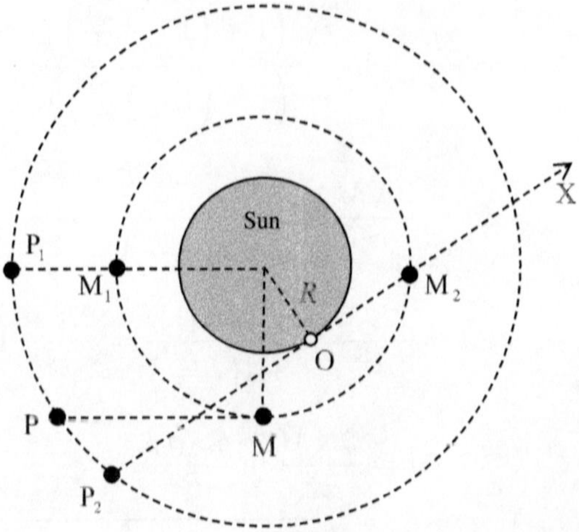

Fig. 41.2

$$n = e^{2\frac{KM}{rc^2}} \approx 1 + 2\frac{KM}{rc^2}; \quad n - 1 = 2\frac{KM}{rc^2};$$

$$r = \sqrt{R^2 + x^2};$$

$$\Delta t = \frac{2KM}{c^3} \int_{-l_1}^{+l_2} \frac{\mathrm{d}x}{\sqrt{R^2 + x^2}};$$

$$\int \frac{\mathrm{d}x}{\sqrt{R^2 + x^2}} = \ln\left(x + \sqrt{R^2 + x^2}\right);$$

$$\Delta t = \frac{2KM}{c^3} \ln \frac{\sqrt{R^2 + l_2^2} + l_2}{\sqrt{R^2 + l_1^2} - l_1};$$

$$l_1 = \sqrt{r_\mathrm{P}^2 - R^2}; \quad l_2 = \sqrt{r_\mathrm{M}^2 - R^2};$$

$$\Delta t = \frac{2KM}{c^3} \ln \frac{r_\mathrm{M} + \sqrt{r_\mathrm{M}^2 - R^2}}{r_\mathrm{P} - \sqrt{r_\mathrm{P}^2 - R^2}} = \frac{2KM}{c^3} \ln \frac{r_\mathrm{M}\left(1 + \sqrt{1 - \frac{R^2}{r_\mathrm{M}^2}}\right)}{r_\mathrm{P}\left(1 - \sqrt{1 - \frac{R^2}{r_\mathrm{P}^2}}\right)};$$

$$R << r_\mathrm{M}; \quad R << r_\mathrm{P};$$

$$\Delta t = \frac{2KM}{c^3} \ln \frac{r_\mathrm{M}\left(1 + \left(1 - \frac{R^2}{r_\mathrm{M}^2}\right)^{1/2}\right)}{r_\mathrm{P}\left(1 - \left(1 - \frac{R^2}{r_\mathrm{P}^2}\right)^{1/2}\right)};$$

$$\Delta t = \frac{2KM}{c^3} \ln \frac{r_{\mathrm{M}}\left(1 + \left(1 - \frac{1}{2}\frac{R^2}{r_{\mathrm{M}}^2}\right)\right)}{r_{\mathrm{P}}\left(1 - \left(1 - \frac{1}{2}\frac{R^2}{r_{\mathrm{P}}^2}\right)\right)};$$

$$\Delta t = \frac{2KM}{c^3} \ln \frac{r_{\mathrm{M}}\left(2 - \frac{1}{2}\frac{R^2}{r_{\mathrm{M}}^2}\right)}{r_{\mathrm{P}}\left(\frac{1}{2}\frac{R^2}{r_{\mathrm{P}}^2}\right)};$$

$$R << r_{\mathrm{M}}; \quad \Delta t = \frac{2KM}{c^3} \ln \frac{2r_{\mathrm{M}}}{r_{\mathrm{P}}\left(\frac{1}{2}\frac{R^2}{r_{\mathrm{P}}^2}\right)} = \frac{2KM}{c^3} \ln \frac{2r_{\mathrm{M}}}{\frac{1}{2}\frac{R^2}{r_{\mathrm{P}}}};$$

$$\Delta t = \frac{2KM}{c^3} \ln \frac{4r_{\mathrm{M}}r_{\mathrm{P}}}{R^2};$$

$$\Delta t = \frac{2KM}{c^3} \ln \left(4\frac{\sqrt{\left(R^2 + r_{\mathrm{P}}^2\right)\left(R^2 + r_{\mathrm{M}}^2\right)}}{R^2}\right);$$

$$R << r_{\mathrm{P}}; \quad R << r_{\mathrm{M}}; \quad \Delta t = \frac{2KM}{c^3} \ln \left(4\frac{r_{\mathrm{P}}r_{\mathrm{M}}}{R^2}\right);$$

$$\Delta t \approx 33 \cdot 10^{-5}\,\mathrm{s}.$$

Problem 42

The Deviation of Light Rays in the Gravitational Field of the Sun

Photons, constituent particles of the light emitted by a certain star, Σ, interact gravitationally with the Sun when they pass through its immediate vicinity and, as a result, the light rays from the star Σ are deflected from their original direction.

a) *Determine*, within the limits of classical mechanics, the angular deviation α_{classic} of the light rays coming from the star Σ, when they pass through the immediate vicinity of the Sun. Compare the calculated angular deviation with that resulting from observational determinations made during solar eclipses (when the light of the stars is not "drowned out" by the light of the Sun), if $\alpha_{\text{obs}} \approx 8.4 \cdot 10^{-6}$ radians.

   Given: $M = 1.99 \cdot 10^{30}\,\text{kg}$ – the mass of the Sun; $R = 6.95 \cdot 10^{8}\,\text{m}$ – the radius of the Sun; $K = 6.67 \cdot 10^{-11}\,\text{Nm}^2\,\text{kg}^2$ – the constant of universal attraction; $c = 3 \cdot 10^{8}\,\text{ms}^{-1}$ – the speed of light in a vacuum.

b) Within the general theory of relativity, it is demonstrated that the differential equation of the relativistic motion of a photon coming from a distant star, Σ, reaching the distance r in the gravitational field of the Sun is

$$\frac{\mathrm{d}}{\mathrm{d}t}(n^2\vec{\mathrm{v}}) = -2\frac{KM}{r^3}\vec{r},$$

where

$$n = e^{2\frac{KM}{rc^2}}$$

is the gravitational index of refraction.

Determine, within the limits of the theory of general relativity, the angular deviation $\alpha_{\text{relativistic}}$ of light rays coming from the star Σ, when they pass through the immediate vicinity of the Sun.

Solution

a)

Method I

The ray of light comes from a distant star, Σ, passes tangentially to the sphere of the Sun and continues on its way to $+\infty$, being deflected by an angle α upon approaching the Sun, as indicated by the drawing in Figure 42.1.

In classical mechanics, the differential equation of motion of a photon coming from a distant star in the gravitational field of the Sun is

$$\frac{\mathrm{d}}{\mathrm{d}t}(\vec{\mathrm{v}}) = -\frac{KM}{r^3}\vec{r}.$$

Therefore, through integration:

$$\mathrm{d}(\vec{\mathrm{v}}) = -\frac{KM}{r^3}\vec{r}\cdot \mathrm{d}t;$$

$$\vec{\mathrm{v}}_{(+\infty)} - \vec{\mathrm{v}}_{(-\infty)} = -KM\int_{-\infty}^{+\infty}\frac{\vec{r}(t)}{r^3(t)}\cdot \mathrm{d}t;$$

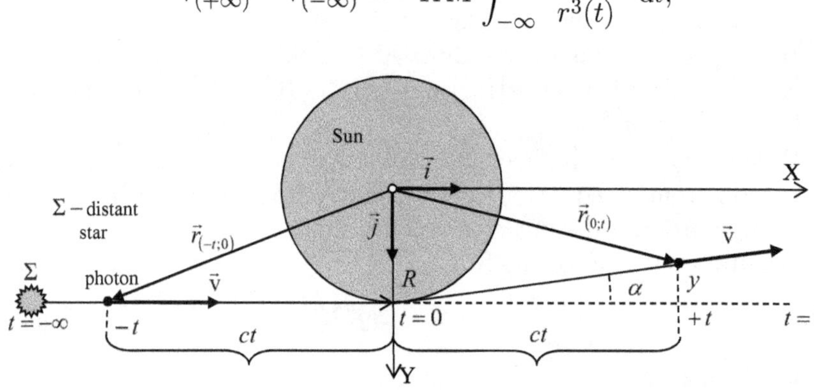

Fig. 42.1

$$\vec{r}(t) = -ct\cdot\vec{i} + R\cdot\vec{j}; \quad r = \sqrt{R^2 + c^2t^2};$$

$$\int_{-\infty}^{+\infty} \frac{\vec{r}(t)}{r^3(t)}\cdot \mathrm{d}t = \int_{-\infty}^{+\infty} \frac{-ct\cdot\vec{i} + R\cdot\vec{j}}{(R^2+c^2t^2)^{3/2}}\cdot \mathrm{d}t$$

$$= -\vec{i}\int_{-\infty}^{+\infty} \frac{ct\mathrm{d}t}{(R^2+c^2t^2)^{3/2}} + R\cdot\vec{j}\int_{-\infty}^{+\infty} \frac{\mathrm{d}t}{(R^2+c^2t^2)^{3/2}};$$

$$I_1 = \int_{-\infty}^{+\infty} \frac{ct\mathrm{d}t}{(R^2+c^2t^2)^{3/2}}; \quad I_2 = \int_{-\infty}^{+\infty} \frac{\mathrm{d}t}{(R^2+c^2t^2)^{3/2}};$$

$$ct = x; \quad t = \frac{x}{c}; \quad c\mathrm{d}t = \mathrm{d}x; \quad \mathrm{d}t = \frac{\mathrm{d}x}{c};$$

$$I_1 = \int_{-\infty}^{+\infty} \frac{x\frac{\mathrm{d}x}{c}}{(R^2+x^2)^{3/2}} = \frac{1}{c}\int_{-\infty}^{+\infty} \frac{x\mathrm{d}x}{(R^2+x^2)^{3/2}};$$

$$I_1 = \frac{1}{c}\int_{-\infty}^{0} \frac{x\mathrm{d}x}{(R^2+x^2)^{3/2}} + \frac{1}{c}\int_{0}^{\infty} \frac{x\mathrm{d}x}{(R^2+x^2)^{3/2}};$$

$$I_1 = -\frac{1}{c}\int_{0}^{\infty} \frac{x\mathrm{d}x}{(R^2+x^2)^{3/2}} + \frac{1}{c}\int_{0}^{\infty} \frac{x\mathrm{d}x}{(R^2+x^2)^{3/2}};$$

$$I_1 = 0;$$

$$I_2 = \int_{-\infty}^{+\infty} \frac{\mathrm{d}t}{(R^2+c^2t^2)^{3/2}}$$

$$= \int_{-\infty}^{0} \frac{\mathrm{d}t}{(R^2+c^2t^2)^{3/2}} + \int_{0}^{+\infty} \frac{\mathrm{d}t}{(R^2+c^2t^2)^{3/2}};$$

$$I_2 = -\int_{0}^{-\infty} \frac{\mathrm{d}t}{(R^2+c^2t^2)^{3/2}} + \int_{0}^{+\infty} \frac{\mathrm{d}t}{(R^2+c^2t^2)^{3/2}};$$

$$I_2 = \int_{0}^{\infty} \frac{\mathrm{d}t}{(R^2+c^2t^2)^{3/2}} + \int_{0}^{+\infty} \frac{\mathrm{d}t}{(R^2+c^2t^2)^{3/2}};$$

$$I_2 = 2\cdot\int_{0}^{\infty} \frac{\mathrm{d}t}{(R^2+c^2t^2)^{3/2}};$$

$$\int_{-\infty}^{+\infty} \frac{\vec{r}(t)}{r^3(t)} \cdot \mathrm{d}t = -\vec{i}\int_{-\infty}^{+\infty} \frac{ct\mathrm{d}t}{(R^2+c^2t^2)^{3/2}}$$

$$+ R\cdot\vec{j}\int_{-\infty}^{+\infty} \frac{\mathrm{d}t}{(R^2+c^2t^2)^{3/2}};$$

$$\int_{-\infty}^{+\infty} \frac{\vec{r}(t)}{r^3(t)} \cdot \mathrm{d}t = -\vec{i}I_1 + R\cdot\vec{j}I_2; \quad I_1 = 0;$$

$$\int_{-\infty}^{+\infty} \frac{\vec{r}(t)}{r^3(t)} \cdot \mathrm{d}t = R\cdot\vec{j}\int_{-\infty}^{+\infty} \frac{\mathrm{d}t}{(R^2+c^2t^2)^{3/2}};$$

$$\int_{-\infty}^{+\infty} \frac{\vec{r}(t)}{r^3(t)} \cdot \mathrm{d}t = 2R\cdot\vec{j}\cdot\int_{0}^{\infty} \frac{\mathrm{d}t}{(R^2+c^2t^2)^{3/2}};$$

$$ct = x; \quad t = \frac{x}{c}; \quad c\mathrm{d}t = \mathrm{d}x; \quad \mathrm{d}t = \frac{\mathrm{d}x}{c};$$

$$\int_{0}^{\infty} \frac{\mathrm{d}t}{(R^2+c^2t^2)^{3/2}} = \int_{0}^{\infty} \frac{\mathrm{d}t}{\left(R^2\left(1+\frac{c^2t^2}{R^2}\right)\right)^{3/2}};$$

$$\int_{0}^{\infty} \frac{\mathrm{d}t}{(R^2+c^2t^2)^{3/2}} = \frac{1}{R^3}\int_{0}^{\infty} \frac{\mathrm{d}t}{\left(1+\frac{c^2t^2}{R^2}\right)^{3/2}};$$

$$\frac{ct}{R} = z; \quad \mathrm{d}t = \frac{R}{c}\cdot\mathrm{d}z;$$

$$\int_{0}^{\infty} \frac{\mathrm{d}t}{\left(1+\frac{c^2t^2}{R^2}\right)^{3/2}} = \int_{0}^{\infty} \frac{\frac{R}{c}\mathrm{d}z}{(1+z^2)^{3/2}} = \frac{R}{c}\cdot\int_{0}^{\infty} \frac{\mathrm{d}z}{(1+z^2)^{3/2}};$$

$$\int_{0}^{\infty} \frac{\mathrm{d}t}{(R^2+c^2t^2)^{3/2}} = \frac{1}{R^3}\cdot\frac{R}{c}\cdot\int_{0}^{\infty} \frac{\mathrm{d}z}{(1+z^2)^{3/2}};$$

$$\int_{0}^{\infty} \frac{\mathrm{d}t}{(R^2+c^2t^2)^{3/2}} = \frac{1}{R^2c}\cdot\int_{0}^{\infty} \frac{\mathrm{d}z}{(1+z^2)^{3/2}};$$

$$z = \mathrm{tg}\ \varphi;$$

$$\mathrm{d}z = \mathrm{d}\left(\mathrm{tg}\ \varphi\right) = \mathrm{d}\left(\frac{\sin\varphi}{\cos\varphi}\right) = \frac{\cos\varphi\cdot\mathrm{d}\left(\sin\varphi\right) - \sin\varphi\cdot\mathrm{d}\left(\cos\varphi\right)}{\cos^2\varphi}$$

$$= \frac{\cos^2\varphi\cdot\mathrm{d}\varphi + \sin^2\varphi\cdot\mathrm{d}\varphi}{\cos^2\varphi}; \quad \mathrm{d}z = \frac{\mathrm{d}\varphi}{\cos^2\varphi};$$

$$(1+z^2)^{3/2} = (1+\mathrm{tg}^2\,\varphi)^{3/2} = \left(\frac{1}{\cos^2\varphi}\right)^{3/2} = \frac{1}{(\cos^2\varphi)^{3/2}} = \frac{1}{\cos^3\varphi};$$

$$\int_0^\infty \frac{\mathrm{d}z}{(1+z^2)^{3/2}} = \int_0^{\pi/2} \frac{\frac{\mathrm{d}\varphi}{\cos^2\varphi}}{\frac{1}{\cos^3\varphi}} = \int_0^{\pi/2} \cos\varphi\cdot\mathrm{d}\varphi$$

$$= \int_0^{\pi/2} \mathrm{d}\,(\sin\varphi) = 1;$$

$$\int_0^\infty \frac{\mathrm{d}t}{(R^2+c^2t^2)^{3/2}} = \frac{1}{R^2c}\cdot\int_0^\infty \frac{\mathrm{d}z}{(1+z^2)^{3/2}} = \frac{1}{R^2c};$$

$$\int_0^\infty \frac{\mathrm{d}t}{(R^2+c^2t^2)^{3/2}} = \frac{1}{R^2c};$$

$$\int_{-\infty}^{+\infty} \frac{\vec{r}(t)}{r^3(t)}\cdot\mathrm{d}t = 2R\cdot\vec{j}\cdot\int_0^\infty \frac{\mathrm{d}t}{(R^2+c^2t^2)^{3/2}};$$

$$\int_{-\infty}^{+\infty} \frac{\vec{r}(t)}{r^3(t)}\cdot\mathrm{d}t = 2R\cdot\vec{j}\cdot\frac{1}{R^2c};$$

$$\int_{-\infty}^{+\infty} \frac{\vec{r}(t)}{r^3(t)}\cdot\mathrm{d}t = \vec{j}\cdot\frac{2}{Rc};$$

$$\vec{\mathrm{v}}_{(+\infty)} - \vec{\mathrm{v}}_{(-\infty)} = -KM\int_{-\infty}^{+\infty} \frac{\vec{r}(t)}{r^3(t)}\cdot\mathrm{d}t;$$

$$\vec{\mathrm{v}}_{(+\infty)} - \vec{\mathrm{v}}_{(-\infty)} = -2\frac{KM}{Rc}\vec{j};$$

$$\vec{\mathrm{v}}_{(-\infty)} = c\cdot\vec{i};\quad \vec{\mathrm{v}}_{(+\infty)} = c\left(\cos\alpha\cdot\vec{i} - \sin\alpha\cdot\vec{j}\right);$$

$$\cos\alpha\approx 1;\quad \sin\alpha\approx\alpha;$$

$$\vec{\mathrm{v}}_{(+\infty)} = c\left(\vec{i} - \alpha\cdot\vec{j}\right);$$

$$\vec{\mathrm{v}}_{(+\infty)} - \vec{\mathrm{v}}_{(-\infty)} = c\left(\vec{i} - \alpha\cdot\vec{j}\right) - \mathrm{v}c\cdot\vec{i};$$

$$\vec{\mathrm{v}}_{(+\infty)} - \vec{\mathrm{v}}_{(-\infty)} = -c\alpha\cdot\vec{j};$$

$$\vec{\mathrm{v}}_{(+\infty)} - \vec{\mathrm{v}}_{(-\infty)} = -2\frac{KM}{Rc}\vec{j};$$

$$-c\alpha \cdot \vec{j} = -2\frac{KM}{Rc}\vec{j};$$

$$\alpha = 2\frac{KM}{Rc^2};$$

$$M = 1.99 \cdot 10^{30}\,\text{kg}; \quad R = 6.95 \cdot 10^8\,\text{m}; \quad K = 6.67 \cdot 10^{-11}\,\text{Nm}^2\text{kg}^2;$$

$$c = 3 \cdot 10^8\,\text{ms}^{-1};$$

$$\alpha_{\text{classic}} = \frac{2KM}{Rc^2} \approx 4.2 \cdot 10^{-6}\,\text{radians};$$

$$\alpha_{\text{obs}} \approx 8.4 \cdot 10^{-6}\,\text{radians};$$

$$\alpha_{\text{obs}} = 2 \cdot \alpha_{\text{classic}}.$$

The difference between α_{obs} and α_{classic} must be explained if one considers the corrections brought by the theory of general relativity.

Method II

It is known that, as shown in the drawing in Figure 42.2, when launched from point P_0 with speed $\vec{v}_0$, at an angle α to the direction of $\vec{r}_0$, a material point with mass m, located under the gravitational action of a planet with radius R and mass M, will evolve on a conic trajectory (a Keplerian trajectory, in which the center of the planet is located), whose equation, in plane polar coordinates, is

$$\frac{1}{r} = \frac{1 + e\cos(\theta - \theta')}{p}.$$

It is known that $\arcsin\left(\frac{KM}{Rc^2}\right) \approx \frac{KM}{Rc^2}$.

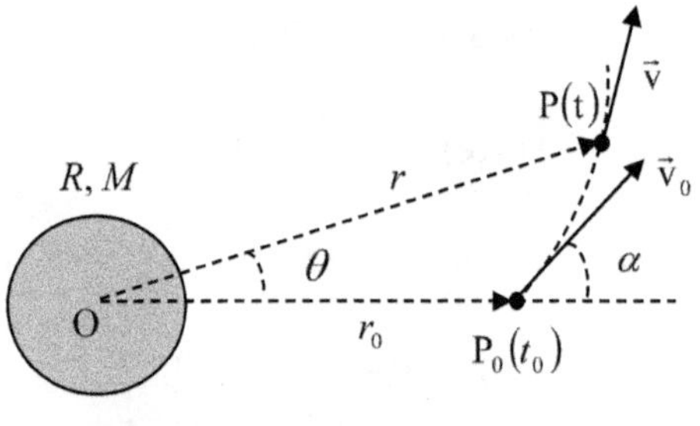

Fig. 42.2

Considering the conditions under which the launch is initiated (the injection of the material point at point P_0), meaning its values r_0 and v_0, the shape of the trajectory of the material point is determined as follows:

$$\frac{1}{r} = \frac{1 + e\cos(\theta - \theta')}{p}; \quad p = \frac{r_0^2 v_0^2 \sin^2\alpha}{KM};$$

$$\mathrm{tg}\,\theta' = \frac{p}{r_0 - p}\mathrm{ctg}\,\alpha;$$

$$e^2 = 1 + \frac{2r_0 v_0^2 \sin^2\alpha}{KM}\left(\frac{r_0 v_0^2}{2KM} - 1\right).$$

The trajectory is an ellipsis if $r_0 v_0^2 < 2KM$, $e < 1$; a parabola if $r_0 v_0^2 = 2KM$, $e = 1$; or a hyperbola if $r_0 v_0^2 > 2KM$, $e > 1$.

In particular, if the injection is carried out along the direction $\alpha = 90°$, it results in:

$$p = \frac{r_0^2 v_0^2}{KM}; \quad \mathrm{ctg}\,\alpha = 0; \quad \mathrm{tg}\,\theta' = 0; \quad r = \frac{p}{1 + e\cos\theta};$$

$$e^2 = 1 + \frac{2r_0 v_0^2}{KM}\left(\frac{r_0 v_0^2}{2KM} - 1\right) = \left(1 - \frac{r_0 v_0^2}{KM}\right)^2.$$

When the conic is a hyperbola and the initial conditions for the injection of the material point on the trajectory are those represented in the drawing in Figure 42.3, the relations of the angle between the asymptotes, 2β, and the polar coordinate θ, as well as the angular deviation, δ, are established using the equation of the conic as follows:

$$\frac{1}{r} = \frac{1 + e\cos(\theta - \theta')}{p};$$

$$\mathrm{tg}\,\theta' = \frac{p}{r_0 - p}\mathrm{ctg}\,\alpha;$$

$$\alpha = 90°, \quad \theta' = 0;$$

$$r \to \infty; \quad 1 + e\cos\theta = 0; \quad \cos\theta = -\frac{1}{e};$$

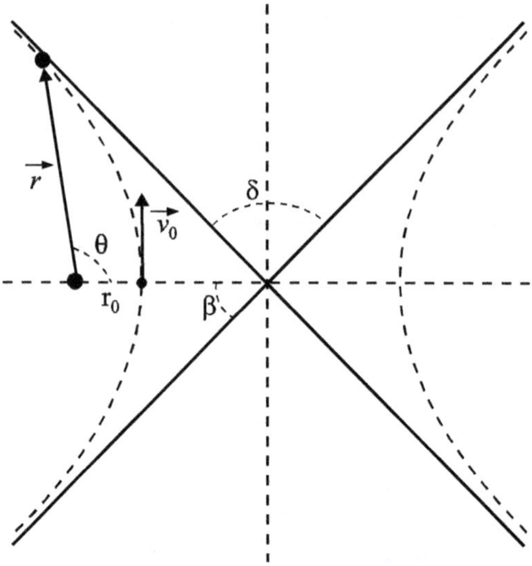

Fig. 42.3

$$\theta = \pm \left[\arccos\left(-\frac{1}{e}\right) \right] + 2k\pi;$$

$$\theta = \pm \left[\frac{\pi}{2} + \arcsin\left(\frac{1}{e}\right) \right] + 2k\pi.$$

Thus, if we limit ourselves to only the values of θ between 0 and π, we have:

$$\theta = \frac{\pi}{2} + \arcsin\left(\frac{1}{e}\right);$$

$$\theta + \beta \approx \pi; \quad \beta = \pi - \theta; \quad \delta = \pi - 2\beta; \quad \delta = 2\arcsin\left(\frac{1}{e}\right);$$

$$e^2 = \left(1 - \frac{r_0 v_0^2}{KM}\right)^2;$$

$$e = \pm\left(1 - \frac{r_0 v_0^2}{KM}\right);$$

$$e = +\left(1 - \frac{r_0 v_0^2}{KM}\right); \quad r_0 v_0^2 > 2KM; \quad r_0 v_0^2 > KM; \quad e < 0;$$

$$e = -\left(1 - \frac{r_0 v_0^2}{KM}\right); \quad r_0 v_0^2 > 2KM; \quad e > 1;$$

$$e = \frac{r_0 v_0^2}{KM} - 1;$$

$$\delta = 2 \arcsin\left(\frac{1}{\frac{r_0 v_0^2}{KM} - 1}\right).$$

Particularizing for a photon from a stellar light ray which passes through the vicinity of the Sun, tangent to its surface, when $r_0 = R$, $v_0 = c$, $\vec{c} \perp \vec{R}$ ($\alpha = 90°$), $Rc^2 = 62.55 \cdot 10^{24}\,\mathrm{m^3 s^{-2}}$ and $2KM = 26.54 \cdot 10^{19}\,\mathrm{m^3 s^{-2}}$, since $Rc^2 \gg 2KM$ and $e > 1$, then the trajectory of the stellar photon, after the gravitational interaction with the Sun, will be an arc of a hyperbola, with the center of the Sun in its focus.

For the eccentricity of this hyperbola, we have:

$$e^2 = \left(1 - \frac{Rc^2}{KM}\right)^2 \approx \left(-\frac{Rc^2}{KM}\right)^2;$$

$$e^2 = \frac{R^2 c^4}{K^2 M^2}; \quad e = \frac{Rc^2}{KM} \gg 1;$$

$$\frac{1}{e} = \frac{KM}{Rc^2} \ll 1; \quad \arcsin\left(\frac{1}{e}\right) \approx \frac{1}{e}.$$

Thus, due to the gravitational action of the Sun, the deviation of the light ray, coming from the star Σ, at infinity, will be:

$$\delta = 2 \arcsin\left(\frac{1}{e}\right) = \frac{2}{e}; \quad e = \frac{Rc^2}{KM}; \quad \frac{1}{e} = \frac{KM}{Rc^2};$$

$$\delta = \frac{2KM}{Rc^2} \approx 4.2 \cdot 10^{-6}\,\text{radians} = \delta_{\text{classic}}.$$

This result does not coincide with the experimental determinations made during solar eclipses (when the light of the stars is not "drowned out" by the light of the Sun):

$$\delta_{\text{obs}} \approx 8.4 \cdot 10^{-6}\,\text{radians} = 2\delta.$$

The discrepancy is resolved by general Relativistic mechanics, which proves that

$$\delta_{\text{relativisty}} = 2\delta = \frac{4KM}{Rc^2} \approx 8.4 \cdot 10^{-6}\text{radians} = 2\delta_{\text{classic}}.$$

b) Within the general theory of relativity, it is demonstrated that the differential equation of the relativistic motion of a photon coming from a distant star, Σ, reaching the distance r in the gravitational field of the Sun is

$$\frac{\mathrm{d}}{\mathrm{d}t}(n^2\vec{\mathrm{v}}) = -2\frac{KM}{r^3}\vec{r},$$

where

$$n = e^{2\frac{KM}{rc^2}}$$

is the gravitational index of refraction.

The integration of the photon's motion equation, under the conditions specified by the drawing in Figure 42.4, results in:

$$\vec{\mathrm{v}}(-\infty) = c \cdot \vec{i};$$

$$\vec{\mathrm{v}}(+\infty) = c \cdot \left(\vec{i}\cos\alpha + \vec{j}\sin\alpha\right) \approx c \cdot \left(\vec{i} + \alpha \cdot \vec{j}\right);$$

$$n(-\infty) = n(+\infty) = 1;$$

$$\mathrm{d}\left(n^2\vec{\mathrm{v}}\right) = -2\frac{KM}{r^3}\vec{r}\mathrm{d}t;$$

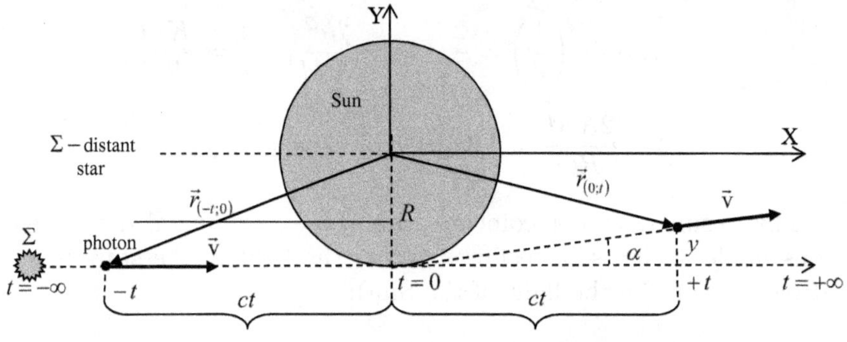

Fig. 42.4

$$\left(n^2\vec{v}\right)_{+\infty} - \left(n^2\vec{v}\right)_{-\infty} = -2KM \int_{-\infty}^{+\infty} \frac{\vec{r}(t)}{r^3(t)}\mathrm{d}t;$$

$$c\alpha \cdot \vec{j} = -2KM \int_{-\infty}^{+\infty} \frac{\vec{r}}{r^3}\mathrm{d}t;$$

$$t \in (-\infty, 0]\,;$$

$$\vec{r} = -ct\vec{i} - R\vec{j}; \quad r^2 = R^2 + c^2t^2;$$

$$t \in [0, +\infty);$$

$$\vec{r} = ct\vec{i} - (R - y)\,\vec{j} = ct\vec{i} - (R - ct \cdot \mathrm{tg}\,\alpha)\,\vec{j};$$

$$\mathrm{tg}\,\alpha \approx \alpha;$$

$$\vec{r} = ct\vec{i} - (R - ct \cdot \alpha)\,\vec{j};$$

$$ct \cdot \alpha \ll R;$$

$$\vec{r} = ct\vec{i} - R\vec{j}; \quad r^2 = R^2 + c^2t^2;$$

$$c\alpha \cdot \vec{j} = -2KM \left(\int_{-\infty}^{0} \frac{-ct\vec{i} - R\vec{j}}{r^3}\mathrm{d}t + \int_{0}^{\infty} \frac{ct\vec{i} - R\vec{j}}{r^3}\mathrm{d}t \right);$$

$$c\alpha \cdot \vec{j} = -2KM \left(\int_{-\infty}^{0} \frac{-ct\vec{i}}{r^3}\mathrm{d}t + \int_{0}^{\infty} \frac{ct\vec{i}}{r^3}\mathrm{d}t + \int_{-\infty}^{0} \frac{-R\vec{j}}{r^3}\mathrm{d}t \right.$$

$$\left. + \int_{0}^{+\infty} \frac{-R\vec{j}}{r^3}\mathrm{d}t \right);$$

$$-t \Rightarrow t; \quad -\vec{i} \Rightarrow \vec{i};$$

$$\int_{-\infty}^{0} \frac{-ct\vec{i}}{r^3}\mathrm{d}t = -\int_{0}^{-\infty} \frac{-ct\vec{i}}{r^3}\mathrm{d}t = -\int_{0}^{\infty} \frac{ct\vec{i}}{r^3}\mathrm{d}t;$$

$$\int_{-\infty}^{0} \frac{-R\vec{j}}{r^3}\mathrm{d}t = -\int_{0}^{-\infty} \frac{-R\vec{j}}{r^3}\mathrm{d}t = -\int_{0}^{+\infty} \frac{R\vec{j}}{r^3}\mathrm{d}t;$$

$$c\alpha \cdot \vec{j} = -2KM\left(\int_{-\infty}^{0} \frac{-ct\vec{i}}{r^3}\mathrm{d}t + \int_{0}^{\infty} \frac{ct\vec{i}}{r^3}\mathrm{d}t + \int_{-\infty}^{0} \frac{-R\vec{j}}{r^3}\mathrm{d}t + \int_{0}^{+\infty} \frac{-R\vec{j}}{r^3}\mathrm{d}t\right);$$

$$c\alpha \cdot \vec{j} = -2KM\left(-\int_{0}^{\infty} \frac{ct\vec{i}}{r^3}\mathrm{d}t + \int_{0}^{\infty} \frac{ct\vec{i}}{r^3}\mathrm{d}t - \int_{0}^{\infty} \frac{R\vec{j}}{r^3}\mathrm{d}t - \int_{0}^{+\infty} \frac{R\vec{j}}{r^3}\mathrm{d}t\right);$$

$$c\alpha \cdot \vec{j} = 4KMR \cdot \vec{j}\int_{0}^{+\infty} \frac{\mathrm{d}t}{r^3};$$

$$\alpha = 4\frac{KMR}{c}\int_{0}^{+\infty} \frac{\mathrm{d}t}{\left(\sqrt{R^2 + c^2t^2}\right)^3};$$

$$\alpha = 4\frac{KM}{cR^2}\int_{0}^{+\infty} \frac{\mathrm{d}t}{\left(1 + \frac{c^2t^2}{R^2}\right)^{3/2}};$$

$$u = \frac{ct}{R}; \quad \mathrm{d}u = \frac{c}{R}\mathrm{d}t;$$

$$\alpha = 4\frac{KM}{Rc^2}\int_{0}^{\infty} \frac{\mathrm{d}u}{(1 + u^2)^{3/2}};$$

$$u = \mathrm{tg}\,\varphi; \quad \mathrm{d}u = \frac{\mathrm{d}\varphi}{\cos^2\varphi};$$

$$\int_{0}^{\infty} \frac{\mathrm{d}u}{(1 + u^2)^{3/2}} = \int_{0}^{\pi/2} \mathrm{d}(\sin\varphi) = 1;$$

$$\alpha_{\text{relativist}} = 4\frac{KM}{c^2R} = 2\alpha_{\text{classic}}.$$

Observation: The problem of the curvature of a light ray when passing by a massive star was first posed by Henry Cavendish, who, knowing the corpuscular theory of light (Newton's theory), calculated the deviation of a photon in the gravitational field of the Sun, using the equation of motion from Newtonian physics:

$$\vec{a} = \frac{\mathrm{d}\vec{v}}{\mathrm{d}t} = \frac{\vec{F}}{m_{\mathrm{photon}}} = -K \cdot \frac{M}{r^3},$$

in which, fortunately, the mass of the photon does not appear explicitly, so he obtained:

$$\alpha_{\mathrm{classic}} = 2\frac{KM}{c^2 R}.$$

Problem 43

Gravitational Refractive Index

The deflection of light rays emitted by a distant star, when passing through the immediate vicinity of the Sun on their way to Earth, proves that photons, in addition to an inertial mass, have a gravitational mass which, according to the principle of equivalence, is equal to the inertial mass.

This gravitational "refraction" of the light ray, like optical refraction, is associated in the theory of general relativity with a gravitational index of refraction.

$$n = \frac{c}{\mathrm{v}},$$

where c is the speed of light in a vacuum measured in a coordinate system far from the gravitational influence of the Sun ($r \to \infty$), and v is the speed of light measured in a coordinate system at a distance r from the center of the Sun, for which the following expression holds:

$$n(r) = e^{2\frac{KM}{rc^2}},$$

where K is the constant of universal attraction, and M is the mass of the Sun.

a) *Determine* v (r) in the approximation $KM \ll rc^2$.

b) *Determine* the relationship between the frequency v_0 of a free photon (in the absence of any gravitational interaction) and the frequency v of a photon at a distance r from the Sun.

Consider that $KM \ll rc^2$, and that

$$e^z \approx 1 + z, \quad \text{if } z \ll 1.$$

Solution

a)

$$x \ll 1; \quad e^x \approx 1 + x; \quad n(r) \approx 1 + 2\frac{KM}{rc^2};$$

$$\mathrm{v}(r) = c\left(1 - 2\frac{KM}{rc^2}\right) \approx c\left(1 - \frac{KM}{rc^2}\right)^2.$$

b) Applying the law of conservation of energy, we get:

$$h\nu_0 = h\nu - K\frac{m_f M}{r}; \quad m_f c^2 = h\nu; \quad \nu_0 = \nu\left(1 - K\frac{M}{rc^2}\right).$$

Problem 44

Stellar Gas Density

In a simple model, a star is considered a gaseous sphere in equilibrium in its gravitational field. The stellar gas consists of hydrogen and helium atoms, completely ionized (plasma). *Determine* the approximate value of the stellar gas density, knowing: r – the radius of the star; T – the star's temperature; n – the relative proportion of hydrogen in the mass of the star; μ_{H} – the molar mass of hydrogen; μ_{He} – the molar mass of helium; R – the universal ideal gas constant; K – the constant of universal attraction. The expression for the radiation pressure inside the star is known, $p_{\mathrm{rad}} = \frac{1}{3}aT^4$, where a is a known constant. The rotation of the star is neglected.

Solution

The hydrostatic equilibrium inside the star means that at every point inside the star, the gravitational forces and the pressure forces balance each other. This means that the stellar gas (plasma) does not spread out into space under pressure forces, but it also does not contract or fall inward under the action of gravity. Hydrostatic equilibrium is achieved globally in the star for a long time. The balance can be disturbed locally, in certain areas, and temporarily during certain phases of stellar evolution.

The total pressure of the stellar gas has two components: the pressure determined by the disordered motions of the particles that make up the stellar gas (p_{gas}) and the pressure due to the radiation emitted by the gas particles (p_{rad}), such that:

$$p_{\text{total}} = p_{\text{gas}} + p_{\text{rad}};$$

$$p_{\text{gas}} = p_{\text{H}} + p_{\text{He}};$$

$$p_{\text{H}}V = \frac{M_{\text{H}}}{\mu_{\text{H}}}RT; \quad p_{\text{He}}V = \frac{M_{\text{He}}}{\mu_{\text{He}}}RT;$$

$$p_{\text{gas}} = \frac{RT}{V}\left(\frac{M_{\text{H}}}{\mu_{\text{H}}} + \frac{M_{\text{He}}}{\mu_{\text{He}}}\right);$$

$$p_{\text{gas}}V = \left(\frac{M_{\text{H}}}{\mu_{\text{H}}} + \frac{M_{\text{He}}}{\mu_{\text{He}}}\right)RT;$$

$$p_{\text{gas}}V = \frac{M}{\mu}RT = \frac{M_{\text{H}} + M_{\text{He}}}{\mu}RT;$$

$$\frac{M_{\text{H}} + M_{\text{He}}}{\mu} = \frac{M_{\text{H}}}{\mu_{\text{H}}} + \frac{M_{\text{He}}}{\mu_{\text{He}}} = \frac{M_{\text{H}}\mu_{\text{He}} + M_{\text{He}}\mu_{\text{H}}}{\mu_{\text{H}}\mu_{\text{He}}};$$

$$\mu = \frac{(M_{\text{H}} + M_{\text{He}})\mu_{\text{H}}\mu_{\text{He}}}{M_{\text{H}}\mu_{\text{He}} + M_{\text{He}}\mu_{\text{H}}};$$

$$p_{\text{gas}}V = \frac{M}{\mu}RT; \quad p_{\text{gas}} = \frac{\frac{M}{V}}{\mu}RT = \frac{\rho}{\mu}RT;$$

$$p_{\text{gas}} = \rho\frac{M_{\text{H}}\mu_{\text{He}} + M_{\text{He}}\mu_{\text{H}}}{(M_{\text{H}} + M_{\text{He}})\mu_{\text{H}}\mu_{\text{He}}}RT;$$

$$M_{\text{H}} = nM; \quad M_{\text{He}} = (1-n)M;$$

$$p_{\text{gas}} = \rho\frac{nM\mu_{\text{He}} + (1-n)M\mu_{\text{H}}}{(nM + (1-n)M)\mu_{\text{H}}\mu_{\text{He}}}RT;$$

$$p_{\text{gas}} = \rho\frac{n\mu_{\text{He}} + (1-n)\mu_{\text{H}}}{(n + (1-n))\mu_{\text{H}}\mu_{\text{He}}}RT;$$

$$p_{\text{gas}} = \rho\frac{n\mu_{\text{He}} + (1-n)\mu_{\text{H}}}{\mu_{\text{H}}\mu_{\text{He}}}RT;$$

$$p_{\text{rad}} = \frac{1}{3}aT;$$

$$p_{\text{total}} = p_{\text{gas}} + p_{\text{rad}};$$

$$p_{\text{total}} = \rho\frac{n\mu_{\text{He}} + (1-n)\mu_{\text{H}}}{\mu_{\text{H}}\mu_{\text{He}}}RT + \frac{1}{3}aT.$$

For the calculation of the gravitational pressure, let us consider, as shown in the drawing in Figure 44.1, a gas cylinder along the radius of the star with a very small cross-sectional area, ΔS. If the gravitational attraction force exerted by the entire star on this gas cylinder is $\vec{F}_{\mathrm{g}}$, then the gravitational pressure determined by this column is $p_{\mathrm{grav}} = F_{\mathrm{grav}}/\Delta S$.

For the calculation of the gravitational attraction force exerted by the entire star on the gas cylinder, we imagine the cylinder divided into n identical cylindrical sectors, each with a height Δr and mass Δm. Assuming the homogeneity of the star, it follows that:

$$F_1 = K\frac{\Delta m \cdot M_1}{\left(r - \frac{\Delta r}{2}\right)^2} = K\frac{\Delta m \cdot M_1}{r^2\left(1 - \frac{\Delta r}{2r}\right)^2} = K\frac{\Delta m \cdot M_1}{r^2}\left(1 - \frac{\Delta r}{2r}\right)^{-2};$$

$$\frac{M}{r^3} = \frac{M_1}{(r-\Delta r)^3}; \quad M_1 = M \cdot \frac{(r-\Delta r)^3}{r^3} = M \cdot (1 - \frac{\Delta r}{r})^3;$$

$$F_1 = K\frac{\Delta m \cdot M}{r^2}\left(1 - \frac{\Delta r}{r}\right)^3 \cdot \left(1 - \frac{\Delta r}{2r}\right)^{-2};$$

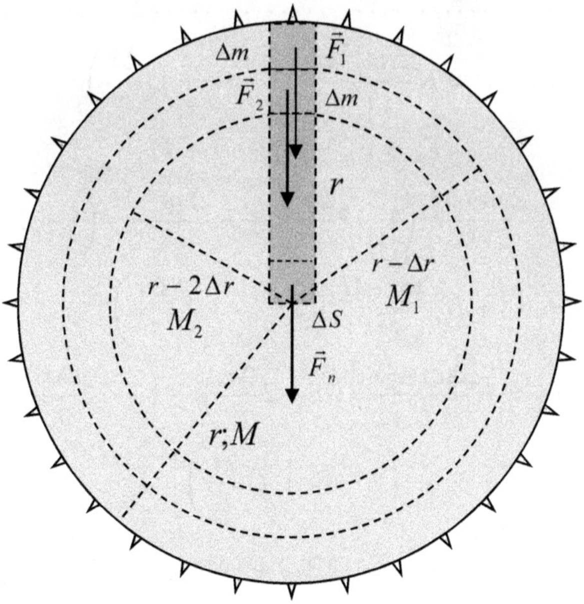

Fig. 44.1

$$F_1 \approx K\frac{\Delta m \cdot M}{r^2}\left(1 - 3\frac{\Delta r}{r}\right) \cdot \left(1 + \frac{\Delta r}{r}\right);$$

$$F_1 \approx K\frac{\Delta m \cdot M}{r^2}\left(1 - 2\frac{\Delta r}{r}\right);$$

$$F_2 = K\frac{\Delta m \cdot M_2}{\left(r - \frac{3}{2}\Delta r\right)^2} = K\frac{\Delta m \cdot M_2}{r^2\left(1 - \frac{3}{2}\frac{\Delta r}{r}\right)^2}$$

$$= K\frac{\Delta m \cdot M_2}{r^2}\left(1 - \frac{3}{2}\frac{\Delta r}{r}\right)^{-2};$$

$$\frac{M}{r^3} = \frac{M_2}{(r - 2 \cdot \Delta r)^3}; \quad M_2 = M \cdot \frac{(r - 2 \cdot \Delta r)^3}{r^3} = M \cdot \left(1 - 2\frac{\Delta r}{r}\right)^3;$$

$$F_2 = K\frac{\Delta m \cdot M}{r^2}\left(1 - 2\frac{\Delta r}{r}\right)^3 \cdot \left(1 - \frac{3}{2}\frac{\Delta r}{r}\right)^{-2};$$

$$F_2 \approx K\frac{\Delta m \cdot M}{r^2}\left(1 - 6\frac{\Delta r}{r}\right) \cdot \left(1 + 3\frac{\Delta r}{r}\right);$$

$$F_2 \approx K\frac{\Delta m \cdot M}{r^2}\left(1 - 3\frac{\Delta r}{r}\right);$$

....

$$F_n \approx K\frac{\Delta m \cdot M}{r^2}\left(1 - (n+1)\frac{\Delta r}{r}\right);$$

$$F_g = F_1 + F_2 + \cdots + F_n;$$

$$F_g = K\frac{\Delta m \cdot M}{r^2}\left(1 - 2\frac{\Delta r}{r}\right) + K\frac{\Delta m \cdot M}{r^2}\left(1 - 3\frac{\Delta r}{r}\right)$$

$$+ \cdots + K\frac{\Delta m \cdot M}{r^2}\left(1 - (n+1)\frac{\Delta r}{r}\right);$$

$$F_g = K\frac{\Delta m \cdot M}{r^2}\left[\left(1 - 2\frac{\Delta r}{r}\right) + \left(1 - 3\frac{\Delta r}{r}\right)\right.$$

$$\left. + \cdots + \left(1 - (n+1)\frac{\Delta r}{r}\right)\right];$$

$$F_g = K\frac{\Delta m \cdot M}{r^2}\left[n - \frac{\Delta r}{r}(2 + 3 + \cdots + (n+1))\right]; \quad n = \frac{r}{\Delta r};$$

$$F_g = K\frac{\Delta m \cdot M}{r^2}\left[n - \frac{1}{n}(1+2+3+\cdots+n)\right]; \quad n = \frac{m}{\Delta m};$$

$$1+2+3+\cdots+n = \frac{(1+n)n}{2};$$

$$F_g = K\frac{m \cdot M}{nr^2}\left[n - \frac{1}{n}\frac{(1+n)n}{2}\right]; \quad F_g = K\frac{m \cdot M}{r^2}\left[1 - \frac{(1+n)}{2n}\right];$$

$$F_g = K\frac{m \cdot M}{r^2}\frac{n-1}{2n}; \quad \frac{n-1}{2n} \approx \frac{1}{2};$$

$$F_g = K\frac{m \cdot M}{2r^2}.$$

The same result is reached if we work with the equivalence presented in the drawing in Figure 44.2 (where the mass of the gas in the radial cylinder, m, is considered the mass of a material point located at the depth $r/2$ inside the star).

In this way, the analyzed gravitational interaction is reduced to the gravitational interaction of the core of the star, having the radius

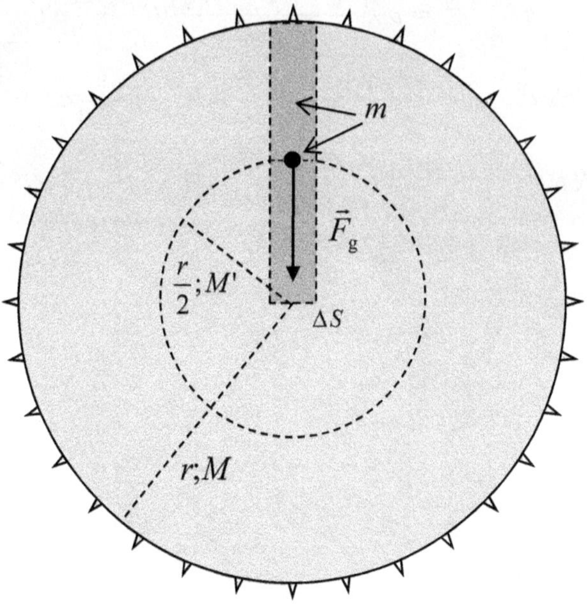

Fig. 44.2

$r/2$ and mass M', with the material point having the mass m located at the distance $r/2$ from the center of the star. The gravitational effect of the spherical shell on the material point inside it is zero.

This results in:

$$F = K\frac{mM'}{\left(\frac{r}{2}\right)^2}; \quad M' = M\frac{\left(\frac{r}{2}\right)^3}{r^3} = \frac{M}{8};$$

$$F = K\frac{m\frac{M}{8}}{\left(\frac{r}{2}\right)^2} = K\frac{mM}{2r^2};$$

$$m = \rho \cdot r \cdot \Delta S; \quad M = \rho\frac{4\pi r^3}{3};$$

$$p_{\text{g}} = \frac{F_{\text{g}}}{\Delta S} = \frac{1}{\Delta S}\frac{\rho \cdot r \cdot \Delta S \cdot \rho \cdot \frac{4\pi r^3}{3}}{2r^2};$$

$$p_{\text{g}} = \frac{2}{3}\pi K r^2 \rho^2;$$

$$p_{\text{total}} = p_{\text{gravitational}};$$

$$\frac{2}{3}\pi K r^2 \rho^2 = \rho\frac{n\mu_{\text{He}} + (1-n)\mu_{\text{H}}}{\mu_{\text{H}}\mu_{\text{He}}}RT + \frac{1}{3}aT;$$

$$\frac{2}{3}\pi K r^2 \cdot \rho^2 - \frac{n\mu_{\text{He}} + (1-n)\mu_{\text{H}}}{\mu_{\text{H}}\mu_{\text{He}}}RT \cdot \rho - \frac{1}{3}aT = 0;$$

$$\frac{2}{3}\pi K r^2 = A; \quad \frac{n\mu_{\text{He}} + (1-n)\mu_{\text{H}}}{\mu_{\text{H}}\mu_{\text{He}}}RT = B; \quad \frac{1}{3}aT = C;$$

$$A\rho^2 - B\rho - C = 0;$$

$$\rho = \frac{B + \sqrt{B^2 + 4AC}}{2A}.$$

Problem 45

The Astronomical Coordinates of the Moon, Earth and Sun

Determine the astronomical coordinates of the Moon, Earth and Sun.

Solution

a) *Equatorial coordinates*

From the spherical triangle EMS, highlighted in the drawing in Figure 45.1, representing the geocentric celestial sphere, which is the analog of the standard spherical triangle ABC shown in the drawing in Figure 45.2, in accordance with the theorems of sines and cosines, we obtain:

$$\angle \mathrm{E} = \alpha_{\mathrm{S}} - \alpha_{\mathrm{M}};$$

$$\sin a \cdot \sin B = \sin b \cdot \sin A;$$

$$\sin(\mathrm{arcMS}) \cdot \sin \mathrm{M_{eq}} = \sin(90^\circ - \delta_{\mathrm{S}}) \cdot \sin(\alpha_{\mathrm{S}} - \alpha_{\mathrm{M}});$$

$$\sin(\mathrm{arcMS}) \cdot \sin \mathrm{M_{eq}} = \cos \delta_{\mathrm{S}} \cdot \sin(\alpha_{\mathrm{S}} - \alpha_{\mathrm{M}}); \qquad (1)$$

$$\cos b = \cos c \cdot \cos a + \sin c \cdot \sin a \cdot \cos B;$$

$$\cos(90^\circ - \delta_{\mathrm{S}}) = \cos(90^\circ - \delta_{\mathrm{M}}) \cdot \cos(\mathrm{arcMS}) + \sin(90^\circ - \delta_{\mathrm{M}}) \cdot \sin(\mathrm{arcMS}) \cdot \cos M_{\mathrm{eq}};$$

$$\sin \delta_{\mathrm{S}} = \sin \delta_{\mathrm{M}} \cdot \cos(\mathrm{arcMS}) + \cos \delta_{\mathrm{M}} \cdot \sin(\mathrm{arcMS}) \cdot \cos M_{\mathrm{eq}};$$

$$\sin(\mathrm{arcMS}) \cdot \cos M_{\mathrm{eq}} = \frac{\sin \delta_{\mathrm{S}} - \sin \delta_{\mathrm{M}} \cdot \cos(\mathrm{arcMS})}{\cos \delta_{\mathrm{M}}}; \qquad (2)$$

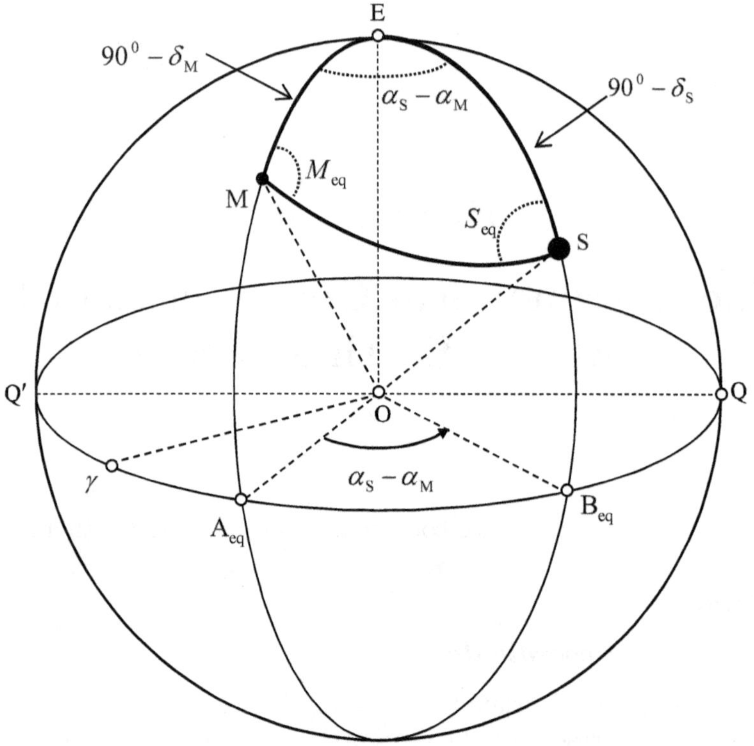

Fig. 45.1

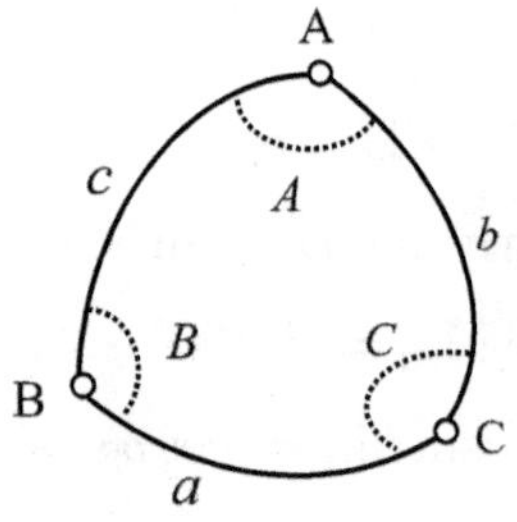

Fig. 45.2

$$\cos a = \cos b \cdot \cos c + \sin b \cdot \sin c \cdot \cos A;$$

$$\cos(\text{arcMS}) = \cos(90^\circ - \delta_S) \cdot \cos(90^\circ - \delta_M) + \sin(90^\circ - \delta_S) \cdot \sin(90^\circ - \delta_M) \cdot \cos(\alpha_S - \alpha_M);$$

$$\cos(\text{arcMS}) = \sin\delta_S \cdot \sin\delta_M + \cos\delta_S \cdot \cos\delta_M \cdot \cos(\alpha_S - \alpha_M). \quad (3)$$

From (2) and (3):

$$\sin(\text{arcMS}) \cdot \cos M_{\text{eq}} = \frac{\sin\delta_{\text{S}} - \sin\delta_{\text{M}} \cdot \cos(\text{arcMS})}{\cos\delta_{\text{M}}};$$

$$\sin(\text{arcMS}) \cdot \cos M_{\text{eq}} = \frac{\sin\delta_{\text{S}} - \sin\delta_{\text{M}} \cdot (\sin\delta_{\text{S}} \cdot \sin\delta_{\text{M}} + \cos\delta_{\text{S}} \cdot \cos\delta_{\text{M}} \cdot \cos(\alpha_{\text{S}} - \alpha_{\text{M}}))}{\cos\delta_{\text{M}}};$$

$$\sin(\text{arcMS}) \cdot \cos M_{\text{eq}} = \frac{\sin\delta_{\text{S}} - \sin\delta_{\text{S}} \cdot \sin^2\delta_{\text{M}} - \cos\delta_{\text{S}} \cdot \sin\delta_{\text{M}} \cdot \cos\delta_{\text{M}} \cdot \cos \cdot (\alpha_{\text{S}} - \alpha_{\text{M}})}{\cos\delta_{\text{M}}};$$

$$\sin(\text{arcMS}) \cdot \cos M_{\text{eq}} = \frac{\sin\delta_{\text{S}} - \sin\delta_{\text{S}} \cdot \sin^2\delta_{\text{M}}}{\cos\delta_{\text{M}}} - \frac{\cos\delta_{\text{S}} \cdot \sin\delta_{\text{M}} \cdot \cos\delta_{\text{M}}}{\cos\delta_{\text{M}}} \cdot \cos(\alpha_{\text{S}} - \alpha_{\text{M}});$$

$$\sin(\text{arcMS}) \cdot \cos M_{\text{eq}} = \frac{\sin\delta_{\text{S}} \cdot (1 - \sin^2\delta_{\text{M}})}{\cos\delta_{\text{M}}} - \cos\delta_{\text{S}} \cdot \sin\delta_{\text{M}} \cdot \cos(\alpha_{\text{S}} - \alpha_{\text{M}});$$

$$\sin(\text{arcMS}) \cdot \cos M_{\text{eq}} = \frac{\sin\delta_{\text{S}} \cdot \cos^2\delta_{\text{M}}}{\cos\delta_{\text{M}}} - \cos\delta_{\text{S}} \cdot \sin\delta_{\text{M}} \cdot \cos(\alpha_{\text{S}} - \alpha_{\text{M}});$$

$$\sin(\text{arcMS}) \cdot \cos M_{\text{eq}} = \sin\delta_{\text{S}} \cdot \cos\delta_{\text{M}} - \cos\delta_{\text{S}} \cdot \sin\delta_{\text{M}} \cdot \cos(\alpha_{\text{S}} - \alpha_{\text{M}}). \tag{4}$$

From (1) and (4):

$$\sin(\text{arcMS}) \cdot \sin M_{\text{eq}} = \cos\delta_{\text{S}} \cdot \sin(\alpha_{\text{S}} - \alpha_{\text{M}});$$

$$\frac{\sin(\text{arcMS}) \cdot \sin M_{\text{eq}}}{\sin(\text{arcMS}) \cdot \cos M_{\text{eq}}} = \frac{\cos\delta_{\text{S}} \cdot \sin(\alpha_{\text{S}} - \alpha_{\text{M}})}{\sin\delta_{\text{S}} \cdot \cos\delta_{\text{M}} - \cos\delta_{\text{S}} \cdot \sin\delta_{\text{M}} \cdot \cos(\alpha_{\text{S}} - \alpha_{\text{M}})};$$

$$\tan M_{\text{eq}} = \frac{\cos\delta_{\text{S}} \cdot \sin(\alpha_{\text{S}} - \alpha_{\text{M}})}{\sin\delta_{\text{S}} \cdot \cos\delta_{\text{M}} - \cos\delta_{\text{S}} \cdot \sin\delta_{\text{M}} \cdot \cos(\alpha_{\text{S}} - \alpha_{\text{M}})}. \tag{5}$$

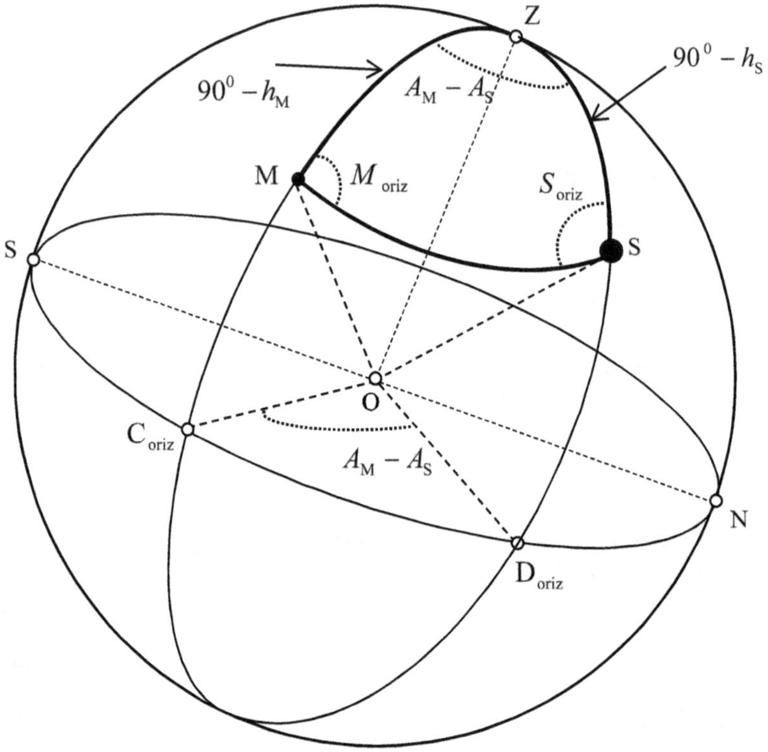

Fig. 45.3

b) *Horizontal coordinates*

Similarly, using the drawing in Figure 45.3, working in horizontal coordinates, the existence of the following relationship is demonstrated:

$$\tan M_{\mathrm{horiz}} = \frac{\cos\ h_{\mathrm{S}} \cdot \sin(A_{\mathrm{M}} - A_{\mathrm{S}})}{\sin h_{\mathrm{S}} \cdot \cos h_{\mathrm{M}} - \cos\ h_{\mathrm{S}} \cdot \sin h_{\mathrm{M}} \cdot \cos(A_{\mathrm{M}} - A_{\mathrm{S}})}. \quad (6)$$

Problem 46

Relativistic Motion in a Central Gravitational Field

Newtonian mechanics tells us that the movement of a material point in the field of the central gravitational force is planar.

Relating the movement of a satellite, S, to a fixed planet, P, when the studied mechanical assembly is associated with the axis systems XY and X′Y′ represented in the drawing in Figure 46.1, using the polar coordinates r and θ, it is known that:

$$\vec{r} = r\vec{\rho}; \quad \vec{\mathrm{v}} = \dot{r}\vec{\rho} + r\dot{\theta}\vec{n}; \quad \vec{a} = (\ddot{r} - r\dot{\theta}^2)\vec{\rho} + (2\dot{r}\dot{\theta} + r\ddot{\theta})\vec{n}; \quad \vec{F} = -F\vec{\rho}.$$

It is also known that the relativistic dynamic effect of force $\vec{F}$ is acceleration:

$$\vec{a} = \frac{1}{m}\vec{F} - \frac{\vec{F}\vec{\mathrm{v}}}{mc^2}\vec{\mathrm{v}},$$

having a collinear component with $\vec{F}$ (as happens in Newtonian mechanics) and a collinear component with $\vec{\mathrm{v}}$ (without a classical correspondent).

Under these conditions, from the fundamental law of dynamics, meaning the differential equation of motion, it follows that:

$$\frac{\mathrm{d}\vec{p}}{\mathrm{d}t} = \vec{F}; \quad \frac{\mathrm{d}(m\vec{\mathrm{v}})}{\mathrm{d}t} = \vec{F}; \quad m = \frac{m_0}{\sqrt{1 - \frac{\mathrm{v}^2}{c^2}}}; \quad m = \frac{m_0}{\sqrt{1 - \frac{\dot{r}^2 + r^2\dot{\theta}^2}{c^2}}},$$

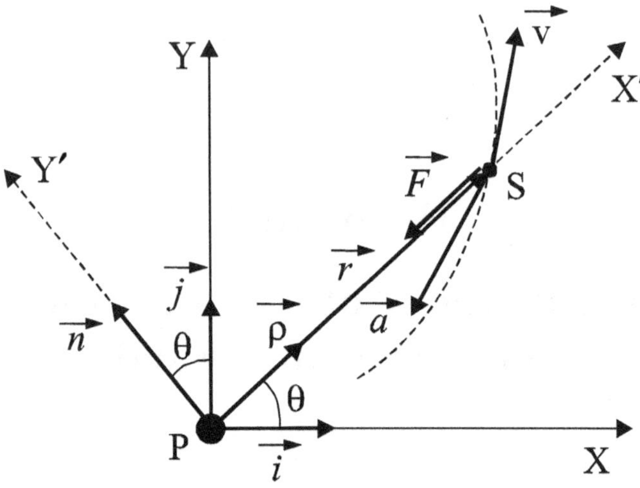

Fig. 46.1

where m_0 is the rest mass of the satellite, and c is the speed of light in a vacuum;

$$m\frac{\mathrm{d}\vec{v}}{\mathrm{d}t} + \vec{v}\frac{\mathrm{d}m}{\mathrm{d}t} = \vec{F}.$$

Demonstrate that the differential equation of the relativistic motion of the satellite, in plane polar coordinates, from which, by integration, the equation of the relativistic trajectory of the satellite in relation to the fixed planet is obtained, is

$$\frac{\mathrm{d}^2}{\mathrm{d}\theta^2}\left(\frac{1}{r}\right) + \left(1 - \frac{K^2 M_0^2}{c^2 C^2}\right)\frac{1}{r} = \frac{K M_0}{C^2}\left(1 + \frac{E}{m_0 c^2}\right),$$

where

$$C = \frac{C_0}{m_0}, \quad C_0 = mr^2\dot{\theta} = \frac{m_0 r^2 \dot{\theta}}{\sqrt{1 - \frac{\dot{r}^2 + r^2\dot{\theta}^2}{c^2}}} = \text{constant}.$$

M_0 is the mass of the planet P.

Particular case: $c \to \infty$.

Solution

Under these conditions, from the fundamental law of dynamics, meaning the differential equation of motion, it follows that:

$$\frac{\mathrm{d}\vec{p}}{\mathrm{d}t} = \vec{F}; \quad \frac{\mathrm{d}(m\vec{v})}{\mathrm{d}t} = \vec{F}; \quad m = \frac{m_0}{\sqrt{1 - \frac{\nu^2}{c^2}}},$$

where m_0 is the rest mass of the satellite;

$$m = \frac{m_0}{\sqrt{1 - \frac{\dot{r}^2 + r^2\dot{\theta}^2}{c^2}}};$$

$$m\frac{\mathrm{d}\vec{v}}{\mathrm{d}t} + \vec{v}\frac{\mathrm{d}m}{\mathrm{d}t} = \vec{F};$$

$$m(\ddot{r} - r\dot{\theta}^2)\vec{\rho} + m(2\dot{r}\dot{\theta} + r\ddot{\theta})\vec{n} + \frac{\mathrm{d}m}{\mathrm{d}t}\dot{r}\vec{\rho} + \frac{\mathrm{d}m}{\mathrm{d}t}r\dot{\theta}\vec{n} = -F\vec{\rho};$$

$$m(\ddot{r} - r\theta^2) + \frac{\mathrm{d}m}{\mathrm{d}t}r = -F;$$

$$m(2\dot{r}\dot{\theta} + r\ddot{\theta}) + \frac{\mathrm{d}m}{\mathrm{d}t}r\dot{\theta} = 0.$$

From the second equation of the system, it follows that:

$$\frac{m}{r}\frac{\mathrm{d}}{\mathrm{d}t}(r^2\dot{\theta}) + \frac{\mathrm{d}m}{\mathrm{d}t}r\dot{\theta} = 0;$$

$$m\frac{\mathrm{d}}{\mathrm{d}t}(r^2\dot{\theta}) + \frac{\mathrm{d}m}{\mathrm{d}t}r^2\dot{\theta} = 0;$$

$$\frac{\mathrm{d}}{\mathrm{d}t}(mr^2\dot{\theta}) = 0;$$

$$mr^2\dot{\theta} = \frac{m_0 r^2\dot{\theta}}{\sqrt{1 - \frac{\dot{r}^2 + r^2\theta^2}{c^2}}} = C_0;$$

$$\frac{C_0}{m_0} = C; \quad m = \frac{C_0}{r^2\dot{\theta}};$$

$$\frac{r^2\dot{\theta}}{\sqrt{1 - \frac{\dot{r}^2 + r^2\dot{\theta}^2}{c^2}}} = C; \quad \dot{\theta} = \frac{C}{r}\frac{\sqrt{1 - \frac{\dot{r}^2}{c^2}}}{\sqrt{r^2 + \frac{C^2}{c^2}}};$$

$$\dot{r} = \frac{\mathrm{d}r}{\mathrm{d}t} = \frac{\mathrm{d}r}{\mathrm{d}e}\dot{\theta} = \frac{\mathrm{d}r}{\mathrm{d}\theta}\frac{C}{r}\frac{\sqrt{1-\frac{\dot{r}^2}{c^2}}}{\sqrt{r^2+\frac{C^2}{c^2}}};$$

$$r^2 = cC\frac{\mathrm{d}r}{\mathrm{d}\theta}\left[C^2\left(\frac{\mathrm{d}r}{\mathrm{d}\theta}\right)^2 + c^2r^2\left(r^2+\frac{C^2}{c^2}\right)\right]^{-1/2}.$$

Under these conditions, the first equation of the previous system becomes:

$$m\frac{\mathrm{d}\dot{r}}{\mathrm{d}t} - mr\dot{\theta}^2 + \frac{\mathrm{d}m}{\mathrm{d}t}\dot{r} = -F;$$

$$\frac{\mathrm{d}}{\mathrm{d}t}(m\dot{r}) - mr\dot{\theta}^2 = -F; \quad m = \frac{C_0}{r^2\dot{\theta}},$$

where C_0 is the direction perpendicular to the plane of the satellite's motion.

It is known as the meaning of the satellite's angular momentum in relation to the planet.

Indeed, from the definition of the angular momentum, it follows that:

$$\vec{M} = \vec{r} \times \vec{p};$$

$$\vec{r} = r\vec{\rho}, \quad \vec{\mathrm{R}} = m\vec{\mathrm{v}} = m\dot{r}\vec{\rho} + mr\dot{\theta}\vec{n};$$

$$\vec{M} = \begin{vmatrix} \vec{\rho} & \vec{n} & \vec{k} \\ r & 0 & 0 \\ m\dot{r} & mr\dot{\theta} & 0 \end{vmatrix} = mr^2\dot{\theta}\vec{k},$$

where $\vec{k}$ is the direction perpendicular to the plane of the satellite's motion.

It is known, however, that for movement under the action of a central force, the angular momentum is conserved. Thus, we have:

$$|\vec{M}| = mr^2\dot{\theta} = C_0 = \text{constant}; \quad C_0 = \frac{m_0}{\sqrt{1-\frac{\mathrm{v}^2}{c^2}}}r^2\dot{\theta};$$

$$\beta = \frac{\mathrm{v}}{c}, \quad \gamma = \frac{1}{\sqrt{1-\beta^2}}; \quad C_0 = m_0\gamma r^2\dot{\theta}.$$

Under these conditions, the first equation of the previous system becomes:

$$m\frac{\mathrm{d}\dot{r}}{\mathrm{d}t} - mr\dot{\theta}^2 + \frac{\mathrm{d}m}{\mathrm{d}t}\dot{r} = -F;$$

$$\frac{\mathrm{d}}{\mathrm{d}t}(m\dot{r}) - mr\dot{\theta}^2 = -F; \quad m = \frac{C_0}{r^2\dot{\theta}};$$

$$\frac{\mathrm{d}}{\mathrm{d}t}\left(\frac{\dot{r}}{r^2\dot{\theta}}\right) - \frac{\dot{\theta}}{r} = -\frac{F}{C_0};$$

$$\dot{r} = \dot{\theta}\frac{\mathrm{d}r}{\mathrm{d}\theta}; \quad \frac{\dot{r}}{\dot{\theta}} = \frac{\mathrm{d}r}{\mathrm{d}\theta};$$

$$\frac{\mathrm{d}}{\mathrm{d}t}\left(\frac{\dot{r}}{r^2\dot{\theta}}\right) = \frac{\mathrm{d}}{\mathrm{d}t}\left(\frac{1}{r^2}\frac{\mathrm{d}r}{\mathrm{d}\theta}\right) = \frac{\mathrm{d}}{\mathrm{d}t}\left[-\frac{\mathrm{d}}{\mathrm{d}\theta}\left(\frac{1}{r}\right)\right] = -\dot{\theta}\frac{\mathrm{d}^2}{\mathrm{d}\theta^2}\left(\frac{1}{r}\right);$$

$$\frac{\mathrm{d}^2}{\mathrm{d}\theta^2}\left(\frac{1}{r}\right) + \frac{1}{r} = \frac{F}{C_0\dot{\theta}}; \quad \dot{\theta} = \frac{C_0}{m_0\gamma r^2};$$

$$\frac{\mathrm{d}^2}{\mathrm{d}\theta^2}\left(\frac{1}{r}\right) + \frac{1}{r} = \frac{Fm_0\gamma r^2}{C_0^2};$$

$$\gamma = \frac{1}{\sqrt{1-\beta^2}};$$

$$\beta^2 = \frac{\mathrm{v}^2}{c^2} = \frac{\dot{r}^2 + r^2\dot{\theta}^2}{c^2};$$

$$\dot{r} = \dot{\theta}\frac{\mathrm{d}r}{\mathrm{d}\theta} = \frac{C_0}{m_0\gamma r^2}\frac{\mathrm{d}r}{\mathrm{d}\theta}; \quad \dot{\theta} = \frac{C_0}{m_0\gamma r^2};$$

$$\beta^2 = \frac{C_0^2}{m_0^2\gamma^2c^2}\left[\left(\frac{1}{r^2}\frac{\mathrm{d}r}{\mathrm{d}\theta}\right)^2 + \frac{1}{r^2}\right];$$

$$\left(\frac{1}{r^2}\frac{\mathrm{d}r}{\mathrm{d}\theta}\right)^2 = \left[-\frac{\mathrm{d}}{\mathrm{d}\theta}\left(\frac{1}{r}\right)\right]^2 = \left[\frac{\mathrm{d}}{\mathrm{d}\theta}\left(\frac{1}{r}\right)\right]^2;$$

$$\beta^2 = \frac{C_0^2}{m_0^2\gamma^2c^2}\left\{\left[\frac{\mathrm{d}}{\mathrm{d}\theta}\left(\frac{1}{r}\right)\right]^2 + \frac{1}{r^2}\right\};$$

$$\gamma^2 = \frac{1}{1-\beta^2} = 1 + \frac{\beta^2}{1-\beta^2} = 1 + \gamma^2\beta^2;$$

$$\gamma = \sqrt{1 + \gamma^2 \beta^2};$$

$$\gamma = \sqrt{1 + \frac{C_0^2}{m_0^2 c^2} \left\{ \left[\frac{\mathrm{d}}{\mathrm{d}\theta} \left(\frac{1}{r} \right) \right]^2 + \frac{1}{r^2} \right\}};$$

$$\frac{\mathrm{d}^2}{\mathrm{d}\theta^2} \left(\frac{1}{r} \right) + \frac{1}{r} = \frac{F m_0 r^2}{C_0^2} \sqrt{1 + \frac{C_0^2}{m_0^2 c^2} \left\{ \left[\frac{\mathrm{d}}{\mathrm{d}\theta} \left(\frac{1}{r} \right) \right]^2 + \frac{1}{r^2} \right\}},$$

representing Binet's equation for the relativistic case.

Observation: for $c \to \infty$, from the previous equation, it follows that

$$\frac{\mathrm{d}^2}{\mathrm{d}\theta^2} \left(\frac{1}{r} \right) + \frac{1}{r} = \frac{F m_0 r^2}{C_0^2},$$

representing Binet's equation, which is known from non-relativistic (classical) mechanics.

On the other hand, the evolution of the satellite around the planet, following the law of conservation of total mechanical energy, results in:

$$E_{\mathrm{C}} + E_{\mathrm{p}} = E = \text{constant};$$

$$E_{\mathrm{c}} = mc^2 - m_0 c^2 = m_0 c^2 (\gamma - 1);$$

$$E_{\mathrm{p}} = -K \frac{m_0 M_0}{r};$$

$$m_0 c^2 (\gamma - 1) + E_{\mathrm{p}} = E; \quad \gamma = 1 + \frac{E - E_{\mathrm{p}}}{m_0 c^2}.$$

Thus, Binet's relativistic equation can be written in an equivalent form:

$$\frac{\mathrm{d}^2}{\mathrm{d}\theta^2} \left(\frac{1}{r} \right) + \frac{1}{r} = \frac{F m_0 r^2}{C_0^2} \left(1 + \frac{E - E_{\mathrm{p}}}{m_0 c^2} \right);$$

$$F = K \frac{m_0 M_0}{r^2};$$

$$\frac{\mathrm{d}^2}{\mathrm{d}\theta^2} \left(\frac{1}{r} \right) + \frac{1}{r} = \frac{K m_0^2 M_0}{C_0^2} \left(1 + \frac{E + K \frac{m_0 M_0}{r}}{m_0 c^2} \right);$$

$$\frac{d^2}{d\theta^2}\left(\frac{1}{r}\right)+\left(1-\frac{K^2m_0^2M_0^2}{c^2C_0^2}\right)\frac{1}{r}=\frac{Km_0^2M_0}{C_0^2}\left(1+\frac{E}{m_0c^2}\right).$$

This represents the differential equation of the relativistic movement of the satellite in plane polar coordinates, from which, by integration, the equation of the relativistic trajectory of the satellite in relation to the fixed planet is obtained:

$$C=\frac{C_0}{m_0};$$

$$\frac{d^2}{d\theta^2}\left(\frac{1}{r}\right)+\left(1-\frac{K^2M_0^2}{c^2C^2}\right)\frac{1}{r}=\frac{KM_0}{C^2}\left(1+\frac{E}{m_0c^2}\right).$$

This is a linear differential inhomogeneous equation of the second degree with constant coefficients, whose general solution is obtained by summing the general solution of the homogeneous equation associated with a particular solution of the inhomogeneous equation (the free term of the inhomogeneous equation being a constant).

This results in:

$$\frac{d^2}{d\theta^2}\left(\frac{1}{r}\right)+\left(1-\frac{K^2M_0^2}{c^2C^2}\right)\frac{1}{r}=0;$$

$$\frac{1}{r}=z_0;\quad 1-\frac{K^2M_0^2}{c^2C^2}=\omega^2;$$

$$z_0''+\omega^2z_0=0;$$

$$z_0=e^{k\theta};\quad k^2+\omega^2=0;$$

$$k_1=i\omega;\quad k_2=-i\omega;$$

$$z_0=C'e^{i\omega\theta}+C''e^{-i\omega\theta};$$

$$z_0=C'(\cos\omega\theta+i\sin\omega\theta)+C''(\cos\omega\theta-i\sin\omega\theta);$$

$$z_0=(C'+C'')\cos\omega\theta+i(C'-C'')\sin\omega\theta;$$

$$C'+C''=C_1;\quad i(C'-C'')=C_2;$$

$$z_0=C_1\cos\omega\theta+C_2\sin\omega\theta;$$

$$C_1=\lambda^*\cos\omega\theta';\quad C_2=\lambda^*\sin\omega\theta',$$

where θ' is not a particular value of θ;

$$z_0 = \lambda^* \cos \omega(\theta - \theta'),$$

representing the general solution of the homogeneous equation.

Because the free term of the inhomogeneous equation is a constant,

$$A = \frac{KM_0}{C^2}\left(1 + \frac{E}{m_0 c^2}\right),$$

it results in:

$$\frac{\mathrm{d}}{\mathrm{d}\theta^2}\left(\frac{1}{r}\right) + \omega^2 \frac{1}{r} = A;$$

$$\frac{1}{r} = z_\mathrm{p}; \quad z_\mathrm{p} = \frac{A}{\omega^2}.$$

Under these conditions, the general solution of the inhomogeneous equation is:

$$\frac{1}{r} = z_0 + z_\mathrm{p};$$

$$\frac{1}{r} = \lambda^* \cos \omega(\theta - \theta') + \frac{A}{\omega^2};$$

$$\lambda^* = \frac{e^*}{p^*}; \quad \frac{A}{\omega^2} = \frac{1}{p^*}; \quad e^* = \lambda^* \frac{\omega^2}{A};$$

$$r = \frac{p^*}{1 + e^* \cos \omega(\theta - \theta')},$$

representing the equation of the satellite's relativistic trajectory;

$$r = \frac{1 - \frac{K^2 M_0^2}{c^2 C^2}}{\frac{KM_0}{C^2}\left(1 + \frac{E}{m_0 c^2}\right) + \lambda^* \left(1 - \frac{K^2 M_0^2}{c^2 C^2}\right) \cos \sqrt{1 - \frac{K^2 M_0^2}{c^2 C^2}}(\theta - \theta')}.$$

From this, for $c \to \infty$, we find the equation of the satellite's non-relativistic trajectory in the form:

$$r = \frac{1}{\frac{KM_0}{C^2} + \lambda \cos(\theta - \theta')};$$

$$r = \frac{\frac{C^2}{KM_0}}{1 + \lambda \frac{C^2}{KM_0} \cos(\theta - \theta')};$$

$$\frac{C^2}{KM_o} = p; \quad \lambda p = e;$$

$$r = \frac{p}{1 + e\cos(\theta - \theta')}.$$

This indicates that (in the non-relativistic version) the trajectory of the satellite is conic, and that the perihelion of the orbit is fixed in space, so that the position vector of the satellite at the perihelion is always the same.

On the other hand, with the equation of the relativistic trajectory of the satellite being different from the equation of the non-relativistic version, it follows that the relativistic trajectory of the satellite in relation to the fixed planet is no longer conic, and that the perihelion of the orbit is no longer fixed in space, so that only the modulus of the position vector of the satellite at perihelion is always the same, its orientation changing. Therefore, the perihelion describes a circle around the planet, having a radius equal to r_{min}.

Consequently, the relativistic trajectory of a satellite relative to a fixed planet is a planar rosette, as shown in the drawing in Figure 46.2.

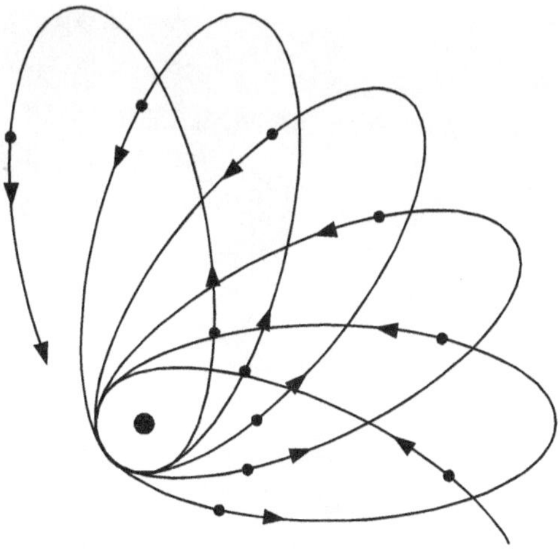

Fig. 46.2

Problem 47

A Flat Cosmic Platform

In the drawings in Figures 47.1 and 47.2, two inertial reference systems are represented: S, an inertial reference system at rest, fixed in relation to a star Σ, and S', a moving inertial reference system, fixed to a rectangular planar cosmic platform of very large dimensions, in rectilinear and uniform motion with the velocity $u < c$ relative to the star Σ. On the surface of the cosmic platform, there is a uniformly distributed system of space shuttles.

At the initial time, $t = t' = 0$, when the origins O and O′ of the two considered inertial reference systems, S and S′, coincide, all spacecraft are simultaneously launched onto the surface of the platform. The laws of motion for each space shuttle relative to the space platform (mobile inertial system, S′) are:

$$x' = \alpha \cdot c \cdot t'; \quad y' = \alpha \cdot a; \quad z' = 0,$$

where: α – a dimensionless parameter associated with the motion of each space shuttle $(-1 \leq \alpha \leq 1)$; a – a positive constant; c – the speed of light in a vacuum.

The number of shuttles for which $0 < \alpha \leq 1$ is equal to the number of shuttles for which $-1 \leq \alpha < 0$. For each value of the parameter α in the system, there is only one shuttle.

a) *Identify* the shape of the starting line, which indicates the placement of the space shuttles in relation to the reference systems S and at the initial time, $t = t' = 0$.

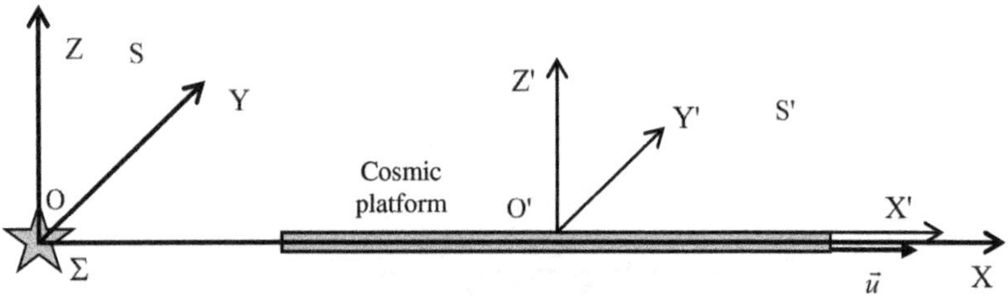

Fig. 47.1

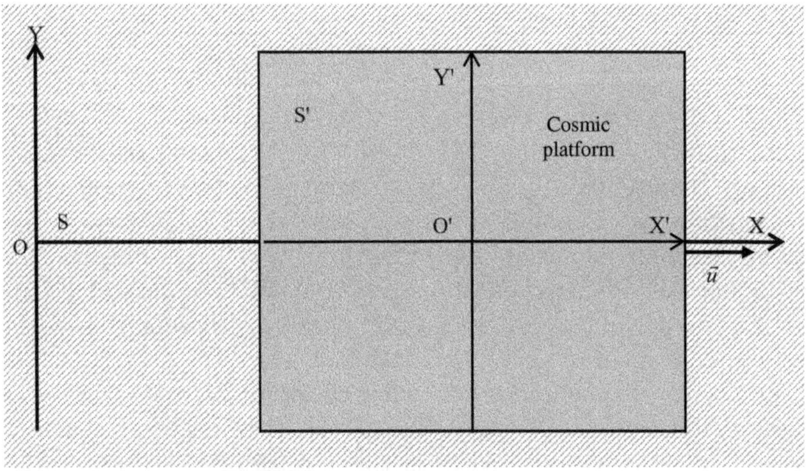

Fig. 47.2

b) *Identify* the shape of the flight line which indicates the relative positions of the space shuttles in relation to the reference system S$'$ at the time t'.

c) *Identify* the shape of the flight line which indicates the relative positions of the space shuttles in relation to the reference system S at the time t.

d) *Determine* the velocities of the space shuttles relative to the reference frame S$'$. *Discuss* their dependence on the values of the parameter α.

e) *Determine* the velocities of the space shuttles relative to the reference frame S. *Discuss* their dependence on the values of the parameter α.

Solution

a) The position coordinates of the space shuttles at the initial moment, in relation to the two reference systems, are determined as follows:

$$x' = \alpha \cdot c \cdot t'; \quad y' = \alpha \cdot a; \quad z' = 0;$$

$$t' = 0;$$

$$x' = 0; \quad y' = \alpha \cdot a; \quad z' = 0;$$

$$x = \frac{x' + ut'}{\sqrt{1 - \frac{u^2}{c^2}}}; \quad y = y'; \quad z = z';$$

$$x = 0; \quad y = \alpha \cdot a; \quad z = 0.$$

The layout of the space shuttle system at the initial time, $t = t' = 0$, in relation to each of the two reference systems (the starting line) is shown in the drawing in Figure 47.3.

The layout relative to system S′ (segment $M'_0N'_0$) is identical to the layout relative to system S (segment M_0N_0).

Relative to the space platform (the inertial system S′), n space shuttles take off in the positive direction of the axis O′X′, n space shuttles take off in the negative direction of the axis O′X′, and one space shuttle, for which $\alpha = 0$, remains at rest on the space platform at the point O′. This happens because $-1 \leq \alpha \leq 1$. In total, $2n + 1$ shuttles take off.

b) From the parametric equations of the space shuttles' motion, written relative to the system S′,

$$x' = \alpha \cdot c \cdot t', \quad y' = \alpha \cdot a, \text{ and } z' = 0,$$

we obtain

$$y' = \frac{a}{ct'}x',$$

which proves that, at any moment t', relative to the observer O′ in the system S′, the line of flight of the space shuttles, as

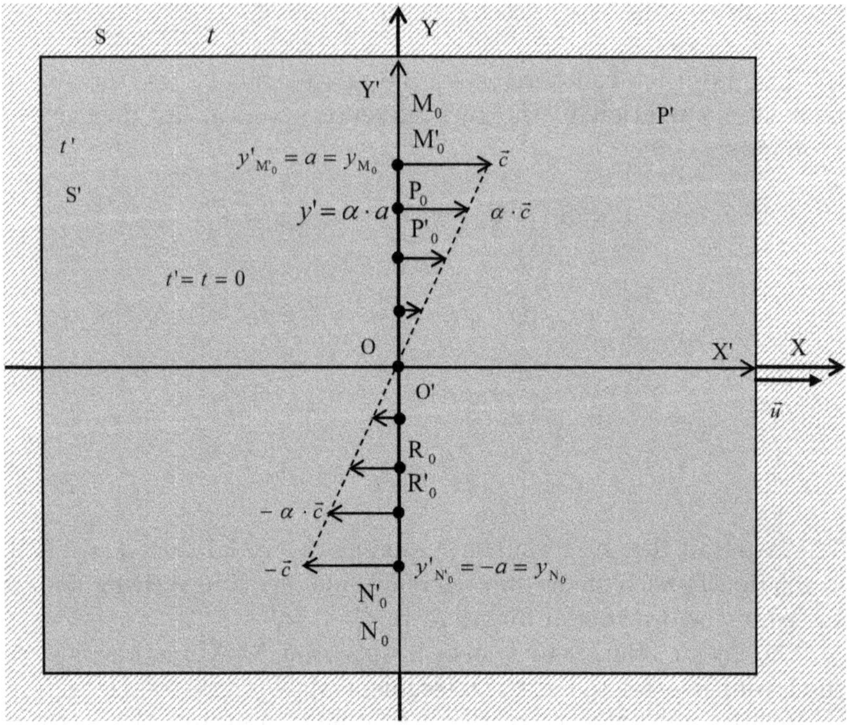

Fig. 47.3

indicated by the drawing in Figure 47.4, is a line segment passing through the origin O′, whose geometric slope is given by the expression

$$\tan\varphi' = \frac{y'}{x'} = \frac{\alpha \cdot a}{\alpha \cdot c \cdot t'} = \frac{a}{c \cdot t'},$$

whose value decreases over time.

On this line, the space shuttles are uniformly distributed on the line segment M′N′ that passes through the origin O′, the position coordinates of the extremities of this segment being:

$$\mathrm{M}'(x'_{\mathrm{M}'} = c \cdot t'; y'_{\mathrm{M}'} = a; z'_{\mathrm{M}'} = 0);$$
$$\mathrm{N}'(x'_{\mathrm{N}'} = -c \cdot t'; y'_{\mathrm{N}'} = -a; z'_{\mathrm{N}'} = 0).$$

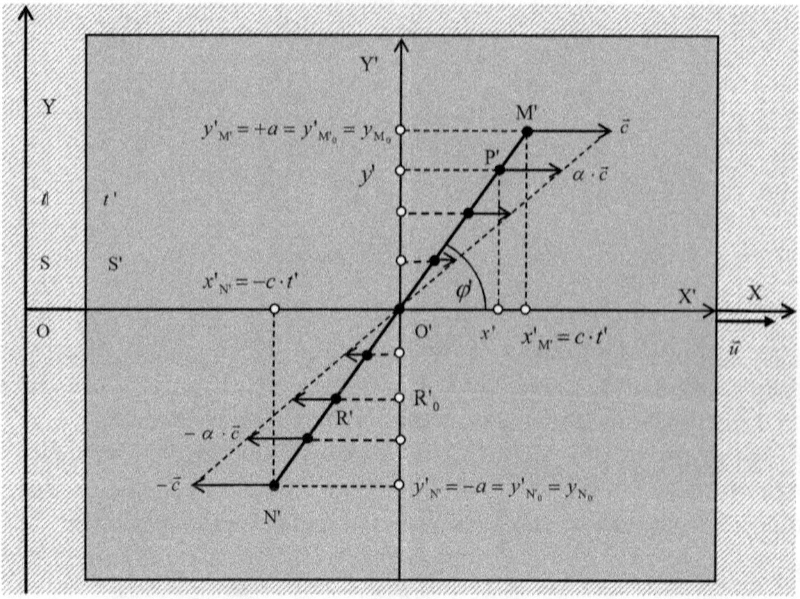

Fig. 47.4

Thus, the distance between the farthest shuttles, the length of the segment M′N′, at time t' is:

$$\mathrm{M'N'} = d' = \sqrt{(x'_{\mathrm{N'}} - x'_{\mathrm{M'}})^2 + (y'_{\mathrm{N'}} - y'_{\mathrm{M'}})^2 + (z'_{\mathrm{N'}} - z'_{\mathrm{M'}})^2};$$

$$\mathrm{M'N'} = d' = 2 \cdot \sqrt{c^2t'^2 + a^2}.$$

c) Using Lorentz transformations, it follows that:

$$x' = \frac{x - ut}{\sqrt{1 - \frac{u^2}{c^2}}}; \quad y' = y; \quad z' = z = 0; \quad t' = \frac{t - \frac{u}{c^2}x}{\sqrt{1 - \frac{u^2}{c^2}}};$$

$$y' = \frac{a}{ct'}x'; \quad y' = \frac{a}{ct'} \cdot \frac{x - ut}{\sqrt{1 - \frac{u^2}{c^2}}};$$

$$y' = \frac{a}{c \cdot \frac{t - \frac{u}{c^2}x}{\sqrt{1-\frac{u^2}{c^2}}}} \cdot \frac{x - ut}{\sqrt{1 - \frac{u^2}{c^2}}} = \frac{a}{c} \cdot \frac{\sqrt{1 - \frac{u^2}{c^2}}}{\sqrt{1 - \frac{u^2}{c^2}}} \frac{x - ut}{t - \frac{u}{c^2}x};$$

$$y' = \frac{a}{c} \cdot \frac{x - ut}{t - \frac{u}{c^2}x} = y;$$

$$y = \frac{ax - aut}{ct - \frac{u}{c}x} = y(x),$$

representing the equation of the flight line (the equation of a hyperbola) on which the space shuttles are arranged at any time t during the flight, in relation to the inertial system S (the star Σ).

The graph of this flight line is the hyperbola arc MN, represented in the drawing in Figure 47.5. The coordinates of its extremities, M and N, in relation to the system S are determined as follows:

$$y_{\mathrm{M}} = y_{\mathrm{M}_0} = y'_{\mathrm{M}'_0} = y'_{\mathrm{M}'} = a;$$

$$y_{\mathrm{M}} = \frac{ax_{\mathrm{M}} - aut}{ct - \frac{u}{c}x_{\mathrm{M}}} = a;$$

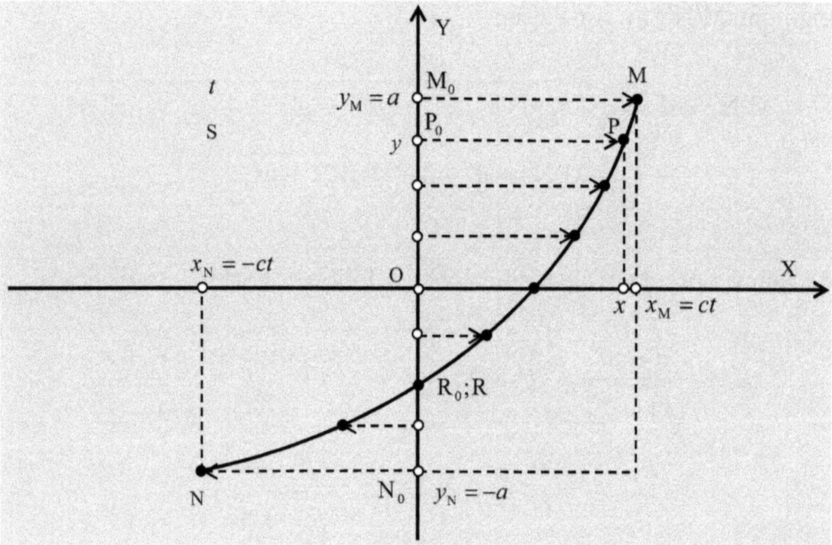

Fig. 47.5

$$x_{\rm M} = ct > 0; \quad y_{\rm M} = a > 0; \quad z_{\rm M} = 0;$$

$$y_{\rm N} = y_{\rm N_0} = y'_{\rm N'_0} = y'_{\rm N'} = -a;$$

$$y_{\rm N} = \frac{ax_{\rm N} - aut}{ct - \frac{u}{c}x_{\rm N}} = -a;$$

$$x_{\rm N} = -ct < 0; \quad y_{\rm N} = -a < 0; \quad z_{\rm N} = 0;$$

$$x_{\rm N} = -x_{\rm M}; \quad y_{\rm N} = -y_{\rm M}.$$

Thus, the distance between the farthest shuttles, the length of the segment MN at time t, is:

$$\mathrm{MN} = d = \sqrt{(x_{\rm N} - x_{\rm M})^2 + (y_{\rm N} - y_{\rm M})^2 + (z_{\rm N} - z_{\rm M})^2};$$

$$\mathrm{MN} = d = 2 \cdot \sqrt{c^2t^2 + a^2}.$$

d) The movements of all space shuttles in relation to the space platform (the inertial system) are uniform rectilinear movements executed in the plane X′O′Y′, in directions parallel to the O′X′ axis and falling into a symmetrical sector with the width $2a$, included between the points $\mathrm{M}'(y'_{\rm M'} = a)$ and $\mathrm{N}'(y'_{\rm N'} = -a)$. The components of their velocities are:

$$\vec{v}' \begin{cases} v'_{X'} = \frac{dx'}{dt'} = \frac{d(\alpha ct')}{dt'} = \alpha c; \\ v'_{Y'} = \frac{dy'}{dt'} = \frac{d(\alpha a)}{dt} = 0; \\ v'_{Z'} = \frac{dz'}{dt'} = 0, \end{cases}$$

with values dependent on the movement parameter $(-1 \leq \alpha \leq 1)$, so that these values are in the range

$$-c \leq v'_{X'} \leq c,$$

corresponding to the points M′, N′ and O′. Thus:

$$v'_{X'(M')} = c; \quad v'_{X'(N')} = -c; \quad v'_{X'(O')} = 0,$$

which means that relative to the space platform (system S′), some of the space shuttles (n) move in the positive direction of the axis

O′X′, others (n) move in the negative direction of the axis O′X′ and 1 space shuttle is stationed at the point O′ on the surface of the cosmic platform.

Given that $\mathrm{v}'_{\mathrm{X}'} = \alpha c$ and $-1 \le \alpha \le 1$, so that $-c \le \mathrm{v}'_{\mathrm{X}'} \le +c$, there must be a value of the parameter α for which $\mathrm{v}'_{\mathrm{X}'} = -u$, which is possible for $\alpha = -\frac{u}{c}$. The location of this space shuttle, on the starting line along the axis O′Y′, is at the point R'_0 of coordinate $y'_{\mathrm{R}'_0} = \alpha a = -\frac{u}{c}a$.

The expression $\mathrm{v}'_{\mathrm{X}'} = -u$ refers to the orientation of the vector component $\vec{\mathrm{v}}'_{\mathrm{X}'}$, as the inverse of the orientation of the axis O′X′. So, we write $|\mathrm{v}'_{\mathrm{X}'}| = u$.

e) The movements of all space shuttles in relation to the star Σ (the inertial system S), are uniform rectilinear movements executed in the plane XOY, in directions parallel to the axis OX and falling into a symmetrical sector with the width $2a$, included between the points $\mathrm{M}(y_{\mathrm{M}} = a)$ and $\mathrm{N}(y_{\mathrm{N}} = -a)$. The velocity components are dependent on the parameter $-1 \le \alpha \le 1$, as evidenced by the following relations:

$$\vec{\mathrm{v}}\begin{cases} \mathrm{v_X} = \frac{\mathrm{d}x}{\mathrm{d}t}; \\ \mathrm{v_Y} = \frac{\mathrm{d}y}{\mathrm{d}t}; \\ \mathrm{v_Z} = \frac{\mathrm{d}z}{\mathrm{d}t}; \end{cases}$$

$$x = \frac{x' + ut'}{\sqrt{1-\frac{u^2}{c^2}}} = \frac{\alpha ct' + ut'}{\sqrt{1-\frac{u^2}{c^2}}} = \frac{(\alpha c + u)t'}{\sqrt{1-\frac{u^2}{c^2}}}; \quad \mathrm{d}x = \frac{(\alpha c + u)\cdot \mathrm{d}t'}{\sqrt{1-\frac{u^2}{c^2}}};$$

$$t = \frac{t' + \frac{u}{c^2}x'}{\sqrt{1-\frac{u^2}{c^2}}} = \frac{t' + \frac{u}{c^2}\alpha ct'}{\sqrt{1-\frac{u^2}{c^2}}} = \frac{\left(1+\frac{\alpha u}{c}\right)t'}{\sqrt{1-\frac{u^2}{c^2}}}; \quad \mathrm{d}t = \frac{\left(1+\frac{\alpha u}{c}\right)\cdot dt'}{\sqrt{1-\frac{u^2}{c^2}}};$$

$$y = y' = \alpha a; \quad \mathrm{d}y = \mathrm{d}y' = \mathrm{d}(\alpha a) = 0;$$

$$z = z' = 0; \quad \mathrm{d}z = \mathrm{d}z' = 0;$$

$$\vec{\mathrm{v}}\begin{cases} \mathrm{v_X} = \frac{\mathrm{d}x}{\mathrm{d}t} = \frac{\alpha c+u}{1+\frac{\alpha u}{c}}; \\ \mathrm{v_Y} = \frac{\mathrm{d}y}{\mathrm{d}t} = 0; \\ \mathrm{v_Z} = \frac{\mathrm{d}z}{\mathrm{d}t} = 0; \end{cases}$$

$$\vec{v}\begin{cases} v_X = \frac{dx}{dt} = \frac{v'_{X'}+u}{1+\frac{v'_{X'}u}{c^2}} = \frac{\alpha c+u}{1+\frac{\alpha cu}{c^2}} = \frac{\alpha c+u}{1+\frac{\alpha u}{c}}; \\ v_Y = \frac{dy}{dt} = \frac{v'_{Y'}}{1+\frac{v'_{Y'}u}{c^2}} \cdot \sqrt{1-\frac{u^2}{c^2}} = 0; \\ v_Z = \frac{dz}{dt} = \frac{v'_{Z'}}{1+\frac{v'_{Z'}u}{c^2}} \cdot \sqrt{1-\frac{u^2}{c^2}} = 0. \end{cases}$$

Under these conditions, due to the values of the movement parameter α $(-1 \le \alpha \le 1)$, for the component v_X, the variants $v_X = 0$, $v_X > 0$ and $v_X < 0$ are possible.

1) If $v_X = 0$, the spacecraft is at rest with respect to the star Σ. This happens when

$$v_X = \frac{\alpha c + u}{1 + \frac{\alpha u}{c}} = 0 \text{ and } \quad \alpha = -\frac{u}{c},$$

which is true for the stationary spacecraft, relative to the star Σ, on the axis OY at the point $R_0 \equiv R'_0$, where the graph of the function $y = y(x)$ intersects the OY axis, as indicated by the drawing in Figure 47.5. From this:

$$y = \frac{ax - aut}{ct - \frac{u}{c}x}; \quad x = 0;$$

$$y = -a\frac{u}{c} = y_{R_0} = y'_{R'_0}.$$

Corresponding to this state, identifying the elements of the given system according to the notation in the drawing in Figure 47.6, we obtain:

$$\vec{v}_1 = \vec{v}_{1,\Sigma} = \vec{v}_{\text{Cosmic platform},\,\Sigma} = \vec{u};$$

$$\vec{v}_2 = \vec{v}_{2,\Sigma} = \vec{v}_{\text{Spacecraft},\,\Sigma} = \vec{v}_X;$$

$$\vec{v}_{21} = \vec{v}_{\text{Spacecraft, Cosmic platform}} = \vec{v}_2 - \vec{v}_1 = \vec{v}_X - \vec{u} = \vec{v}'_{X'};$$

$$\vec{v}_2 = \vec{v}_{21} + \vec{v}_1; \quad v_2 = \sqrt{v_{21}^2 + v_1^2 + 2v_{21}v_1\cos(180°)} = v_{21} - v_1;$$

$$\vec{v}_X = \vec{v}'_{X'} + \vec{u};$$

$$v_X = v'_{X'} - u; \quad v'_{X'} = |v'_{X'}|;$$

$$v_X = |v'_{X'}| - u;$$

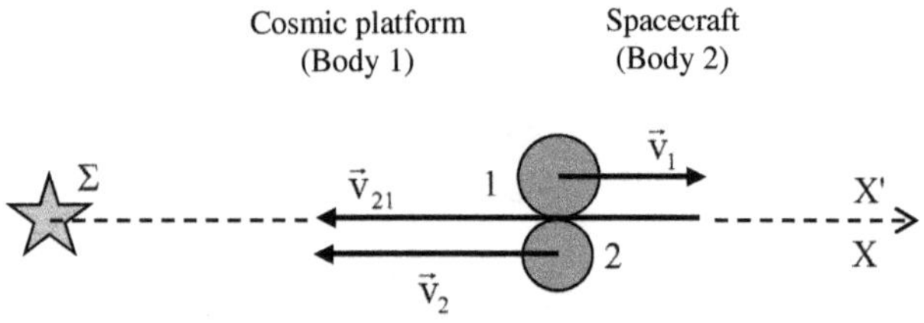

Fig. 47.6

$$v'_{X'} = -u; \quad |v'_{X'}| = u;$$
$$v_X = 0.$$

2) If $v_X > 0$, the orientation of the component $\vec{v}_X$ is the same as the orientation of the OX axis of the system S, and the direction of the spacecraft's motion is the same as the positive direction of the OX axis, which implies that the spacecraft is moving away from the star Σ, so that:

$$v_X = \frac{\alpha c + u}{1 + \frac{\alpha u}{c}};$$
$$\alpha = 1; \quad v_X = c;$$
$$\alpha = 0; \quad v_X = u;$$
$$\alpha = -\frac{u}{c}; \quad v_X = 0;$$
$$0 \le v_X \le c,$$

as the drawing in Figure 47.7 indicates.

3) If $v_X < 0$, the orientation of the component $\vec{v}_X$ is opposite to the orientation of the OX axis of the system S, so that the direction of the spacecraft's motion is opposite to the positive direction of the OX axis, which implies that the spacecraft is approaching the star Σ, such that:

$$v_X = \frac{\alpha c + u}{1 + \frac{\alpha u}{c}};$$
$$\alpha = -1; \quad v_X = -c;$$
$$\alpha = 0; \quad v_X = u;$$

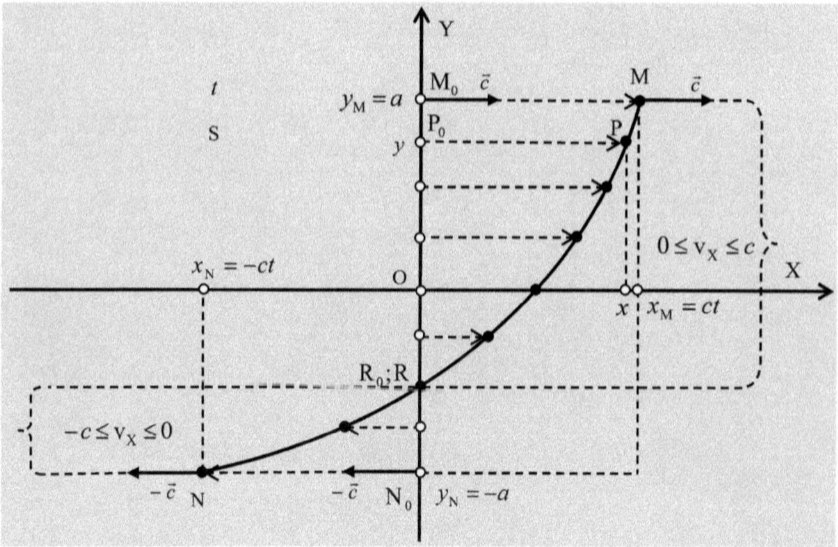

Fig. 47.7

$$\alpha = -\frac{u}{c}; \quad v_X = 0;$$

$$-c \le v_X \le 0,$$

as the drawing in Figure 47.7 indicates.

Problem 48

Attenuation of Gravitational Attraction

A spaceship consisting of two identical space capsules, each with mass $m/2$, evolves around the Earth on the same circular orbit with radius r, in the variants represented in Figures 48.1 and 48.2: either side by side (tangent and connected) or arranged (fixed) at the ends of a very light, sufficiently long bar (like a "barbell"). Therefore, the angle at the center between the two cosmic capsules is α.

a) *Identify* the effect of the existence of a distance between the two spherical capsules on the gravitational attraction exerted by the Earth on the entire spacecraft, compared to the gravitational attraction exerted by the Earth on the same two spherical capsules in contact when the distances of the capsules from the center of the Earth are, at all times, r.

The following are known: the mass of the Earth, M; the constant of gravitational attraction, K.

b) *Determine* the speeds of the two cosmic capsules in each of the two variants. The following are known: the mass of the Earth, M; the constant of gravitational attraction, K.

c) The two cosmic capsules can slide without friction along the bar, but between them is an elastic spring whose elasticity constant is k. *Determine* the length of the spring when it is not on the bar between the two cosmic capsules, l_0.

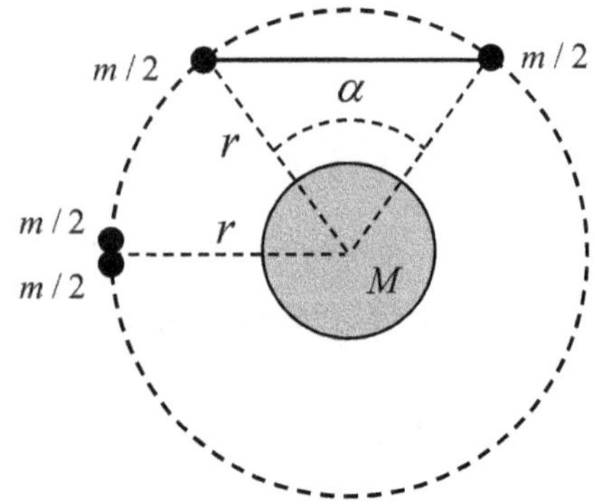

Fig. 48.1

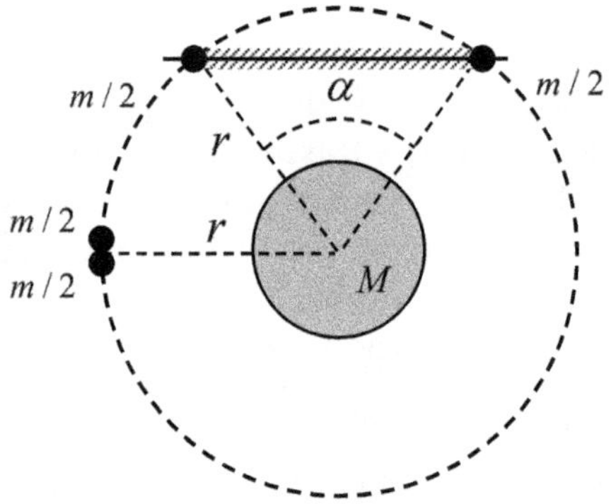

Fig. 48.2

Solution

a) The external forces acting on each capsule, in each of the two variants, are those represented in the drawing in Figure 48.3.

Corresponding to the first variant, the resultant of the two gravitational attraction forces acting on the spacecraft is

$$\vec{F}_{\mathrm{r},0} = 2\vec{F}_0,$$

and it follows that

$$F_{\mathrm{r},0} = 2F_0 = 2 \cdot K\frac{\frac{m}{2}M}{r^2} = K\frac{mM}{r^2}.$$

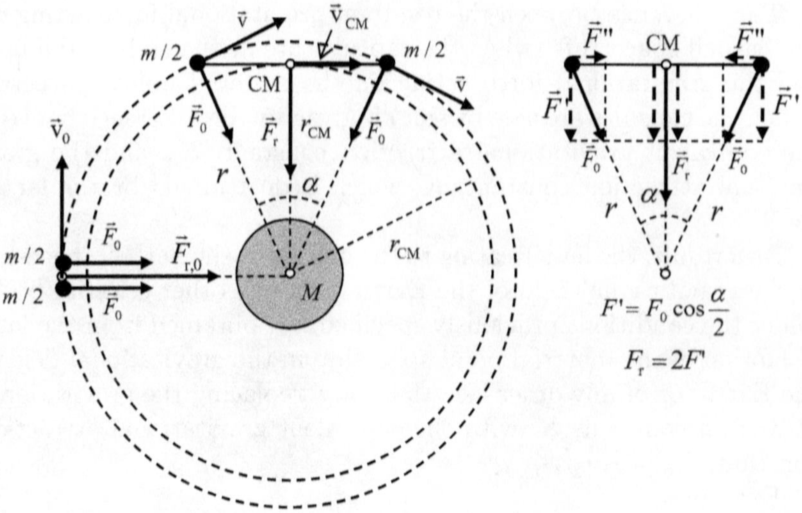

Fig. 48.3

For the second variant, when the barbell-shaped spaceship is equivalent to a single space capsule, having mass m, located at the center of mass, CM, of the barbell, the resultant gravitational force acting on the spaceship is:

$$F_r = 2F' = 2F_0 \cos\frac{\alpha}{2} = 2 \cdot K \cdot \frac{\frac{m}{2}M}{r^2} \cdot \cos\frac{\alpha}{2};$$

$$F_r = K\frac{mM}{r^2}\cos\frac{\alpha}{2}.$$

This results in:

$$F_r = K\cos\frac{\alpha}{2} \cdot \frac{mM}{r^2}; \quad K_e = K\cos\frac{\alpha}{2}; K_e < K,$$

where K_e is the "constant of equivalent gravitational attraction".

$$F_r = K_e\frac{mM}{r^2}; \quad F_{r,0} = K\frac{mM}{r^2}; \quad F_r < F_{r,0}.$$

The resultant gravitational force acting on the barbell spacecraft is less than the resultant gravitational force acting on the material point spacecraft when the capsules that make up the spacecraft are at the same distance from the center of the Earth.

The difference between the resultant gravitational force acting on the barbell spacecraft (when the capsules are at the ends of the bar) and the gravitational force acting on the material point spacecraft (when the capsules are side by side) is given by the difference between the equivalent gravitational attraction constant, K_e, and the gravitational attraction constant, K, which is determined by the factor $\cos\frac{\alpha}{2}$.

As a result, the laws relating to the motion of the barbell spaceship in the gravitational field of the Earth or of any other celestial body, under the conditions previously specified, are obtained from the laws of motion of the material point spaceship in the gravitational field of the Earth, or of any other celestial body, replacing the gravitational attraction constant, K, with the equivalent gravitational attraction constant, $K_e = K\cos\frac{\alpha}{2}$.

Examples:

1) The gravitational potential energy of the barbell spacecraft in the Earth's gravitational field is

$$E_{pg} = -K_e\frac{mM}{r} = -K\cos\frac{\alpha}{2}\cdot\frac{mM}{r}.$$

2) Kepler's laws:

- 1st law: the orbit of the CM of the barbell spacecraft, under the specified conditions, is an ellipse whose focus is the Earth (as indicated by the drawing in Figure 48.3, from the statement of the problem);
- 2nd law: the position vector of the CM of the barbell spacecraft describes flat surfaces with equal areas in equal time intervals;
- 3rd law:

$$T^2 = \frac{4\pi^2}{K_e M}a^3 = \frac{4\pi^2}{K\cos\frac{\alpha}{2}\cdot M}a^3.$$

At the moment of the passage of the material point spaceship through the perigee of the initial elliptical orbit, with the two cosmic capsules in contact, when the distance between the material point spaceship and the center of the Earth is

$$r_{min} = a(1-e),$$

where a and e are the semi-minor axis and the eccentricity of the initial elliptical orbit, respectively, the gravitational potential energy of the system consisting of the material point spaceship with mass m and the Earth, with mass M, is

$$E_{\mathrm{p,g,0}} = -K\frac{mM}{r_{\min}},$$

and the kinetic energy of the material point spaceship is:

$$E_{\mathrm{c,0}} = \frac{m\mathrm{v}_{\max}^2}{2},$$

where $\mathrm{v}_{\max}$ is the speed of the material point spaceship at the moment of passing through the perigee of its initial elliptical orbit;

$$\mathrm{v}_{\max} = \sqrt{KM\frac{1+e}{1-e}}.$$

As a result of the rapid angular separation, α, of the two cosmic capsules, which now represent the barbell spaceship, the gravitational potential energy of the system consisting of the barbell spaceship, with mass m, and the Earth, with mass M, is:

$$E_{\mathrm{p,g}} = -K\frac{mM}{r} = -K\frac{mM}{\frac{r_{\min}}{\cos\frac{\alpha}{2}}} = -K\cos\frac{\alpha}{2}\cdot\frac{mM}{r_{\min}} = -K_{\mathrm{e}}\frac{mM}{r_{\min}};$$

$$E_{\mathrm{p,g,0}} = -K\frac{mM}{r_{\min}};\quad E_{\mathrm{p,g}} = -K_{\mathrm{e}}\frac{mM}{r_{\min}};$$

$$K_{\mathrm{e}} = K\cos\frac{\alpha}{2} < K;$$

$$E_{\mathrm{p,g}} > E_{\mathrm{p,g,0}}.$$

At the moment of the passage of the material point spaceship through the apogee of the initial elliptical orbit, with the two cosmic capsules in contact, when the distance between the material point spaceship and the center of the Earth is maximum,

$$r_{\max} = a(1+e),$$

where a and e are the semi-minor axis and the eccentricity of the initial elliptical orbit, respectively, the gravitational potential energy of the system consisting of the material point spaceship, with mass m, and the Earth, with mass M, is

$$E_{\mathrm{p,g,}} = -K\frac{mM}{r_{\mathrm{max}}},$$

and the kinetic energy of the material point spaceship is

$$E_{\mathrm{c}} = \frac{m\mathrm{v}_{\mathrm{min}}^2}{2},$$

where $\mathrm{v}_{\mathrm{min}}$ is the speed of the material point spaceship at the time of passing through the apogee of its initial elliptical orbit;

$$\mathrm{v}_{\mathrm{min}} = \sqrt{KM\frac{1-e}{1+e}}.$$

As a result of the rapid angular separation, α, of the two space capsules, which now represent the barbell spaceship, the gravitational potential energy of the system consisting of the barbell spaceship, with mass m, and the Earth, with mass M, is

$$E_{\mathrm{p,g,}} = -K_{\mathrm{e}}\frac{mM}{r_{\mathrm{max}}} = -K\cos\frac{\alpha}{2}\cdot\frac{mM}{r_{\mathrm{max}}}.$$

b)

$$F_{0,\mathrm{r}} = K\frac{mM}{r^2} = \frac{m\mathrm{v}_0^2}{r};$$

$$\mathrm{v}_0 = \sqrt{K\frac{M}{r}};$$

$$F_{\mathrm{r}} = K_{\mathrm{e}}\frac{mM}{r^2} = K\cos\frac{\alpha}{2}\cdot\frac{mM}{r^2} = K\frac{mM}{r^2}\cdot\cos\frac{\alpha}{2} = \frac{m\mathrm{v}_{\mathrm{CM}}^2}{r_{\mathrm{CM}}};$$

$$r_{\mathrm{CM}} = r\cdot\cos\frac{\alpha}{2};$$

$$K\frac{mM}{r^2}\cdot\cos\frac{\alpha}{2} = \frac{m\mathrm{v}_{\mathrm{CM}}^2}{r\cdot\cos\frac{\alpha}{2}}; \quad K\frac{M}{r}\cdot\cos\frac{\alpha}{2} = \frac{\mathrm{v}_{\mathrm{CM}}^2}{\cos\frac{\alpha}{2}};$$

$$\mathrm{v}_{\mathrm{CM}}^2 = K\frac{M}{r}\cdot\cos^2\frac{\alpha}{2};$$

$$\mathrm{v_{CM}} = \cos\frac{\alpha}{2} \cdot \sqrt{K\frac{M}{r}} = \mathrm{v_0}\cos\frac{\alpha}{2};$$

$$T = \frac{2\pi \cdot r_{\mathrm{CM}}}{\mathrm{v_{CM}}} = \frac{2\pi \cdot r}{\mathrm{v}}; \quad \frac{r_{\mathrm{CM}}}{\mathrm{v_{CM}}} = \frac{r}{\mathrm{v}};$$

$$\mathrm{v} = \frac{r}{r_{\mathrm{CM}}}\mathrm{v_{CM}} = \frac{r}{r \cdot \cos\frac{\alpha}{2}}\cos\frac{\alpha}{2} \cdot \sqrt{K\frac{M}{r}};$$

$$\mathrm{v} = \sqrt{K\frac{M}{r}} = \mathrm{v_0}.$$

c)

$$F_{\mathrm{e}} = k \cdot \Delta l = F'' = F_0 \cdot \sin\frac{\alpha}{2} = K\frac{mM}{r^2} \cdot \sin\frac{\alpha}{2} = k(l_0 - l);$$

$$l - 2r\sin\frac{\alpha}{2}; \quad l_0 - l = \frac{K}{k} \cdot \frac{mM}{r^2} \cdot \sin\frac{\alpha}{2};$$

$$l_0 = \left(2r + \frac{K}{k} \cdot \frac{mM}{r^2}\right) \cdot \sin\frac{\alpha}{2}.$$

Other Books by Mihail Sandu

1. **PROBLEME DE FIZICĂ PENTRU GIMNAZIU**, *Editura Didactică şi Pedagogică*, Bucharest, 1977, 327 p.
2. **FIZICĂ, MANUAL PENTRU CLASA a VI-a**, coauthor Emanuel Nichita, *Editura Didactică şi Pedagogică*, Bucharest, 1978, 160 p.
3. **PROBLEME DE FIZICĂ PENTRU GIMNAZIU**, coauthors Emanuel Nichita and Tudorel Ştefan, *Editura Didactică şi Pedagogicş*, Bucharest, 1982, 223 p.
4. **CAIET METODIC CU PROBLEME DE FIZICĂ PENTRU GIMNAZIU**, *Casa Personalului Didactic*, Rm. Vâlcea, 1983, 181 p.
5. **GHIDUL PROFESORULUI ŞI AL ELEVULUI PENTRU CERCURILE DE FIZICĂ, VOL. I**, *Casa de Cultură a Ştiinţei şi Tehnicii pentru tineret*, Rm. Vâlcea, 1984, 498 p.
6. **GHIDUL PROFESORULUI ŞI AL ELEVULUI PENTRU CERCURILE DE FIZICĂ, VOL. II**, *Casa de Cultură a Ştiinţei şi Tehnicii pentru tineret*, Rm. Vâlcea, 1984, 405 p.
7. **CULEGERE DE PROBLEME DE FIZICĂ, VOL. I**, *Societatea de Ştiinţe Fizice şi Chimice din România*, Bucharest, 1986, 338 p.

8. **CULEGERE DE PROBLEME DE FIZICĂ, VOL. II**, *Societatea de Ştiinţe Fizice şi Chimice din România*, Bucharest, 1986, 118 p.

9. **PROBLEME DE FIZICĂ**, *Editura "Scrisul Românesc"*, Craiova, 1988, 277 p.

10. **GHID PENTRU CERCURILE DE FIZICĂ**, *Editura Academiei Române*, Bucharest, 1991, 325 p.

11. **PROBLEME DE FIZICĂ PENTRU GIMNAZIU**, *Editura Didactică şi Pedagogică*, Bucharest, 1991, 208 p.

12. **500 PROBLEME DE FIZICĂ**, *Editura Tehnică*, Bucharest, 1991, 283 p.

13. **PROBLEME DE PERFORMANŢĂ ÎN FIZICĂ**, *Editura Tehnică*, Bucharest, 1992, 340 p.

14. **PROBLEME DE FIZICĂ PENTRU GIMNAZIU**, *Editura "Lumina"*, Chişinău, Republic of Moldova, 1993, 272 p.

15. **CAIET METODIC CU PROBLEME DE FIZICĂ PENTRU GIMNAZIU**, *Editura "Lumina"*, Chişinău, Republic of Moldova, 1993, 179 p.

16. **TEME ŞI PROBLEME PENTRU CERCURILE DE FIZICĂ**, *Editura "Hyperion XXI"*, Bucharest, 1993, 254 p.

17. **PROBLEME DE FIZICĂ DIN REVISTA "KVANT", VOL. I**, *Editura Didactică şi Pedagogică*, Bucharest, 1993, 182 p.

18. **PROBLEME DE FIZICĂ PENTRU LICEU**, *Editura "Ex Libris"*, Rm. Vâlcea, 1993, 378 p.

19. **PROBLEME DE FIZICĂ DIN REVISTA "KVANT", VOL. II**, *Editura Didactică şi Pedagogică*, Bucharest, 1994, 194 p.

20. **CULEGERE DE PROBLEME DE FIZICĂ, VOL. I**, *Editura "TipCim"*, Chişinău, Republic of Moldova, 1995, 338 p.

21. **CULEGERE DE PROBLEME DE FIZICĂ, VOL. II**, *Editura "TipCim"*, Chişinău, Republic of Moldova, 1995, 117 p.

22. **GHIDUL PROFESORULUI ŞI AL ELEVULUI PENTRU CERCURILE DE FIZICĂ, VOL. I**, *Editura "TipCim", Chişinău*, Republic of Moldova, 1995, 498 p.

23. **PROBLEME DE FIZICĂ PENTRU GIMNAZIU**, *Editura Didactică şi Pedagogică*, Bucharest, 1995, 202 p.

24. **GHIDUL PROFESORULUI ŞI AL ELEVULUI PENTRU CERCURILE DE FIZICĂ, VOL. II**, *Editura "TipCim", Chişinău*, Republic of Moldova, 1996, 400 p.

25. **PROBLEME DE FIZICĂ PENTRU GIMNAZIU**, *Editura "ALL"*, Bucharest, 1996, Ediţia I, 503 p.

26. **PROBLEME DE FIZICĂ DIN REVISTA "KVANT", VOL. III**, *Editura Didactică şi Pedagogică*, Bucharest, 1996, 208 p.

27. **PROBLEME DE FIZICĂ DIN REVISTA "KVANT", VOL. IV**, *Editura Didactică şi Pedagogică*, Bucharest, 1996, 229 p.

28. **PROBLEME DE FIZICĂ PENTRU GIMNAZIU**, *Editura "ALL"*, Bucharest, 1997, Ediţia a II-a, 503 p.

29. **PROBLEME DE FIZICĂ – CONCURSURI ŞI OLIMPIADE**, *Editura "Petrion"*, Bucharest, 1998, 335 p.

30. **PROBLEME DE FIZICĂ PENTRU GIMNAZIU**, *Editura "ALL"*, Bucharest, 1998, Ediţia a III-a, 503 p.

31. **MECANICĂ FIZICĂ** (university course), *Editura Didactică şi Pedagogică*, Bucharest, 2002, 462 p.

32. **MECANICĂ TEORETICĂ** (university course), *Editura Didactică şi Pedagogică*, Bucharest, 2002, 476 p.

33. **ASTRONOMIE** (university course), *Editura Didactică şi Pedagogică*, Bucharest, 2003, 400 p.

34. **FIZICĂ. ÎNVĂŢĂMÂNTUL LICEAL ŞI CERCETAREA ŞTIINŢIFICĂ. MONOGRAFIA JUDEŢULUI VÂLCEA**, *Editura Conphys*, Rm. Vâlcea, 2004, 406 p.

35. **IN MEMORIAM – PROFESORUL LIVIU TĂTAR**, *Editura Conphys*, Rm. Vâlcea, 2004.

36. **PROBLEME DE FIZICĂ PENTRU GIMNAZIU**, *Editura "ALL"*, Bucharest, 2005, Ediţia a IV-a, 416 p.

37. **TEORIA RELATIVITĂŢII** (university course), *Editura Didactică şi Pedagogică*, Bucharest, 2005, 341 p.

38. **EVRIKA! PROBLEME DE FIZICĂ**, *Editura Didactică şi Pedagogică*, Bucharest, 2005, 600 p.

39. **CONCURSUL DE SELECŢIE PENTRU OLIMPIADA INTERNAŢIONALĂ DE FIZICĂ – 2005**, coauthor, *Editura Conphys*, Rm. Vâlcea, 2006, 262 p.

40. **PROBLEME DE FIZICĂ PENTRU GIMNAZIU**, *Editura "ALL"*, Bucharest, 2007, Ediţia a V-a, 416 p.

41. **ACTIVITĂŢI OLIMPICE LA C.P.P.P. CĂLIMĂNEŞTI-CĂCIULATA, 2004 – 2005 – 2006, ASTRONOMIE ŞI ŞTIINŢE**, *Editura Conphys*, Rm. Vâlcea, 2007, 611 p.

42. **ACTIVITĂŢI OLIMPICE LA C.P.P.P. CĂLIMĂNEŞTI-CĂCIULATA, 2004 – 2005 – 2006, FIZICĂ**, *Editura Conphys*, Rm. Vâlcea, 2007, 530 p.

43. **TOP-FIZ, PROBLEME DE FIZICĂ**, *Editura Didactică şi Pedagogică*, Bucharest, 2009, 833 p.

44. **LUCRĂRI EXPERIMENTALE DE FIZICĂ**, *Editura Conphys*, Rm. Vâlcea, 2009, 156 p.

45. **ŞTIINŢE PENTRU JUNIORI. CHIMIE – FIZICĂ – BIOLOGIE**, coauthor, *Editura MISTRAL*, Bucharest, 2009, 358 p.

46. **CONCURSUL INTERJUDEŢEAN DE FIZICĂ "LIVIU TĂTAR"**, coauthor, *Editura Universitaria*, Craiova, 2011, 236 p.

47. **ŞTIINŢE PENTRU JUNIORI. CHIMIE – FIZICĂ – BIOLOGIE**, coauthor, *Editura MISTRAL*, Ediţia a 2 – a, Bucharest, 2011, 358 p.

48. **PROBLEME ŞI EXPERIMENTE DE FIZICĂ**, *Editura Didactică şi Pedagogică*, Bucharest, 2012, 820 p.

49. **INTERNATIONAL OLYMPIAD ON ASTRONOMY AND ASTROPHYSICS, PROBLEMS, 2007–2013, SOLUTIONS, COMMENTS, DETAILS, EXTENSIONS**, *Editura CYGNUS*, SUCEAVA, 2014, 600 p.

50. **OLIMPIADA INTERNAŢIONALĂ DE ASTRONOMIE Ş ASTROFIZICĂ, PROBLEME, VOL. 1**, *Editura CYGNUS*, Suceava, 2015, 400 p.

51. **OLIMPIADA INTERNAŢIONALĂ DE ASTRONOMIE ŞI ASTROFIZICĂ, PROBLEME, VOL. 2**, *Editura CYGNUS*, Suceava, 2015, 460 p.

52. **CONCURSUL INTERJUDEŢEAN DE FIZICĂ "LIVIU TĂTAR"**, coauthor, *Editura else*, Craiova, 2016, 280 p.

53. **INTERNATIONAL OLYMPIAD ON ASTRONOMY AND ASTROPHYSICS, PROBLEMS, VOL. I**, *Trisula Adisakti*, Jakarta, Indonesia; Amazon, USA, 2016, 500 p.

54. **INTERNATIONAL OLYMPIAD ON ASTRONOMY AND ASTROPHYSICS, PROBLEMS, VOL. II**, *Trisula Adisakti*, Jakarta, Indonesia; Amazon, USA, 450 p.

55. **ASTRONOMIE Ę ASTROFIZICĂ – EVENIMENTE OLIMPICE – 1996–2010, PROBLEME**, *Editura Didactică şi Pedagogică*, Bucharest, 2018, 972 p.

56. **EXPERIMENTE DE FIZICĂ**, *Editura Didactică şi Pedagogică*, Bucharest, 2018, 476 p.

57. **ASTRONOMIE ŞI ASTROFIZICĂ. TEME ŞI PROBLEME PENTRU OLIMPIADE Ę CONCURSURI**, *Editura Didactică şi Pedagogică*, Bucharest, 2019.

58. **ASTRONOMY AND ASTROPHYSICS. THEMES AND PROBLEMS FOR OLYMPIADS AND COMPETITIONS**, *Editura Didactică şi Pedagogică*, Bucharest, 2019.

59. **ASTRONOMIE ŞI ASTROFIZICĂ – EVENIMENTE OLIMPICE – 2011–2013, PROBLEME, VOL. II**, *Editura Didactică şi Pedagogică*, Bucharest, 2020, 500 p.

60. **ASTRONOMIE ŞI ASTROFIZICĂ – EVENIMENTE OLIMPICE – 2013–2015, PROBLEME, VOL. III**, *Editura Didactică şi Pedagogică*, Bucharest, 2020, 500 p.

61. **ASTRONOMIE ŞI ASTROFIZICĂ – EVENIMENTE OLIMPICE – 2016–2017, PROBLEME, VOL. IV**, *Editura Didactică şi Pedagogică*, Bucharest, 2020, 500 p.

62. **ASTRONOMIE ŞI ASTROFIZICĂ – EVENIMENTE OLIMPICE – 2017–2018, PROBLEME, VOL. V**, *Editura Didactică şi Pedagogică*, Bucharest, 2020, 500 p.

63. **ASTRONOMIE ŞI ASTROFIZICĂ – EVENIMENTE OLIMPICE – 2019, PROBLEME, VOL. VI**, *Editura Didactică şi Pedagogică*, Bucharest, 2020, 548 p.

64. **INTERNATIONAL ASTRONOMY OLYMPIAD, VOL. I**, *Editura CYGNUS,* Suceava, 2021, 464 p.

65. **INTERNATIONAL ASTRONOMY OLYMPIAD, VOL. II,** *Editura CYGNUS,* Suceava, Romania, 2021, 484 p.

66. **INTERNATIONAL OLYMPIAD ON ASTRONOMY AND ASTROPHYSICS, VOL. III,** *Editura CYGNUS,* Suceava, Romania, 2021, 589 p.

67. **ASTRONOMY AND ASTROPHYSICS. ALL MY PROBLEMS FOR INTERNATIONAL OLYMPIADS, VOL. I,** *Editura CYGNUS,* Suceava, Romania, 2021, 344 p.

68. **ASTRONOMY AND ASTROPHYSICS. ALL MY PROBLEMS for INTERNATIONAL OLYMPIADS, VOL. II,** *Editura CYGNUS,* Suceava, Romania, 2021, 314 p.

69. **"SUB CERUL BUCOVINEI", VOL. I, CONCURSUL INTERJUDEŢEAN DE FIZICĂ "CYGNUS", CONCURSUL NAŢIONAL DE FIZICĂ "MARIN DACIAN BICA", Clasele VI–IX, 2015–2019,** *Editura CYGNUS,* Suceava, 2021, 300 p.

70. **"SUB CERUL BUCOVINEI", Vol. II, CONCURSUL INTERJUDEŢEAN DE FIZICĂ "CYGNUS", CONCURSUL NAŢIONAL DE FIZICĂ "MARIN DACIAN BICA", CLASELE X–XII, 2015–2019,** *Editura CYGNUS,* Suceava, 2021, 326 p.

71. **"SUB CERUL BUCOVINEI", VOL. III, CONCURSUL NAŢIONAL DE ASTRONOMIE ŞI ASTROFIZICĂ "MARIN DACIAN BICA" FIZICĂ, 2015–2019, CONCURSUL INTERDISCIPLINAR "ŞTIINŢELE PĂMÂNTULUI", OLIMPIADA NAŢIONALĂ DE ASTRONOMIE ŞI ASTROFIZICĂ",** *Editura CYGNUS,* Suceava, 2021, 388 p.

72. **PROBLEME DE FIZICĂ, VOL. I, EVENIMENTE OLIMPICE, SELECŢII PENTRU IPhO 1995–2009,** *Editura Didactică şi Pedagogică,* Bucharest, 2021, 340 p.

73. **PROBLEME DE FIZICĂ, VOL. II, EVENIMENTE OLIMPICE, SELECŢII PENTRU IPhO 2010–2021,** *Editura Didactică şi Pedagogică,* Bucharest, 2021, 340 p.

74. **ASTRONOMIE ŞI ASTROFIZICĂ, EVENIMENTE OLIMPICE, 2020–2021, PROBLEME, VOL. VII,** *Editura Didactică şi Pedagogică,* Bucharest, 2022, 600 p.

75. **PROBLEME DE PERFORMANŢĂ ÎN FIZICĂ, VOL. I**, *Editura Didactică şi Pedagogică*, Bucureşti, 2023, 664 p.

76. **ASTRONOMIE ŞI ASTROFIZICĂ, EVENIMENTE OLIMPICE, 2020–2022, PROBLEME, VOL. VIII,** *Editura Didactică şi Pedagogică,* Bucharest, 2023, 595 p.

77. **ASTRONOMY AND ASTROPHYSICS – MY NEW PROBLEMS, VOL. I** (for IOAA 2023, Edition XVI, Poland), *Editura Didactică şi Pedagogică*, Bucureşti, 2023, 395 p.

78. **ASTRONOMY AND ASTROPHYSICS – MY NEW PROBLEMS, VOL. II** (for IOAA 2023, Edition XVI, Poland), *Editura Didactică şi Pedagogică*, Bucureşti, 2023, 460 p.

79. **ASTRONOMY AND ASTROPHYSICS – MY NEW PROBLEMS, SECOND ED., VOL. I** (for IOAA Juniors 2023, Edition II, Greece), *Editura Didactică şi Pedagogică*, Bucharest, 2023, 395 p.

80. **ASTRONOMY AND ASTROPHYSICS – MY NEW PROBLEMS, SECOND ED., VOL. II** (for IOAA Juniors 2023, Edition II, Greece), *Editura Didactică şi Pedagogică*, Bucharest, 2023, 460 p.

81. **ASTRONOMY AND ASTROPHYSICS – MY NEW PROBLEMS, THIRD ED., VOL. I** (for IAO 2023, Edition XXVII, China), *Editura Didactică şi Pedagogică*, Bucharest, 2023, 460 p.

82. **ASTRONOMY AND ASTROPHYSICS – MY NEW PROBLEMS, THIRD ED., VOL. II** (for IAO 2023, Edition XXVII, China), *Editura Didactică şi Pedagogică*, Bucharest, 2023, 460 p.

83. **PROBLEME DE PERFORMANŢĂ ÎN FIZICĂ, VOL. II,** *Editura Didactică şi Pedagogică,* Bucharest, 2023, 500 p.

84. **PROBLEME DE PERFORMANŢĂ ÎN FIZICĂ, VOL. III,** *Editura Didactică şi Pedagogică,* Bucharest, 2024, 500 p.

85. INTERNATIONAL PRE-OLYMPIC CONTEST OF PHYSICS, ROMANIA, HUNGARY, MOLDOVA, *Editura Didactică şi Pedagogică,* Bucharest, 2024, 350 p.

86. ŞTIINŢE PENTRU JUNIORI ÎN ROMÂNIA 2004–2023, PROBLEMELE MELE DE FIZICĂ, *Editura Didactică şi Pedagogică,* Bucharest, 2024, 480 p.

87. SCIENCE FOR JUNIORS IN ROMANIA 2004–2023, MY PROBLEMS OF PHYSICS, *Editura Didactică şi Pedagogică,* Bucharest, 2024, 480 p.

88. ŞTIINŢE PENTRU JUNIORI ÎN ROMÂNIA, 2024, PROBLEMELE MELE DE FIZICĂ, *Editura Didactică şi Pedagogică,* Bucharest, 2024, 200 p.

89. SCIENCE FOR JUNIORS IN ROMANIA, 2024, MY PROBLEMS OF PHYSICS, *Editura Didactică şi Pedagogică,* Bucharest, 2024, 200 p.

90. CONCURSUL NAŢIONAL DE FIZICĂ ŞI CHIMIE "IMPULS PERPETUUM!" AL ELEVILOR DE GIMNAZIU DIN MEDIUL RURAL, PROBLEMELE MELE DE FIZICĂ, 2011, 2013–2020, *Editura CONPHYS,* Rm. Vâlcea, 2025, 200 p.

91. PROBLEMELE MELE DE FIZICĂ PENTRU EVENIMENTE OLIMPICE, VOL. I, 2022–2023, *Editura Didactică şi Pedagogică,* Bucharest, 2025, 300 p.

92. PROBLEMELE MELE DE FIZICĂ PENTRU EVENIMENTE OLIMPICE, VOL. II, 2023–2024, *Editura Didactică şi Pedagogică,* Bucharest, 2025, 350 p.

93. ASTRONOMIE ŞI ASTROFIZICĂ, PENTRU ELEVI OLIMPICI, STUDENŢI ŞI PROFESORI, *Editura Didactică şi Pedagogică,* Bucharest, 2025, 450 p.

94. ASTRONOMY AND ASTROPHYSICS OLYMPIAD: TOPICS, PROBLEMS AND SOLUTIONS, VOL. I, *World Scientific,* London, 2025, 596 p.

95. ASTRONOMY AND ASTROPHYSICS: PROBLEMS AND SOLUTIONS, VOL. II, *World Scientific*, London, 2025, 432 p.

96. ASTRONOMY AND ASTROPHYSICS: PROBLEMS AND SOLUTIONS, VOL. III, *World Scientific,* London, 2025, 456 p.

97. PHYSICS OLYMPIAD: PROBLEMS AND SOLUTIONS, *World Scientific,* London, 2025, 396 p.

www.ingramcontent.com/pod-product-compliance
Lightning Source LLC
Chambersburg PA
CBHW071443161025
34052CB00008B/27
* 9 7 8 1 8 0 0 6 1 6 9 3 6 *